The world of nature is not one-dimensional like a landscape painting, however beautiful to the eye. It is composed of many elements and layers, some seen, most hidden. In this marvellous book, Simon Grove highlights them all, leaving no stone—or seashell—unturned in his quest to present a well-rounded natural history of Tasmania. Travelling through the seasons, he also overcomes life-threatening illness on his remarkable journey.

—Donald Knowler
'On the Wing' columnist
Sunday Tasmanian, Hobart

Whether naturalist or scientist, Simon's formidable knowledge of the Tasmanian environment shines in this powerfully woven narrative on the significance of small things and the bigger picture. Deeply personal and introspective, it explores the powerful connection between nature, home, and us.

—Dr Sally Bryant AM
Wildlife scientist

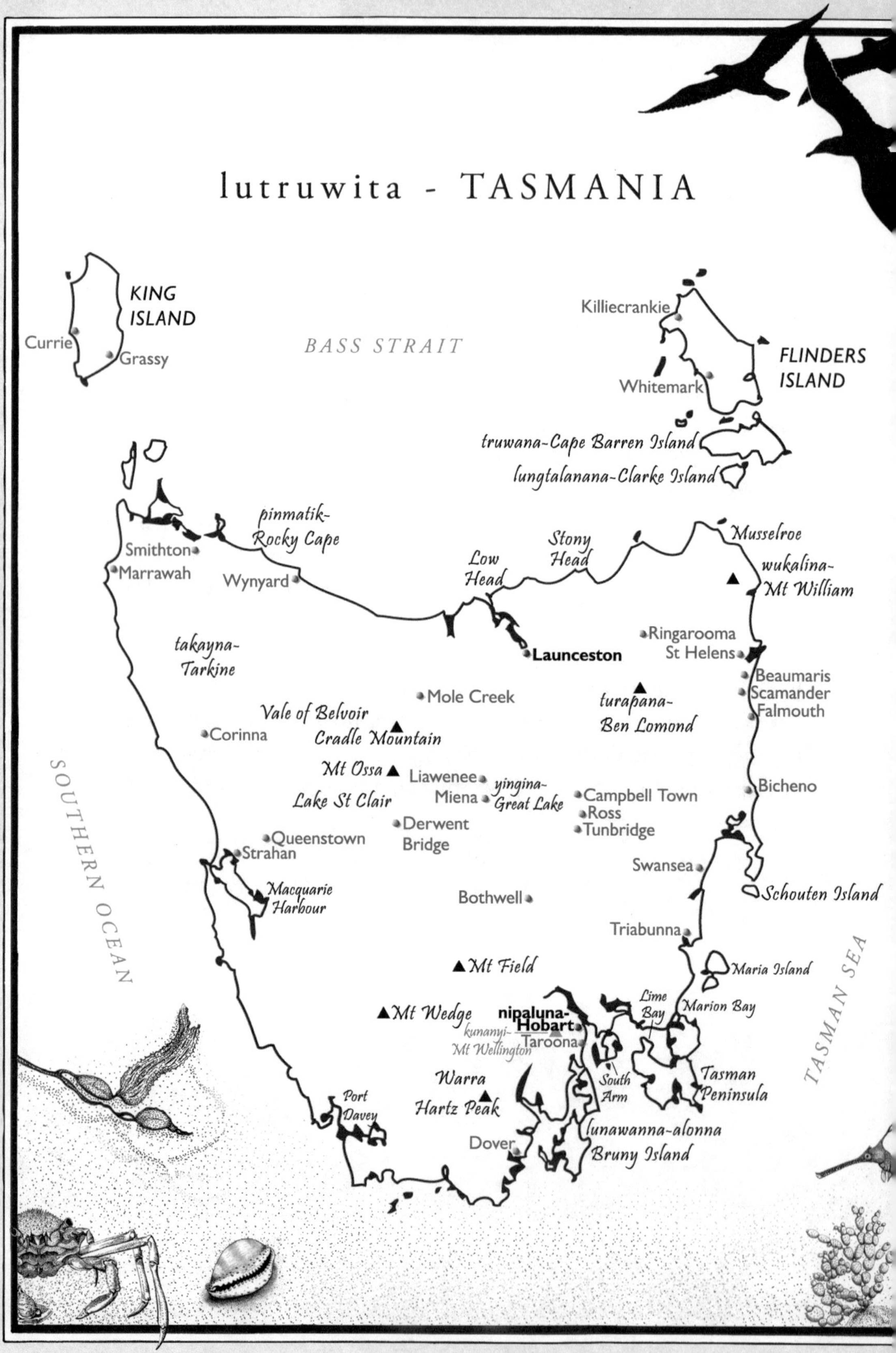
lutruwita - TASMANIA
KING ISLAND
Currie
Grassy
BASS STRAIT
Killiecrankie
FLINDERS ISLAND
Whitemark
truwana-Cape Barren Island
lungtalanana-Clarke Island
pinmatik-Rocky Cape
Smithton
Marrawah
Wynyard
Low Head
Stony Head
Musselroe
wukalina-Mt William
Ringarooma
St Helens
Launceston
takayna-Tarkine
Beaumaris
Scamander
Falmouth
Mole Creek
turapana-Ben Lomond
Vale of Belvoir
Cradle Mountain
Corinna
Mt Ossa
Liawenee
yingina-Great Lake
Miena
Campbell Town
Ross
Tunbridge
Bicheno
Lake St Clair
Derwent Bridge
Queenstown
Strahan
Swansea
SOUTHERN OCEAN
Macquarie Harbour
Bothwell
Schouten Island
Triabunna
Maria Island
TASMAN SEA
Mt Field
Lime Bay
Marion Bay
Mt Wedge
nipaluna-Hobart
kunanyi-Mt Wellington
Taroona
South Arm
Tasman Peninsula
Warra
Hartz Peak
Port Davey
Dover
lunawanna-alonna Bruny Island

Seasons in the South

A TASMANIAN NATURALIST'S JOURNEY OF DISCOVERY —*AND RECOVERY*

Simon Grove

Illustrations by
Keith Davis

Glass House Books
Brisbane

2nd edition,

ISBN: 9781923435247 (PB); 9781923435254 (eBook)

1st edition: Forty South Publishing Pty Ltd
fortysouth.com.au

Digital artwork p. 226 by Andi Magnusson

NATIONAL LIBRARY OF AUSTRALIA
A catalogue record for this book is available from the National Library of Australia

Contents

January

February

Out of time *an interlude*

March

April

May

June

July

August

Author's note

This is my Tasmanian story, and also the story of the island that I have come to call home. It's a story of its flora and (especially) fauna, both on land and off shore; its weather, its landscapes, its natural systems; sometimes of its people, too, albeit mostly those who, like me, call themselves naturalists. Who I am, why I should want to tell you these things, and why you might care (I hope that you will), are things that you will learn more about as you read on.

My stories are presented as a series of vignettes, arranged into a month-by-month account of the Tasmania that I have come to know and love, with its patterns, cycles, twists and surprises. The accounts are based on my own recollections; but where time has made these hazy, I have used some imagination to recreate the ambience. Scientific names have been largely kept out of the narrative because they are known to worry the uninitiated; but an index of some is offered at the end of the book, for those who value them as gateways to knowing about the natural world in a particular way, as I do.

There are other ways of knowing, though, and other ways of articulating what we know. To the First Peoples of this island—Palawa and Pakana—who have occupied it for at least forty thousand years, Tasmania is Lutruwita. Many words in their language, Palawa kani, are reserved for use by Palawa/Pakana; but some place names, including Lutruwita, have been offered to those of us without their heritage to use in appropriate contexts. In this book, I generally default to 'Tasmania' because it's the term that most of us are familiar with; but I value and feel more connected to the island when calling it Lutruwita, so I use this term when it seems appropriate to do so. I have tried to adopt the same approach with other place names, too, but I have generally provided both the Palawa kani and English term on first mention. Notwithstanding all that is good about this island today, I regret the damage that was done when it was colonised. I honour those who remain and are the custodians of special knowledge and ways of knowing. I acknowledge Palawa and Pakana, and their Elders and this Country. I hope that my offering to them, of what is in this book, will be welcome.

—Simon Grove
Taroona, Tasmania
September 2023

Author Simon Grove

SIMON GROVE is Senior Curator of Invertebrate Zoology at the Tasmanian Museum and Art Gallery in Hobart. He hails from England. Following doctoral research in the Daintree rainforests of tropical North Queensland, he moved to Tasmania with his young family in 2001, to work as a Conservation Biologist. A lifelong naturalist, he is author of *The Seashells of Tasmania: A Comprehensive Guide*, and has also published widely on Tasmanian natural history and ecology. Simon regularly chats about Tasmanian invertebrates on local radio, and in 2019 was awarded the Australian Natural History Medallion—the 'Nobel Prize for Australian naturalists'.

Illustrator Keith Davis

KEITH DAVIS has been drawing, sketching and painting nature for five decades, and his artworks have been exhibited widely. He is a qualified artist and natural history illustrator, and also has education and science credentials. He is a passionate nature and sustainability educator, designer of permaculture systems, and author and illustrator of *Seasons of the Far South*, an educational book about the flora and fauna of Tasmania's south.

For my mother Pauline, who took every opportunity to encourage my childhood love of the natural world; my father Tony, whose unenviable commuter lifestyle enabled us to live in the countryside; my wife Chris, and sons James and Ben, who have accepted my naturalist foibles with good humour, and have been there during so many memorable and defining moments in my Tasmanian journey; and my three immuno-compatible siblings, especially my elder brother Jon—donor of top-quality, life-saving stem cells from his bone marrow, during my time of need.

I also dedicate this work to the original custodians of Lutruwita-Tasmania—Elders who lived and breathed this land for tens of thousands of years before British colonists brutally stole it; and to the Palawa/Pakana of today, who safeguard the knowledge and ways of knowing of their predecessors. This land is their Country.

© K.Davis 2022

September

New beginnings

'Did you hear about the twin towers?'

'Yeah—on the News. It's hard to believe.'

I had awoken to those words from a deep slumber. They were spoken by two men, who happened to be strolling past our unit, in Hobart's Battery Point. It was one of our first mornings in Tasmania and my jetlagged brain struggled to comprehend their words. Failing, I drifted back to sleep again. Sometime later, I was again awoken, this time by the crazed chattering of an unfamiliar bird. Despite my tiredness, I needed to know what it was. I jumped up, donned a dressing gown, staggered to the window and drew the curtains, binoculars at the ready. The glare of a crisp, clear Hobart morning momentarily blinded me; but then, in an ornamental cherry tree newly adorned with soft pink blossom, I made out a medium-sized songbird as perky and alert as I was befuddled. It sported a black cap, white nape and belly, and grey wings and tail. There was no mistaking the intent of the hooked tip to its bill, nor the menacing glint in its eye. I had become familiar with black butcherbirds during three years in the rainforests of Queensland's Wet Tropics, and this looked like one of their relatives. Leafing through my well-thumbed Australian bird book, I was startled by the stiff, dry pages almost crackling beneath my touch. It was such an unfamiliar experience after the all-pervading dampness of the tropics. Then I came to it: Bingo—a grey butcherbird! Being able to slot this southern version into a ready-made mental 'pigeon'-hole gave me immense pleasure. It was a dash of familiarity in an otherwise unfamiliar land.

The butcherbird surveyed its domain, seemingly indifferent to incessant mobbing by New Holland honeyeaters which, with good reason, objected to the predatory butcherbird's presence; indeed, its very existence. Eventually the birds departed, and I turned on the kettle for a cuppa, automatically switching on the radio, too. Instantly, I was assaulted by coverage of what has become known as 9/11. How awful the calamity; yet how distant it seemed to me on this bright Hobart morning in September 2001. With my wife, Chris, and our two young sons, James and Ben, I had flown from my native England to Singapore, then on to Melbourne, and, finally, Hobart—just in time, it seemed. There would be no more international flights, no contrails in the sky; not for a while, at least. I had voluntarily left so much behind: my family, my upbringing, so much of my life in Britain; even my recent three-year sojourn in tropical Queensland. Everything seemed a long, long way away from the island I had come to, which I then knew only as 'Tasmania', but which is Lutruwita to the Palawa/Pakana—the survivors of its First Peoples. I had already noticed how the small-town capital city, Hobart, faced pointedly away from the rest of Australia, towards the frigid Southern Ocean. I considered how lucky we were to have found this quiet corner of the world at such a tense time of global upheaval; yet how alien it all seemed. I was daunted by the prospect of starting life afresh here.

But who am I, and why does it matter? Perhaps it doesn't, because this book is primarily about Tasmanian nature; however, I can't escape from the fact that it's my narrative, and you, I believe, may well ask why this English naturalist bloke feels he should write a book like this, especially when he has the audacity to call himself Tasmanian. So I will introduce myself. As for your natural question: the answer, as with so much in life, is rooted in serendipity. Although each of us experiences our timelines as linear affairs, we are all entangled in the lives of others, and those intersections are where magic sometimes happens, opportunities arise, and life may take new turns. Mine is no exception. People noted a naturalist in me from my earliest years, and nudged me along this path. In time it lead to my settling in Tasmania and writing this book. The fact that I continue to travel on the trajectory they guided me towards affirms the value of their efforts. I shan't name them here, but I thank them all.

I might not have known the word at first, but it seems there has never been a time in my life when I did not consider myself a naturalist. One of my earliest memories is of 'saving' earthworms that were drowning on a flooded cricket oval. I had gathered them all into a handy, empty, glass milk bottle, so that I could release them somewhere drier. I also remember my big brother feeding me earthworms; but, while I have my suspicions, I cannot be sure that the two memories are connected. From my early years, I remember being enthralled by the sight and sound of bumblebees on a flowering buddleia bush in our garden, deciding that their interests would be best served if I were to gently entrap them in my cupped hands (they never stung me) and transfer them to individual 'hives' that I had made in our sandpit. Much to my chagrin, they usually escaped through the front entrance at the earliest opportunity, unless I sealed them in with more sand. I remember spending still summer evenings in my pyjamas, outside the back door, entranced by swarms of tiny flying beetles dancing in the air above our dustbins. I recall the musty smell of the pages of my scrapbook, where I glued pressed flower's (yes, I confess: the errant apostrophe appears in my handwritten title on the front cover) that I had collected from the local heathland on early morning dog-walking sorties with neighbours.

Lessons in natural history can be quite blunt. I recall the pain and indignation of my first ant sting while I was sitting on the lawn making daisy chains. The guilt of discovering that the lizards I had brought back from the heathland, and kept in a small glass terrarium, had fallen prey to the ground beetles that I had decided might make good companions for them. More guilt when my younger sister was bitten on the finger by one of the wood mice that I had ensconced in another terrarium, having caught them in my hands and transferred them, without incident, from their home in a patch of woodland beyond the garden fence. And the shock and terror of being surrounded by an angry swarm of wasps, emerging from their nest in this same patch of woodland, after I had disturbed it in one of my dens in the bracken. My parents lost count of the number of stings the wasps inflicted on my torso, but at least I paid more attention to their kind afterwards. I soon learnt to tell them from their harmless, yellow-striped hoverfly mimics, which also frequented our garden.

With a growing library of books on birds, butterflies, beetles, seashells and wildflowers, every walk in the countryside, and every family holiday by the seaside, was an opportunity for new encounters with nature. The mewing calls of herring gulls on the roof of our Cornish seaside holiday cottage are still with me today (I had to get up at dawn to find out what they were); as is the musty, salty smell of the brightly coloured sea urchin tests (exoskeletons) that we bought from a local souvenir shop for use as lampshades. I recall the delight of finding my first cowrie shell (really a ribbed cowrie) on a Cornish beach, not long after I had seen one featured in a wonderful painting of a seashore scene in one of my Ladybird books. In time we amassed a jam jar full of cowries, which were brought out for use as counters and currency in family card games. Other family holidays, to the far corners of Britain, offered more beachcombing, more country walks, more sights, smells and sounds of nature; all of which I still hold dear.

In my last two years of primary school, I developed a love affair with seashells, sparked by receiving a little basket of (supposedly) Polynesian shells from my grandparents, at Christmas. I filled school exercise books with drawings of shells and other marine life, copied their scientific names from a catalogue, and wrote that I aimed to be a marine biologist when I grew up. I developed a card index system for my shell treasures, neatly typed with two fingers on an ancient typewriter. Before long, my bedroom had become a public shell museum, even featuring on a local television series. I was already a regular visitor to the local museum, and I soon ended up as their youngest volunteer. I worked alongside a more senior volunteer (a much-loved mentor), during my school holidays. We were occupied with identifying and displaying the museum's dusty, Victorian-era seashell collection. Later, we moved on to curating their birds and butterflies. One thing led to another, and I was soon the youngest member of the local natural history society (by several decades). There, I was taken under the wing of another mentor, and enjoyed monthly outings to look for birds, butterflies, orchids and fungi. Sometimes, I was made responsible for reporting on our finds in the local newspaper. I saved up my pocket money and bought my first manual single-lens reflex camera and spent many a happy summer's day hunting insect and floral subjects in local nature reserves reached, mostly, by bike rides through the network of country lanes that started at the bottom of our driveway.

Fast forward through the angst-ridden teenage years (sometimes even escaping into nature couldn't lift my mood). Forward, too, through the liberating freedom of student life at a university situated on the picturesque Devonshire coast, with a backdrop of rolling hills and moorland that offered endless opportunities for immersing in nature. Emerging out the other side of these emotionally-charged years, I found myself as a young adult with a first-class degree in nature conservation and marine biology, and a whole wide world awaiting exploration. My first tropical experience was a three-month scientific expedition to the rainforests and coral reefs of Costa Rica. This was followed by six months, the same year, of marine biological research amongst the mangrove forests of the Kenyan coast. Life in England seemed tame after those experiences: I enjoyed several largely field-based jobs in nature conservation, which took me to some of the loveliest parts of the country; but I yearned for more adventure. So I spent three months as a nature guide in the Peruvian Amazon, then signed up for a forestry master's course at Oxford, with the aim of returning to the tropics to help save the rainforests. This was when I met Chris, my future wife, and we spent the following couple of years working together, on a tropical forestry and conservation project in Uganda. It was a formative experience that got me my next job, on a forestry aid project in Indonesia. It also made me start thinking about the nature of sustainable forest management in the tropics; particularly about how it had to be about more than just the trees. I wanted to draw attention to insects as indicators of sustainability, especially those which depended on dead wood, because my experiences in England had taught me that these were amongst the organisms most at risk from forest use and abuse. Therefore, I applied for a Commonwealth scholarship to study for a doctorate at James Cook University, in Cairns. I couldn't believe my luck when it all came together.

For the next three years, Chris and I sweated it out in the Daintree rainforests, and got married, and had two kids. The sheer exuberance of plant and animal life in Queensland's Wet Tropics makes this place wondrous, but it was a tough environment for doing fieldwork during the steamy summers. A mid-study cyclone that left a trail of destruction across my study area added to my discomfort; but there's nothing quite like an expiring visa to focus an academic mind. Somehow, with great relief, I completed my thesis within the allotted time span. Chris likes to summarise the results of my research as '*the more dead wood there is in a tropical forest, the more deadwood insects you get*'. It isn't far from the truth, but at least we now have the evidence.

At the end of my candidature, we were all set to continue our careers in tropical rainforest conservation and management, as I had been offered an academic job in Fiji. Then a Tasmanian colleague sent me a cutting from *The Mercury* (Hobart's daily newspaper), for a position as a conservation biologist with Forestry Tasmania. Chris and I had previously enjoyed a holiday in Tasmania, so I applied, and got the job. We squeezed in a short visit to England before returning, to begin our new life in Tasmania. It was September 2001—snow season, and the world upside down.

Snow season

If you were to invert a map of Tasmania and superimpose it at its corresponding latitude in Europe, it would sit comfortably in the Mediterranean Sea between Barcelona, Marseille, Menorca, and Sardinia—places synonymous with languid breezes gently ruffling palm tree fronds and the hair of animated diners at plazas crammed with outdoor eateries. But Hobart was emphatically not exuding any hint of the Mediterranean as we settled into our first month in Tasmania. Outside work hours, Hobart's streets were deserted; and even during the day, a disconcertingly large proportion of Hobart's citizens were swaddled in puffer jackets (affectionately known as Tasmanian tuxedos—I have since come to appreciate these as *de rigueur* attire both outdoors and in). Like every September since then, the dominant vibe was the sound of chilly air barrelling in from the Southern Ocean, regularly dusting the top third of what locals called, simply, 'The Mountain' with snow. In Hobart—known as Nipaluna in Palawa kani—there's no escaping the looming presence of Kunanyi-Mount Wellington at any time of year; but sudden snowfalls seem to peak in September, regularly imbuing the mountain with a heightened air of grandeur.

More than two decades have passed since 9/11 and our move to Tasmania. Much has changed in the world: wars, politics, society, technology, economy, environment, climate, and to top it all, a pandemic. Tasmania, too, has witnessed a rapid transformation. No longer such a sleepy backwater, we've seen sea-changers, tree-changers, forest wars, the rise of café and foodie culture, arts-, adventure- and nature-based tourism, aurora chasers. We have become uncomfortably aware of the terrible injustices inflicted by colonial culture on Lutruwita's First Peoples, and it is now dawning on us that those deeds reverberate through what we call the natural world—which is also a cultural landscape. We've had a foretaste of the droughts, heatwaves, bushfires and floods that climate experts predict will become increasingly prevalent in the years ahead. Yet *plus ça change, plus c'est la même chose*: butcherbirds still stare down taunting honeyeaters amid the leafy streetscapes of suburban Hobart; and the 250 kilometres of Bass Strait remain a formidable barrier, both physical and psychological, between Tasmania and the Australian mainland. Above all, day-to-day life in the path of the Roaring Forties remains in thrall to fickle weather: a choreography of winds, ceaselessly changing cloud formations and rain bands orchestrated by an endless succession of cold fronts and low-pressure systems streaming in from the west. These battle equally determined high pressure ridges descending from the north, for supremacy in the skies over Tasmania.

While the chill of winter is never far away in September, it is, nevertheless, an auspicious month, a time of promise, of awakening. Around the equinox, the air seems particularly restless, newly energised by the heat of a sun that sweeps across the skies in increasingly confident arcs, rapidly lengthening the days. In our earliest weeks in Hobart, I learnt to seek out sheltered spots outdoors during my lunch hour, letting the sun's gift of warmth seep through my many layers of clothing; layers that were sorely needed whenever clouds skittered in from the south to block the sun. I had come to Tasmania to become Forestry Tasmania's conservation biologist. I had the academic freedom and support at Forestry to develop a research program as I saw fit. I had been recruited because of my background in what I came to call 'deadwoodology'. It tied in with industry moves to develop a market for wood residues other than (or in addition to) woodchips. My employers wanted my research to help them understand the ecological risks of such a move, and how to manage those risks. There were diversions along the way, though: I joined others at Forestry and the University of Tasmania, researching alternatives to clearfelling; and I developed and coordinated a range of research projects aimed at better understanding the ecology of the fire-dependent wet eucalypt forests (on spatial scales from single trees to landscapes, and on timescales from years to centuries). We sampled insects as our study organisms for many of these projects; and, in the process, we built up a vast collection and database of Tasmanian forest insects, especially beetles. Birds, fungi, mosses and vascular plants were also significant parts of this research portfolio.

But all this was still in the future: we had only just arrived, and house-hunting took us from the city to the suburbs and beyond, where an awakening of life seemed particularly evident. Roadsides erupted in a spectacular display of floral optimism: brash yellow wattles and *Crassulas*, delicate flowering cherries and apricots, outlandishly giant blue *Echiums*, to name but a few. We came to understand the well-worn trope about four seasons in a day. Some view it as a reason to stay away. If your only experience of Tasmanian weather is a soaking when you dash from house to car, or billowing umbrellas on a city street, or a stiff southerly sea breeze that, with perfect timing, whips the waves into whitecaps the moment you arrive at the beach for a swim, then such a view is understandable. But here's the challenge Tasmania offers you: embrace the weather, immerse yourself in the climate, breathe in the seasons; tap into their elemental energies and you may find, as I have, that the forces of nature enliven and invigorate you. They also provide the backdrop needed to fully appreciate the quiet lives and dramas playing out all around us, in Tasmania's unique—and I really mean *unique*—natural world. As a naturalist, I had arrived in paradise.

Snow rarely descends to sea level in Tasmania, but it's most likely to do so in September. Though it seldom lasts more than a few hours, it will be a topic for conversation for much longer. For our young sons, snow in the front yard was nothing short of magical. I can still hear their shrieks of delight, as they peered through the window at fluffy white blobs descending silently from the sky. Trips up the mountain, to the snow, enchanted children and adults alike. They remain a popular family pastime in early spring, although it's not often that snowfall corresponds with weekends, or fine weather. On the mountain's plateau-like summit, snow is more often accompanied by howling, bone-chilling winds; and the perilously ice-varnished, boulder-strewn paths challenge even the adventurous. There are rewards for persistence, though. Duck down from the plateau, into the lee of the winds, and you can find yourself enjoying the privilege of feeling as though you have the place all to yourself. Except you don't, of course; nature is all around you.

But this book is not about the Tasmanian weather or its southern seasons; it is about the interplay between these and the *living* world. It is also about my own experiences as a naturalist immersed in this world, and my journey of discovery, spanning twenty-something years on this wonderful island. I spent my first decade with Forestry Tasmania. It was a fine and thoroughly rewarding job. During leisure time, I rekindled my interest in seashores and seashells, and built up a substantial collection. For much of my second decade, and beyond, I have held the position of senior curator of invertebrate zoology at the Tasmanian Museum and Art Gallery. It has proven to be the best job that I could possibly imagine, allowing me to combine my interests in insects, seashells and all the other local invertebrates. The Tasmanian Forest Insect Collection followed me to the museum, as did my Tasmanian seashell collection. Twice, however, I have succumbed to acute myeloid leukaemia. While I was recovering from my first bout, I was kept sane by writing a guide to the seashells of Tasmania. I only survived the second bout thanks to my elder brother, who donated to me a transplant of bone-marrow stem cells. Such experiences, and more, have shaped my life, my outlook and my values; but in times of darkness as well as joy, it is nature that has been my constant companion and guide, my solace, and my inspiration. My life is all the richer for it.

A memorable mountain shrimp

On one of those wintry September days, some years ago, I found myself gingerly inching my way down a treacherously icy track on Kunanyi's eastern flanks. At the foot of the steepest section there's a small watercourse, not much more than a runnel, which eventually flows into the Hobart Rivulet. It offered a good place to stop and regain my composure after the ungainly descent. The wind still howled up on the distant plateau, but down here all was calm. I felt my cheeks flush and blood return to my fingers and toes. A party of black currawongs careened through the leaden skies. These hardy, irrepressibly outgoing birds of Tasmania's wilder reaches seem to take great delight in riding out the cold even while other mountain birds, such as crescent honeyeaters, descend to the suburbs below. Ever the socialites, on this occasion, their communal aerial antics made absorbing viewing, set off perfectly by their discordant, grating, honking calls.

Before me, sheets of translucent ice jutted out over parts of the runnel from its rocky fringes, so that the water beneath flowed in silent, mesmeric pulses, interspersed with low burbling sounds as the runnel periodically gulped air. The water would have been only a degree or two above zero, but two pairs of twitching antennae protruding from a dark nook between some submerged cobblestones signalled that it was warm enough for one of the mountain's strangest inhabitants. I stood motionless as the mountain shrimp tentatively stepped forward. If I hadn't seen it move, I would have missed it, because its olive-green body, about as long as my thumb, was the colour of the alga-covered gravel over which it crawled. Its slender, many-legged, segmented trunk ended in a lobster-like fan of a tail. Two beady, compound eyes watched for danger as the animal continued its traverse of the runnel. It soon disappeared beneath another stone. It was my first-ever encounter with the famed mountain shrimp, and I was immediately enthralled. For a few minutes I waited, hoping it would reappear, but it didn't. Feeling cold, I moved on.

Continuing down the mountain with a spring in my step, I pondered my good fortune in having found a home at the foot of this wild mountain, where the calls of the currawongs echo among dolerite columns, and where, for time immemorial, the shrimps have been simply doing their thing in cold, babbling creeks. The Tasmanian mountain shrimp is one of a suite of odd, freshwater crustaceans that call Tasmania, and nowhere else in the world, home. Most charismatic is the giant freshwater crayfish, a lobster-like denizen of the state's northern rivers, which I have yet to see, and the world's largest freshwater invertebrate. The mountain shrimps can't compete on size or charisma, but in an evolutionary sense they're at least as special. Fossils of very similar-looking marine shrimps date back some 250 million years, well before the dinosaurs and long before other shrimps, prawns and lobsters appear in the fossil record. This is why today's mountain shrimps are sometimes called living fossils (a term that annoys evolutionary biologists no end). In the modern era they eke out their existence in cold, well oxygenated mountain tarns, in peaty pools and runnels and in subterranean rivers in karst country—but only in Tasmania.

At the time that I saw my first, the world knew of five species in the family, only one of which had been given a common name. But recently the eponymous Tasmanian mountain shrimp was found to comprise three closely related species, each occupying a different part of western Tasmania. This had the effect of making the species that I had seen one of the rarest and most vulnerable of the lot, since its entire global range was then thought to comprise Kunanyi's eastern flanks, overlooking the city of Hobart. Yet, intriguingly, the same species has since been detected in small streams flowing into the Tyenna and Florentine Rivers, on the south-western fringe of Mount Field National Park, some sixty kilometres to the northwest of Kunanyi. There, it lives almost alongside its other two sibling species. Disjunct and overlapping distributions like this are supposed to say much about Tasmania's deep history—but this case remains a bit of a mystery, one in which scientific conjecture can sound a bit like a just-so story, at least to my ears. Nevertheless, I am enthralled by the idea that there are so many discoveries still to be made in Tasmania, perhaps because, in the country of my birth and upbringing, it seemed that almost everything about the fauna had been well and truly figured out, long ago.

An Antipodean England, not

To many mainland Australians, the Tasmania they see—one created by colonists—seems almost like another country. That country is England. There's a perception that its houses and settlements, its gentle farming landscapes with winding, hedge-lined roads, and its predominantly Anglo people, have more in common with the Old Country than does the rest of the nation. I dare say there is some truth in this. But would the same apply to the natural world, too? Having arrived from England, I was, at first, keen to see the similarities and push the differences to the back of my mind. Without hard evidence, I had convinced myself that, compared to the Wet Tropics of North Queensland, Tasmania's cool-temperate faunal and floral diversity would prove tractable and accessible for a naturalist like me. I was very much looking forward to getting to know what was what, in the same way as I had developed a pretty comprehensive knowledge of England's plants and animals during my childhood and youth. This was something that I had singularly failed to do in North Queensland, because of the sheer diversity of life there.

And so it seemed, at first. Most of the weeds and grasses that burst forth along road verges and in paddocks—thistles, docks, brambles (blackberries), ragwort, plantain, Yorkshire fog, sweet vernal grass—were imports from Europe and already familiar to me, as were the spring songs of the blackbirds, goldfinches, skylarks and other European songbirds, which the colonists' acclimatisation societies had managed to establish in their new home, easing their homesickness. Garden-dwelling slaters, snails, earwigs, ladybirds, millipedes and spiders, cautiously emerging from winter hibernation, were old acquaintances, too. I later learnt that the garden snail was—can you believe this?—deliberately introduced to southern Tasmania during the late nineteenth century, by a Taroona resident, Charles Beddome, who had a bit of a thing for snails (many local species were named in his honour). He clearly had little interest in gardening. From the oaks and elms in Hobart's parks, to the hawthorn hedges and vanilla-scented, gorse-covered hillsides of the Midlands, I clutched at the similarities to affirm a sense of belonging here. It seemed natural to me that the first couple of many articles I would write for *The Tasmanian Naturalist* would make much of these common elements.

But how wrong could I be! It didn't take long to realise it. One doesn't have to look very deeply into Tasmanian nature to appreciate that this is no England in the South. The similarities are superficial, and largely confined to the Europeanised parts of the island—agricultural and urban land. Even there, the differences far outweigh the similarities. The English countryside isn't, for the most part, home to vineyards, opium poppy fields, or plantations of olives or walnuts; irrigation is seldom warranted; and merino sheep aren't a fixture either—they wouldn't cope with the constant damp. Nor do English paddocks host remnant trees left after forest clearance—the forests there were cleared thousands of years ago.

As for Tasmanian nature—well, I hope this book will demonstrate that our natural environments and the species that dwell in them are exceptional, and not remotely akin to those of a faraway island off the northwest coast of Europe. The climate drivers of the natural world differ, too, in manifold ways; and these are reflected in Tasmania's ever-changing weather and the idiosyncratic march of its seasons. So let me clear this up now: Tasmania is patently *not* an Antipodean England.

Of cabbage gums and noisy miners

September is perhaps the most predictably green month. As warmth returns to the land—hesitantly at first, but with more conviction as the days pass—lawns, verges and pastures sprout emerald coats, making the most of a surfeit of soil moisture while it lasts. It's the only time a road trip through the sheep-grazing country of the Midlands seems inviting, to me. The verdant paddocks appear full of promise and the lambs' fleeces are still fresh and white, having yet to take on the dirty grey or red-brown hue of summer's dust. Though the Midlands look glorious now, the region receives less annual rainfall than anywhere else in Tasmania; and, in dry years, less than some of the world's great deserts.

Whenever the wind switches to a northerly, which it is wont to do with increasing gusto as the month progresses, welcome swallows and tree martins dash over Bass Strait and fan out across the relatively easy terrain of the Midlands, hawking for newly-hatched insects as they swoop and bank over the grasslands. They are joined by black-faced cuckoo-shrikes, whose ventriloquial calls seem the very essence of spring's promise; and by swamp harriers, quartering over the boundless paddocks in search of ground-nesting birds or young rabbits. Starlings have long since ousted most native songbirds from these first-settled parts of Tasmania, and they are widely despised by naturalists; but even starlings can look dapper in September, resplendent in shimmering breeding plumage that glints violet in the bright sunlight (and even more so in ultraviolet light, which is how birds see it). By now they will have already staked out the best nesting hollows in the remnant cabbage gums that still dot the Midlands hillsides.

Those cabbage gums have, of course, been here longer than the starlings, and even in euphoric spring weather, their stark isolation is a sobering reminder of the near-total loss of the previously expansive grassy woodlands that were particularly attractive to the first settlers and their sheep. As with most of Lutruwita's landscapes, these park-like woodlands were a cultural artefact, having been created and maintained for thousands of years through the well-honed burning practices of Lutruwita's First Peoples. Most relict gums have now reached the end of their natural lives but left no offspring because the bare ground affords no protection to their seedlings. Some have succumbed to drought, aided and abetted by livestock compacting the soil around them, and by the introduced pasture grasses that, in combination with the addition of artificial fertilisers and the subtraction of regular fires, disrupt the trees' essential mycorrhizal fungal associations. Others, left exposed in a much-simplified ecological community, have fallen victim to the depredations of brush-tailed possums and leaf-eating and sap-sucking insects. This, too, is ultimately due to livestock, for their droppings over-enrich the soil around the base of trees, leading to the tree channelling the bonanza of nutrients into its foliage and, unwittingly, boosting its palatability and attractiveness to herbivores.

In a further cruel twist, a native honeyeater, the noisy miner, is also responsible for the gum trees' decline. These birds are one of the few species that can outdo starlings in belligerence and thuggery. They seem to particularly thrive in fragmented and degraded woodlands. Like many other native birds in eucalypt country, their energy needs are largely met by consuming lerp, the ornate, sugary, protective housings delicately crafted by psyllid bugs, which make a living sucking the sap from gum leaves. But noisy miners are intolerant of other birds moving in on their patch and will aggressively chase away all comers. Their dog-in-the-manger attitude assures them of an ample supply of lerp all for themselves—usually far more than they can possibly consume. The result is that psyllid populations can explode to a point at which the tree can no longer sustain the loss of vital sap, sending it into a spiral of decline. But let's not be too hasty in vilifying the miners: in a more natural setting, they would be outsiders, confined to woodland edges or areas where the trees were unusually sparse, such as waterlogged or saline hollows or frost pockets. They would have needed every bit of their boldness to keep the birds of the woodland interior from completely ousting them from the landscape. It's just that, over the past two hundred years, the tables have turned rather drastically in favour of miners and their partners in crime, the psyllid bugs.

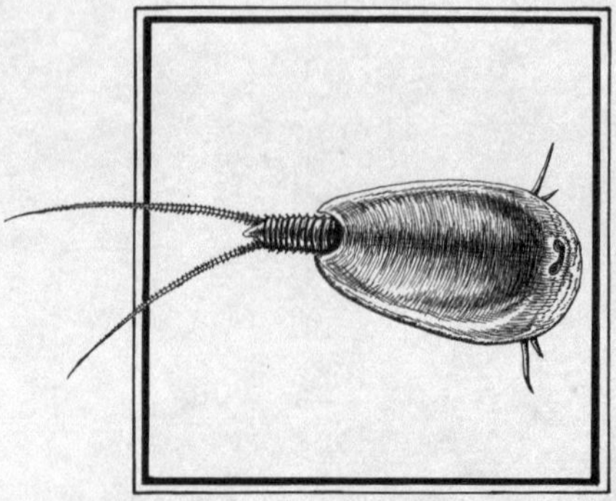

Bill's bountiful billabong

It was on a breezy, showery September day, a couple of years back, that I found myself driving the Midland Highway with my friend Dave, whose mate, Bill, runs a mixed farm not far from Symmons Plains. Dave is a geologist by profession, but takes a keen interest in all sorts of outdoor activities, from mountain-biking to bushwalking and, it now seems, the pursuit of strange crustaceans. Some days earlier, I had mentioned to Dave that I had been hunting, without success, for tadpole shrimps. These odd little crustaceans, which are about the size of a ten-cent piece (remember those?) and look a bit like miniature horseshoe crabs but with forked tails, are said to be not uncommon across Australia. They can be encountered wherever water from winter rains lingers just long enough for them to complete their short lifecycles, but not long enough for predatory fish to establish. The Australian outback is prime country for tadpole shrimps, but they also occur in more settled areas. My attempts to track them down in the Hobart area had proven fruitless though, and I had resigned myself to yet another year of dipping out, to borrow a birding term. But not long afterwards, Dave phoned to report that 'Bill's billabong' had, again, been topped up by floodwaters spilling over from the South Esk River, after an unusually wet winter. The floodplain of the South Esk is dotted with former meanders. Their outlines are clearly visible in aerial photos, although most are now infilled; however, a few still function as temporary wetlands when the river overflows its banks. Bill's billabong, Dave suggested, would be the perfect place to look for tadpole shrimps.

It seemed like a long shot, but I figured that, if we failed to find the shrimps, at least we would have a grand day out, and we were unlikely to return completely empty-handed. I loaded a long-handled dipnet, a kitchen sieve, a pair of waders, and assorted trays and sample bottles into the back of the museum ute, and off we went. On calling in at the farmhouse, it transpired that Bill and Jill were eager to join us in our quest, needing no enticing with the Eccles cakes that I had bought for us all at the Ross bakery, on our way from Hobart.

Together we bumped our way through the bottom paddock and up a slight rise overlooking the billabong, where we parked the ute. I must admit that I felt a certain amount of pressure to deliver on Dave's hunch. With the three of them watching, it was down to me to don the waders first, and step cautiously into the water. Each squelching step sent plump brown tadpoles (real frog tadpoles, rather than the shrimp variety) wriggling off in all directions. I still can't believe what happened next. As I tipped out the contents of my very first dip into a white tray on the open tailgate of the ute, there, before us, was our quarry: a small but perfectly formed tadpole shrimp! Whoops of delight followed as, in turn, we peered into the tray for a closer look. My initial anguish completely evaporated as two more shrimps were captured in further dips. The tray was also teeming with other pond creatures with otherworldly names: seed shrimps, water fleas and cyclopoid copepods; plus an assortment of water-beetles and water-bugs of various shapes and sizes. Initially, I thought of these as 'also-rans' alongside the main quarry. It took the excited chatter of my fellow dip-netters for me to appreciate that these, too, were worth more than a passing glance. The oval seed shrimps came in two sizes. At first it seemed as though they were rotating aimlessly in midwater, until it became clear that they had all magically ended up at the more brightly lit end of the tray. Meanwhile the water fleas, which always remind me of microscopic chooks, calmly bobbed along to their own internal rhythm; and the one-eyed, torpedo-shaped copepods zigzagged maniacally from one side of the tray to the other.

How did they all find themselves in Bill's billabong? The obvious explanation—that they arrived with the floodwater from the river—doesn't stack up because none of these creatures can survive for long in flowing water; nor in the presence of hungry fish. The truth is rather stranger: the beetles and bugs flew in, visually homing in on the shimmering surface of the billabong from a great height. The tadpole shrimps, water fleas and seed shrimps would have hatched from cyst-like eggs. These may have been buried for years in the sediment at the bottom of the dried-up billabong, or they may have blown in with dust on the wind, or they may have been carried from other ponds on the legs of waterbirds. The copepods would have emerged from dormant juveniles rather than cyst-like eggs, but in other respects they follow the same strategy. Then, setting aside how the different creatures got here, all would have to feed up. Those that weren't adults on arrival would shed their skins multiple times, grow, and, eventually, breed; all within a matter of weeks. The warming waters of spring are crucial for making this possible because warmth speeds up their life cycle; but warmth is a double-edged sword: it also hastens evaporation. While any adult beetles and bugs would be able to fly off in search of new ponds, the fate of the crustaceans would be to perish, as the billabong dried out.

Our fate that day was more positive. We returned to Bill and Jill's farmhouse for a celebratory cuppa—with Eccles cakes, of course. It set Dave and me up well for the long journey back to Hobart.

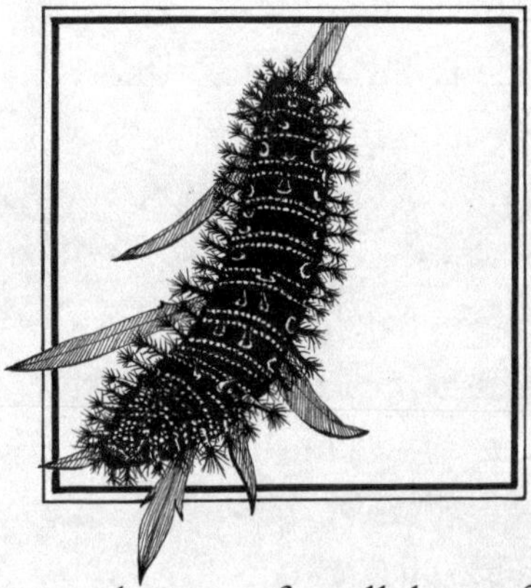

Hairy cats

When contemplating what form this book would take, I decided that it ought to feature a butterfly for every month, since they have been my favourites since childhood—my 'gateway' into the whole wide world of invertebrates. For the summer months I've been spoilt for choice—after all, butterflies are the very essence of summer, even in Tasmania. It's a lot less easy to pick ones that can carry the flag for the cooler months, when most butterflies are at other stages in their life cycles, as eggs, larvae (caterpillars) or pupae (chrysalides). Whilst we might tend to view the adult, butterfly stage as the culmination of all those earlier stages and the only one worth talking about, that would be rash. A butterfly or moth may spend mere days as an adult—its main purpose, then, is to disperse, find a mate and reproduce before dying. In contrast to this, it may spend months as a larva. This is the only stage during which it grows: from a tiny hatchling, through several moults, until it's ready to pupate and undergo its transmogrification into an ethereal butterfly.

A caterpillar is essentially a chewing and digesting machine, whose one purpose in life is to assimilate enough nutrition to ensure that the resultant butterfly (or moth) can have a stab at reproduction. But it must also avoid being eaten—no mean feat when you lack wings for escape and your tiny mind is set on other pressing matters. Some go for camouflage, hiding in plain sight—the leaf-green caterpillars of a cabbage white butterfly come to mind here. Some secrete themselves away, such as those of skipper butterflies, which feed within silken tents woven among the leaf sheaths of sedges. Many sport hairs and bristles though. These may serve purely as a physical deterrent, or they might contain poisons or irritants. It's hairy ones such as these that give caterpillars their name—the word is derived from an Old French term that broadly translates as 'hairy cat'. In a quirky entwining of etymology and entomology, however, the current French word for a caterpillar is derived from an earlier Latin word meaning 'little dog'. Meanwhile, the more sciency term, 'larva', is the Latin word for a ghost or a mask. This name implies that a larval form of an insect masks its true, adult form.

So, to my butterfly—or, rather, caterpillar—of the month. One bright September morning, I headed out for a walk along a coastal path that follows the low cliffs between Lauderdale and Seven Mile Beach, a few kilometres to the east of Hobart. Although the sun shone, there was a wind blowing in from the south, across Frederick Henry Bay, and the air was decidedly chilly. I stopped to watch some gannets fishing in the bay, admiring their seamless and near-instantaneous transition from graceful bird of the air to purposeful, torpedo-like underwater fish-hunter. It is strange to contemplate how these birds may well have flown all the way from breeding colonies around New Zealand to fish our inshore waters (while our own local birds winter further west). The path along this stretch of coast is heavily trampled in places, and the path-side vegetation regularly dies off due to drought, while also being exposed to sea spray. Buckshorn plantain, its rosettes of hairy, multi-lobed leaves hugging the ground, is one of few plant species that can survive in this harsh environment, though even its growth is here stunted by the elements.

I sat myself down on a patch of bare ground and immediately sensed the warmth that it radiated, which contrasted with the chill air just millimetres higher. Casting my gaze landwards, I spotted a bristly black caterpillar about two centimetres long, sitting motionless on the bare earth. I cautiously prodded it with a twig, and it curled into a semicircle, its multi-branched bristles standing proud along its back while its soft underbelly was safely tucked away. Then I spotted another one, and another. I got my eye in and, before long, I had counted fifty or sixty in the space of just a few square metres. It seemed that the caterpillars were all basking on the bare ground to warm up between meals, a feat made easier by their dark colouration. I recognised them as larvae of meadow argus butterflies, for whom plantains are a significant larval foodplant. Meadow argus are found across much of Australia, so they clearly don't mind a bit of heat; but here, in southern Tasmania, both the butterflies and their larvae must seek out warmth. Bare ground is so much better at absorbing the sun's heat than dense vegetation, which would seem to be why there were so many larvae on this otherwise rather unassuming section of clifftop. Lusher grassland just back from the cliff edge, where the plantain appeared to be thriving, seemed much more tempting a feeding place, but it was probably too cold for caterpillars.

I tend to think of the meadow argus as a cheery and faithful denizen of our parks, gardens and nature strips, because these are the places where we go to great lengths to keep the grassy sward short and open, inadvertently enhancing both warmth and plantains. My guess is, however, that these butterflies might once have been a much rarer insect in Tasmania, perhaps confined to an ancestral homeland along clifftops and on the fringes of saltmarshes. They're a beautiful embodiment of the principle that everything we do to the land can give rise to ecological winners, as well as losers. To put it another way, nature abhors a vacuum. The void left by the departure of one set of species (whether at the hand of humanity or otherwise) provides an opportunity for others. The original custodians of Lutruwita knew this well, since it was the basis for their skilled, cyclical, fire-based management of the land, which would have optimised the availability of game animals as well as bush tucker such as yam daisies, berries and seeds, while incidentally providing for all the other species that thrived under their care at different stages in the cycle. Taken to extremes (as is our more recent tendency as a species), the demands our management imposes on the planet could leave us with a world populated by little more than humans, crops, rats and cockroaches; but, in moderation, nature tends to make the most of the opportunities that our activities offer, and, just sometimes, the results can delight us.

The bristles on the meadow argus caterpillar were deterrent enough that I merely prodded one, rather than picking it up. But there's a family of moths whose larval bristles take deterrence to the next level. Their local members are known as cup moths because the emerging moth exits its globular pupal case via a circular opening, leaving a cup-like empty case. From the moth's point of view, and more importantly, that of its potential predators, the grey-brown pupal case bears a remarkable resemblance to the seed capsule of a eucalypt. Indeed, it is usually attached to a eucalypt trunk or twig (the caterpillars having fed on the eucalypt leaves). This enhances the camouflage and gains the pupa protection from predation. But that's the least striking thing about cup moths. Actually, this is not strictly true—the adult moths are indeed less striking, being nondescript, drab-brown affairs; but their larvae assail the senses far more. Rather than a general covering of bristles, the cup moth caterpillar's are arranged in tufts, which are perched on retractable tubercles. At rest, or when feeding, the caterpillar keeps these neatly tucked in, but, if it feels threatened, out they pop, revealing formidable armour to any would-be attacker. This effect is further enhanced by dramatic colouration. Each larval instar (growth stage) sports its own colour scheme, but the basics comprise contrasting regions of pink, yellow, blue, green and black. Birds would recognise such vivid colouring as a warning to back off. Should a predator persist in its attack, it would soon find out that it was no idle warning, since the bristles are poison-tipped and can cause serious pain, even for humans.

The typical cup moth species in Tasmania used to be the pale cup moth, a relatively innocuous member of the eucalypt leaf-feeding guild of insects; but in recent years a second species, the painted cup moth, has become much more prominent. I first became aware of the scale of their impact a few years ago, when I was walking home from the local beach, one day in late September. The ridge behind Taroona is swathed in eucalypt woodland, with white peppermint—a species endemic to south-eastern Tasmania—patchily dominating the canopy. Typically, its blue-green foliage is hard to discern among the other eucalypt species, at a distance, but on this day the ridge was blotched in brown and the usually sharp skyline was a diffuse smudge due to bright light filtering through the sparse canopy along the ridgeline. Intrigued, I resolved to take a closer look, so, later that day, I walked the steep Truganini Track, up towards Mount Nelson. From the wet, ferny gully, with its tall blue gums and brown stringybarks, I emerged into the open woodland of the upper slopes, where white peppermint dominates. It's one of my favourite trees (though I have quite a few favourites), especially in its gnarly, thick-set, old-growth form (an uncommon sight these days). The trees here aren't particularly large or old, but they still imbue the woodland with a sense of timelessness that always draws me in.

Today was different, though. Stopping to catch my breath, I became aware of a gentle pitter-pattering, as though of light rain; yet no rain had been forecast and the sky was clear. Gazing at a patch of leaf litter I soon spotted the source of the sound: frass—that's the technical term for insect poo. Cup moth caterpillar poo was raining down on the woodland floor. Here and there a caterpillar had lost its grip and tumbled to the ground, too; and the longer I stared at the ground, the more of them I managed to pick out, as they attempted a long, slow haul back up the tree trunks to the canopy. A single caterpillar may only consume a few peppermint leaves in its lifetime, but here I was seeing—and hearing—the impacts of thousands upon thousands of them doing so simultaneously. The vague browning that I had observed from Taroona was now discernible as a host of partially-eaten and dying leaves on the peppermints overhead. Those poison-tipped bristles certainly had a lot to answer for.

It's all too easy to conclude that such a significant level of defoliation signals disaster for the affected trees. Newspaper articles duly reported the event as a plague that would lead to the death of thousands of hectares of forest. But, to misquote a Mark Twain misquote, reports of the woodland's death have proven to be greatly exaggerated. There have been several more plague years since that first one, but in non-plague years the trees seem to recover. Perhaps they are assisted by a transient boost to soil fertility, provided by the rain of frass. Yet something has clearly changed in the environment to induce these events. Is a warming climate allowing more of the cup moths' eggs and young larvae to survive the winter? Or is a drier climate reducing the peppermints' rate of growth and their ability to defend themselves against these herbivores? Could it be a delayed response to altered fire regimes since European colonisation? Without knowing the cause, it is difficult to predict what the future might hold for our cherished white peppermint woodlands and their uncomfortable relationship with the cup moths. I don't have any answers yet, and I doubt that anybody else does; but I am left with a feeling of unease, that these outbreaks are a forewarning of worse changes to come in the decades ahead.

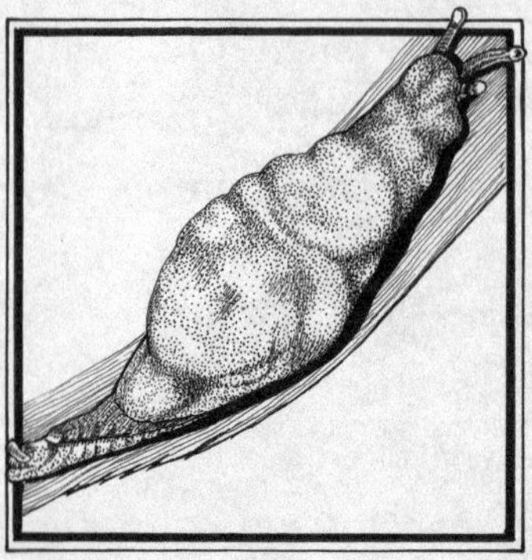

Into the rainforest

When Chris' mum, Norma, died, a few years back, part of her legacy was a second-hand campervan that we bought with Chris's share of the estate. Chris named it Norma in her honour. One recent September, I drove Norma on a solo trip to the Tarkine region, in the far northwest of Tasmania. Ostensibly it was to set up some insect traps in the rainforest. Rainforest remains a relatively under-sampled habitat, and it has lots of potential for unusual and endemic species. My other motivation was to enjoy a spell of solitude and nature therapy.

In contemporary lore, the Tarkine—Takayna—conjures images of wild rainforests, rivers, and coasts. It is billed as 'untouched wilderness', home to the largest remaining tract of intact temperate rainforest on the planet. The reality of the Tarkine is a bit more nuanced. Parts of it have been heavily exploited since European colonisation, at first by mineral prospectors and then through mining and forestry. It is a wilderness, today, in the sense that almost nobody lives there; not even its First Peoples, who once prospered along the coastal margins. Furthermore, nature has a knack of papering-over our infringements, given enough time. These days the Tarkine is, without a doubt, a wonderful part of the island. It comprises rugged, deeply dissected country; and it is remote, unpopulated, and cloaked in a mixture of rainforest, wet eucalypt forest, wet heathland and buttongrass moorland.

I grew up imagining 'rainforest' as a lush and exuberant place of great diversity, where an explosion of life thrived under a tropical sun, sustained by daily torrential downpours. That was, indeed, the first concept of a rainforest, or *Regenwald*, in the original, nineteenth-century German name. During my youthful travels in Costa Rica and Peru, I experienced daily deluges of rain and incredibly abundant biodiversity, which confirmed those initial impressions. Tiny Costa

Rica—significantly smaller than Tasmania—boasted twice the number of bird species of the whole of the US and Canada; while the Tambopata Reserve, in the Peruvian Amazon (where I had worked as a nature guide) sported three times more butterfly species than the entire continent of Australia, despite its total land area being not much bigger than greater Hobart. Over time, however, ecologists have expanded the concept of rainforest. It now includes various temperate forests, within which an abundance of rain remains the overriding determinant of what can or can't live or grow there. But in Australia, on the planet's driest continent, where fire plays a fundamental role in the ecology of most ecosystems, rainforest has come to mean something slightly different. It refers to virtually any forest comprising plant species that lack clear adaptations to fire. Elsewhere in the world, the term 'dry rainforest' would be an oxymoron; not so in Australia, where it applies to dry-climate forest types growing in fire-protected environments. The significant point is that Australian—and Tasmanian—rainforests are defined more by an absence of bushfires than by the presence of abundant rain. This matters in a place like the Tarkine because, here, the rainforest forms a mosaic with wet eucalypt forest, and eucalypts are quintessentially fire-dependent.

The Tarkine may get a lot of rain, but it's not enough to exclude fire. How this plays out is a product of the lie of the land and occasional lightning strikes, and, it is likely, past cultural burning practices. Aboriginal burning certainly maintained the open heathlands and moorlands towards the coast, but any role it may have had in maintaining inland wet eucalypt forests remains unclear. As a rule, rainforest in the Tarkine occupies the more sheltered, inland parts of the landscape, while eucalypts dominate most of the ridgelines, especially on northerly aspects where the forest dries out quite readily. Eucalypts are highly flammable, and although the adult trees are often killed in bushfires, their seedlings regenerate profusely in the aftermath of fire, generally from seed shed by long-lasting and fire-resistant capsules remaining on any branches or twigs that weren't consumed by the flames. In the Tarkine, it is usual for regrowing wet eucalypt forest to eventually host an understorey of slow-growing rainforest tree species such as myrtle and sassafras—

we call these 'mixed forests'. Rainforest trees such as these will have progressively colonised the already-established eucalypt forest over the decades or centuries since the last fire, arriving from windblown or bird-dropped seed. Rainforest trees grow slowly compared to eucalypts; but, over the years, as the eucalypts die of old age, they could potentially reclaim the ridgelines entirely, if it were not for the periodic intervention of fire, for rainforest trees have little natural resistance to this. The corollary also applies: little by little, the eucalypts could lay claim to the valleys and lower slopes through gradually encroaching on the rainforest edge with each successive fire. Their chances of doing so are slim, though, both because of the reluctance of rainforest to burn (though a warming climate may change this) and because a eucalypt's seeds don't travel far from the parent tree, since they are essentially gravity-dispersed. Their seedlings, too, are light-demanding. In shade they fail to thrive, often succumbing to pathogenic fungi or to browsing animals. Thus, these two radically different forest formations appear to maintain a dynamic equilibrium—an uneasy truce—over space and time: individual battles may be won by each side, but the war is never over because each side is only really competitive within its own stronghold, its own ecological envelope, as dictated by topography, aspect and the vagaries of fire weather over the course of a century or more.

Whilst eucalypt forests very visibly wax and wane over the years (and sometimes over mere hours as a bushfire sweeps through), our temperate rainforests, when unadulterated by eucalypts, are the epitome of timelessness. With soft, springy ground accommodating one's footsteps without audible complaint, it's easy to walk through a rainforest in noble silence. Some call such an experience 'forest bathing'. It offers exhilaration without undue energy expenditure. Amble gently, mindfully, and let timelessness pervade your senses. You will be amongst trees of all ages and sizes, because no fire has ever reset the forest back to time-zero. Instead, individual trees typically age gracefully—sometimes majestically, sometimes grotesquely—and eventually collapse or topple, at which point there will already be plenty of shade-tolerant seedlings and saplings of the same tree species waiting patiently in the wings, ready to reach for the light and fill the gap.

Sometimes honey fungus or myrtle wilt will take out a whole clump of trees in one hit; but that's generally about as big a disturbance event as you will get in the rainforest.

Our rainforests are timeless in another sense, too: they don't seem to change much from season to season. It took me a while to realise this, perhaps because it's harder to notice an absence of seasonal signatures than to observe their presence. The Tarkine rainforests occupy the part of Tasmania with the most equable climate, so temperature isn't a reliable clue to the season. The mosses and ferns that tend to dominate the understorey don't have flowers, so there is no obvious peak flowering season at ground level, and no riot of spring colour. While sassafras and leatherwood trees have discrete flowering seasons (early spring and midsummer, respectively), their flowers are usually well above head height and often only noticed as browning, soggy petals on the ground, once detached by wind and rain. As for myrtles, they develop a fiery-orange flush of new growth in spring, but their flowers, which come a bit later in the year, are far from showy. Autumn can bring a riot of colour if you look closely enough, thanks to the many fungi that send up fruiting bodies at that time of year. But it seems to me that insects may be the clearest marker of the seasons: more mosquitoes than gnats? It's probably summer. More gnats than mosquitoes? It could well be winter.

Back to my campervan trip then. The eight-hour drive (with tea-brewing, leg-stretching and fossicking stops, of course) is about as far as one can travel by road from Hobart. For the most part the scenery is pastoral: sheep country through the Midlands shifting into cattle country as you press on into the northwest. Irrigation schemes are progressively transforming low-productivity grazing land into intensively managed expanses of bright yellow canola and rich green ryegrass. In the northwest, much of the farming land was blessed with natural fertility in the form of deep, red, basalt soils or the silt and peat of long-drained swamps. Where herds of dairy cows now graze beneath overhead irrigation lines, European pioneers once cleared dense forests of blackwood and paperbark. The words 'clean' and 'green' are often used together, as though one complements the other; but despite the phrase having become part of Tasmania's brand identity, I can't help but feel that the adjectives have been irrevocably decoupled in these intensively managed parts of the landscape. Still, although most of the forest has long been cleared from the agricultural heartlands, the distant hills and mountains, which rise abruptly from the plains, remain reassuringly wooded and provide a stunning backdrop to much of the drive.

I arrived at dusk. I made a cuppa on the gas stove and allowed myself to unwind into the sights and sounds of the rainforest and a babbling creek, just a few metres from where I had parked. A lone black currawong began its discordant yet slightly mournful evening call from the top of a dead myrtle. I placed its tone somewhere between a trumpeting and a honking, and imagined that it signified a sort of winding-down routine, bookending the day in preparation for roosting. In about ten hours' time, it would be greeting the early morning with a similar vocalisation, perhaps with more of a hint of optimism for the day ahead, or triumph at having conquered another night.

At some point, which I only discerned a minute or two after its last call, the currawong decided enough was enough. Then the only sound was the creek, swollen by recent rains and continuing its babbling unabated. As the night closed in around me, my mind played unsettling tricks, turning the murmuring of water over well-worn pebbles into snatches of speech, laughter, singing and crying. How easy it would be to attribute agency and animacy to the mysterious source of these sounds. Maybe I should? The spirits of the ancestors, perhaps; or the creek itself, striking up conversation with the trees, or with me. It would have plenty of tales to tell, I'm sure. All things seem possible when you're alone in a dark, timeless, immense rainforest.

I clasped a bowl of steaming soup as the stars appeared, dimly and in ones and twos at first, as if they were not quite sure that it was their time to shine. Continuing to peer skywards I found that, the longer I did so, the more my eyes adjusted to the darkness and the more stars seemed to pop into existence. Before long, the heavens were aglow with the combined light from countless celestial bodies: distant suns, galaxies, and nebulae, perfectly framed by the silhouetted canopy of the myrtles. I felt both very, very insignificant, and rather special, as I contemplated the incomprehensibly vast spans of time and space over which these photon streams had travelled, anticipating no endpoint to their journeys, yet all converging on my retinas at this one moment. The works or manifestations of more ancestral spirits? I wasn't sure whether I found the thought comforting or unsettling. On the one hand, I liked the notion that I might not be alone out here; but on the other, the ancestors would have every right to feel aggrieved with the likes of me, given the brutality with which European colonists—my people—had usurped the land and culture of Lutruwita's First Peoples. Would they make their anger known to me during the night, and if so in what way? Thumping on the roof of the van? Scratching their fingernails down the windows? Knocking over pots and pans left outside? Cackling and moaning from the edge of the forest? I would hate to end up like one of those hapless kids in *The Blair Witch Project*.

I was in a jittery frame of mind, then, as I donned my puffer jacket and a headtorch and ventured out for a post-dinner wander down a path that winds its way alongside the creek, through the rainforest. Shadows jumped out at me as the beam from my torch intercepted the towering trunks of ancient treeferns, and it occurred to me that this would be a good explanation for their alternative name of 'manferns'. More spirits out to get me. Yet for all my self-spooking, the forest at night can be a captivating place, as various animals emerge to feed and meet under the cover of darkness. And, of course, this is why I ventured out in the first place. I took control of my breathing and felt my mind and body relax as I conquered my fears and entered happy bug-seeker mode.

My first encounter was with a snail bearing a shiny, slightly translucent black shell and two pairs of tentacles at the front of its orange-tinged body. Milligan's carnivorous snail is a predator that relentlessly hunts down other snails, albeit at a snail's pace. Capturing one, it envelops the victim in its expandable 'foot' before tucking into the flesh with a rasping set of teeth known, in the trade, as a radula. Endemic to north-western Tasmania, its empty shell is commonly encountered along rainforest paths—perhaps after the snail has fallen victim, itself, to predatory ground beetles. I could see by the intermittent trail of slime glinting in my torchlight that this one had spent the day in leaf litter at the foot of a treefern. I wondered whether it was a creature of habit, like some other snails: always homing in on the same patch of leaf litter at the end of its nocturnal hunting forays.

Speaking of which, another trail of slime now gleamed in my torchlight. This one headed straight up the trunk of a sassafras tree. The slime-maker was at head height: a delightfully corpulent bicoloured native slug, endemic to western Tasmania. Despite its name (which I admit to having just contrived, based on its scientific name), these grainy-textured slugs tend towards either grey-brown or a diaphanous peppermint green, depending on what sort of forest they inhabit. Rainforest ones are usually green, like the one in front of me. Bicoloured native slugs are one of rather few land-dwelling animals that undertake what's known as 'diel migration', a term more usually applied to plankton that rises towards the ocean surface by night and descends to the depths by day. In the slugs' case, they slink determinedly up tree trunks when humidity rises at dusk, probably to graze on lichens or algae growing on the bark; and slide back down again before dawn, to wait out the drier daylight hours hunched-up in the leaf litter. They are not totally set in their ways, since in rainy weather they also venture out in daytime.

Next was a two-streaked darkling beetle—yes, another contrived name (you'd better get used to these), and another rainforest creature endemic to western Tasmania. These beetles hide away in rotting logs by day. By night they seem to spend a lot of time just hanging out. They make no attempt to flee when caught in the spotlight. They're a handsome beetle with an elongate body and with sharply delineated wing cases that look as if they are made of beaten copper and then engraved or etched. From this one's rear end two worm-like, writhing protuberances emerged, which the animal extended further when I gave it a gentle prod. Darkling beetles use these 'eversible abdominal glands' as a means of chemical defence, wafting noxious secretions (benzoquinones) at would-be predators such as quolls or antechinus, and overly familiar entomologists. These defences save the beetles from the bother of having to flee from every possible enemy encounter, leaving them more time and energy to graze on algae and lichens, or to find mates.

I left the path and made my way down a steep, earthy bank to the margins of the creek. Silty deposits either side of the main creek bed were pockmarked with hollows and with the clay chimneys of burrowing crayfish. Tasmania (and southeast Australia) is one of two global hotspots for terrestrial crayfish diversity (the other is the south-eastern United States). Calling them terrestrial is a bit of an exaggeration, since nearly all species are aquatic (like other crayfish) and have gill chambers to extract oxygen from water. The difference is that they mostly live in places where they have to excavate their own waterholes by burrowing down to the water table, so that the cavern they have made—sometimes more than a metre from the surface—part-fills with water. They then mould the excavated clay into 'chimneys'—some tall and proud, others more like the collapsed creations of novice potters. These extend their burrows aerially, sometimes by up to 30 centimetres. Whether this is just a convenient means of disposal of surplus clay, or whether it helps to prevent flooding of the burrow, is unclear; but, with all the burrowing effort involved, I find myself wondering whether the transition to land-living was worth it. In an evolutionary and ecological sense, it clearly was, but for the individual crayfish condemned to a life of digging drudgery, I remain unconvinced.

The crayfish generally stay hidden within their waterholes by day, but at night they emerge to seek out tasty bits of decomposing vegetation to drag back home. One such animal was out tonight, caught in the act of tearing up a fallen sassafras leaf with its oversized blue-grey claws. Though I don't know one burrowing crayfish species from the next by sight, each of Tasmania's native species—of which there are over thirty (all but one being endemic to the island)—has its own restricted range. Where these ranges overlap, the species tend to segregate by habitat. So, if you know where you are and what habitat you're in, you can sometimes deduce your local burrowing crayfish (or *vice versa*, I suppose).

By the way, the word 'crayfish' has nothing to do with fish, etymologically speaking. It's related to the modern French word for shrimp or prawn (*crevette*), and shares a root in Old French *crevice* (related to German *krebs*, English crab), which went on to become *creveis* in Middle English. Once people in England were no longer required to understand Norman French (or, it would seem, pay much heed to evolutionary biology), *creveis* eventually transmuted into crayfish.

With bleary eyes, I decided that it was time to turn van-wards. I could just discern the distinctive call of a Tasmanian treefrog over the noise of the creek. I headed towards the source of the sound, which is not unlike the conversational quacking of a Pacific black duck, but the vocalist had clearly decided that I wasn't the sort of creature that he wanted to attract, because that was the last I heard of him. I couldn't pinpoint him without his call, which was a shame because the Tasmanian treefrog is a handsome, tricoloured animal: a pale fawn base colour overlaid with rich chocolate-brown and bright green marbling, giving an overall effect not unlike that of lush lichens on rainforest logs. The Tasmanian treefrog, too, is endemic to western Tasmania's rainforests and buttongrass moorlands and is at the very northern end of its range in the Tarkine.

The night was mercifully uneventful, with no malevolent spirits appearing; and the morning brought clear skies and the prospect of some springtime warmth. The blowflies seemed to sense it and were up and buzzing before the sun had crept over the ridgeline, settling on and probing my rucksack, in which I had my insect traps packed and ready to go (the irony was lost on the blowflies, I suspect). A kookaburra gave a half-hearted chuckle from its perch on the dead branch of a myrtle overlooking my camping spot, as though aware that anything throatier would come across as inappropriate in this Tasmanian rainforest setting where, by rights, it really didn't belong at all. A fat-bottomed pademelon skulked among the sedges at the edge of the clearing, pretending to graze but really just biding its time in the vain hope that I might throw it a tastier morsel. The murmuring creek had completely lost its nocturnal air of malevolence. I watched its tea-stained waters for a while as I ate my breakfast. Dollops of foam sailing by told of rapids upstream, but here the creek's flow was merely turbulent and, to my mind, deeply relaxing.

But there was work to be done. I donned my pack and followed the track down a muddy bank into the rainforest proper. While I paused for my eyes to adjust to the dimness of the forest interior, I inhaled deep breaths of humid air and savoured its earthy aroma. There should be many more words for the varied smells of soil and forest, but earthy will do quite well in this case. It conjures up the combined odours of damp leaf litter, rotting wood, moss, spores, humus and mud. The forest was almost completely silent, as it often is, but it was also serene and rather lovely: not quite parklike, but with widely-spaced, gnarly old myrtles. Their trunks were intricately scaly-barked, and many bore massive bracket (shelf) fungi. The forest veterans were interspersed with younger myrtles, sassafras, treeferns and the odd blackwood.

Tasmanian rainforests all comprise much the same limited set of tree species, but they come in three structural types. This one was my favourite, a callidendrous rainforest. The word means 'beautifully treed' and applies to rainforests where the trees are large and well-spaced, and where one can walk amongst them unimpeded—which explains why I had chosen to set my insect traps here. By contrast, implicate rainforest is almost impossible to walk through because of sprawling thickets of horizontal—an aptly-named scrambling, scandent tree or shrub that seems to prefer growing horizontally rather than vertically (although, to make things awkward for the bushwalker, it does a bit of both). You tend to get implicate rainforest where the ground is less conducive to the growth of big trees, such as where there is a lot of subsurface water movement. Then there's thamnic rainforest, which usually comprises dense thickets of small myrtles or other rainforest trees and grows in places where poor soils or a harsh climate (or both) restrict their size.

What this particular callidendrous rainforest lacked in tree species richness, it made up for with an impressive array of ferns and mosses. The forest floor bore a veritable shag-pile carpet of emerald-green mosses and leafy liverworts, and every log and stump was similarly draped. I am no bryologist, but, without trying too hard, I could see a dozen distinct types of moss. I was delighted to find Dawson's superb moss growing here. At up to 30 centimetres high, it is the world's tallest freestanding moss species. I had longed to see it, since hearing of its existence soon after moving to Tasmania. Treeferns, too, were festooned with mosses, as well as epiphytic filmy ferns and tassel ferns (the latter are neither ferns nor mosses, but members of a similarly ancient lineage called clubmosses).

Spring is by no means peak season for fungal fruiting, but there were plenty of fruiting bodies around, in various shades of pink, yellow and orange. These are the above-ground, visible manifestations of massively connected underground networks of fungal filaments, which do much of the decomposition and nutrient cycling in the forest (with a little help from the invertebrates). We have only recently come to understand the extent to which fungal networks act as couriers or go-betweens for the forest's trees, shunting, trading, and exchanging sugars, amino-acids, and other nutrients from one tree's root system to the next for the benefit of all. They don't go as far as eliminating competition among trees, but they do provide an alternative way for trees to interact with each other, possibly enabling a bit more negotiation, more give-and-take, more live-and-let-live. I'd drink to that.

To the pleasant acoustic accompaniment of jingling pink robins and the liquid, warbling purr of Tasmanian thornbills emboldened by a rising sun, I scouted for a spot to set up my first trap out of sight of the path. I soon found what I was looking for: a sheltered glade hemmed in by shrubs and treeferns, where flying insects would be funnelled my way. The trap in question is a tentlike contraption called a Malaise trap—and therein lies a story. Dr Malaise was a Swedish medical doctor practising in the early decades of the twentieth century, with a surname that really should have led him towards some other career—entomology, perhaps. Why a Swedish doctor bore a French surname meaning discomfort or sickness, I haven't a clue. I hope his patients didn't understand French, but being Swedes, they probably did. Then again, our school nurse was a Mrs D'Eath, and she never did us any harm. I have since learnt that D'Eath is an anglicisation of a mediaeval Belgian family name originating from the town of Aeth, but that would have gone right over my head as a schoolboy.

But I digress. Entomology actually was Dr Malaise's passion. He would spend his summer holidays camping in the Alps and collecting insects. The story goes that, one day, he left his tent flaps open while he was out collecting. On his return, he noticed that the flies and other insects that had inadvertently flown into his tent had all congregated around a small hole at the tent's highest point. They were trying to escape where light streamed through. Malaise's eureka moment was to surmise that flying insects could be collected passively in a tent-like trap with a hole at the top that led to a collecting chamber charged with a liquid preservative, such as diluted ethanol. The trap that he subsequently designed (and first used in the jungles of Burma) primarily consisted of a vertical screen of fine, dark netting, invisible to flying insects; this was suspended beneath an integral tent-like roof of bright, white netting, higher at one end than the other. At the higher end, a hole led to the collecting chamber. The whole apparatus was mounted on a set of poles and guyed and pegged out, preferably in an insect flightpath. The design was a roaring success. It could be left sampling for weeks, during which time hundreds or thousands of small flying insects might succumb to the deception and end up dropping into the ethanol. Malaise went on to become the world expert in Burmese sawflies, while his eponymous trap for flying insects became the go-to standard for entomologists the world over.

I eventually found spots for three further Malaise traps. Mission accomplished, I retraced my steps to the van, tuning in to a distant Bassian thrush which had struck up its fluty song—it was not unlike that of a blackbird but rather more subdued and, I thought, soulful. I wondered if a soulful disposition explained why Bassian thrushes like to lurk in the darkest recesses of the rainforest and seem reluctant to burst into song (compared with their cocky suburban cousins). I feel like that sometimes, too. Fortunately, my traps would need servicing every few weeks throughout the summer. This set me up with a perfect excuse to revisit, regularly, the enchanted rainforest on the other side of the island.

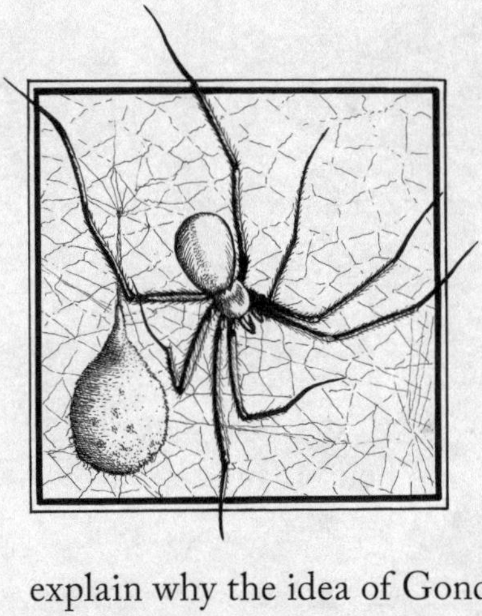

Gondwana origins

As a naturalist or bushwalker, or reader of books such as this one, you won't get far without learning of Tasmania's special Gondwanan heritage. So I thought I should take a little time to familiarise you with that heritage and explain why the idea of Gondwana is relevant to so much that is special in Tasmanian nature. And, because our Gondwanan connections hark back to the supercontinent's cool, southerly latitudes at the time of its disintegration, I figured that I should write about it in a cool month.

Tasmania's chapter in the Gondwanan story starts some two hundred million years ago. Gondwana had existed long before then, but for a couple of hundred million years it had been subsumed into an even larger supercontinent called Pangaea—basically, rather than the planet being home to multiple, separate continents, Pangaea was just one big landmass. But by two hundred million years ago, Gondwana was a supercontinent unto itself, separated from another to the north, called Laurasia, and centred pretty much on the South Pole, although it extended all the way to the tropics. Today's southern continents—Antarctica, Africa, South America, and Australia (including New Guinea)—plus India, Madagascar and New Zealand, were all clumped together within Gondwana.

The supercontinent derives its name from the forested lands of the Gond people in central India, where a sedimentary rock series was first described as Gondwanan some 150 years ago (it actually dates to the Permian-Triassic period, before the breakup of Gondwana). A few years after that, but before theories of continental drift had been formulated, a Swiss geologist, Eduard Suess, hypothesised the former existence of a supercontinent, which he dubbed Gondwanaland, based on shared fossils of Permian-Triassic seed-ferns in those same sediments in India, as well as in Africa and South America. Suess had the mechanism wrong—he postulated land bridges that had since sunk beneath the ocean—but he was onto something, and the name stuck. Those seed-fern fossils, by the way, are now also known from Antarctica and Australia, including Tasmania.

Gondwana was far from homogeneous because it encompassed such a vast range of latitudes, but the temperature gradient from the tropics to the poles was much less steep than today. The polar portion was largely ice-free, and there were fewer biogeographic barriers within the supercontinent than between its modern-day fragments. It's hard to imagine, say, a spider lineage evolving in South America and then somehow colonising Australia across thousands of kilometres of inhospitable ocean (or *vice versa*), whereas it's easier to imagine a spider lineage coming to occupy the parts of Gondwana that eventually *became*

Australia and South America.

About 180 million years ago, Gondwana began to fragment. First to be cast off was Africa, along with Madagascar, followed some sixty million years later by India. That left South America, Australia and New Zealand (or rather, Zealandia) still attached to Antarctica. Zealandia separated about eighty-four million years ago and may have largely sunk beneath the waves before eventually re-emerging as New Zealand (and several other islands). Australia had made half-hearted attempts to separate at the same time as Africa, but it didn't finally part company with Antarctica until about thirty-three million years ago. Signs of those earlier attempts abound in Tasmania, in the form of our wonderful dolerite landforms (think the Organ Pipes on Kunanyi, or the high cliffs of capes Pillar, Hauy and Raoul, on the Tasman Peninsula). Dolerite is the igneous rock derived from upwellings of lava as the crust began to split apart during the Jurassic Period, some 180 million years ago. Tellingly, the only other parts of the world with similar dolerite landforms are Antarctica and South Africa, where it occurs for similar reasons.

South America finally broke free about twenty-three million years ago, with the opening of the Drake Passage. When it did so, the Antarctic Circumpolar Current formed, for the first time imposing an effective barrier to the transfer of heat from the warmer oceans and continents to its north to an increasingly frigid Antarctica, in the south. Ever since, icy Antarctica has more or less stayed put and nearly all of its Gondwanan species have long since perished. In contrast, the cast-off continents have continued to head northwards, further widening the Southern Ocean and ensuring that only their southernmost parts remain within temperate climatic zones.

What does this have to do with nature in today's Tasmania? Quite a lot, really: while all of Australia was once part of Gondwana, Tasmania's present, southerly geographical position means that its climate has remained more similar to the conditions that would have prevailed across Gondwana at the time when Australia-Tasmania split off and began its northward drift. For that reason, Tasmania shares more biological similarities with temperate southern South America and New Zealand than northern Australia shares with tropical South

America or Africa. Botanical examples abound: myrtle beeches, leatherwoods, waratahs and various conifers, for example. My interests are more with the small animals, so here I might mention cave spiders, nannochoristid scorpionflies and moss bugs, to name but a few. What these plants and animals have in common is a liking for the cooler, wetter parts of Tasmania. Indeed, they prefer similar environments in southern South America and, in some instances, New Zealand, New Caledonia, the eastern slopes of the Great Dividing Range of Australia, and the highlands of New Guinea. Generally, they love rain, and they don't like fire. How very un-Australian, you might say—yet these traits would have served them well across much of the continent in past times, and across much of Gondwana before that. It's just that, over the past thirty million years, Australia's climate has become steadily hotter and drier as it drifted north. Some Gondwanan lineages have evolved to cope with these conditions—eucalypts, banksias and she-oaks are Gondwanan (in fact, the earliest-known fossil eucalypt is from South America); but invasions from the tropical reaches of Eurasia have filled in the gaps where other Gondwanan species became extinct in Australia.

Gondwana fires up an austral imagination. I revel in the thought that so much of our nature traces its ancestry back to this ancient southern supercontinent. It's worth remembering, however, that we in the south don't have a monopoly on deep history. Laurasia gave rise to Eurasia and North America, and at least as many plants and animals across the globe have Laurasian origins. And, of course, many can trace their ancestry further back in time to Pangaea. So, while our Gondwanan species are special, their pedigrees are, on average, no more ancient than those of the non-Gondwanan species that can be found in Tasmania today.

Notwithstanding what I have just written, while pretty much everywhere can make claims about how special their natural heritage is, I think Tasmania's tangible links with its Gondwanan past—biological, climatological and geological—do make it *especially* distinct. They certainly contribute to the sense of deep time one feels when immersed in the quiet, dark rainforest. It may be cold, and raining, but as you stumble over damp, mossy dolerite boulders, alongside a gushing, fern-lined creek, it's not hard to feel as though time has stood still for millions of years. Gondwana is gone, but, in Tasmania, its spirit still hangs in the air, feeds the rivers, sculpts the land, and nurtures nature.

©K.DAVIS 2022

October

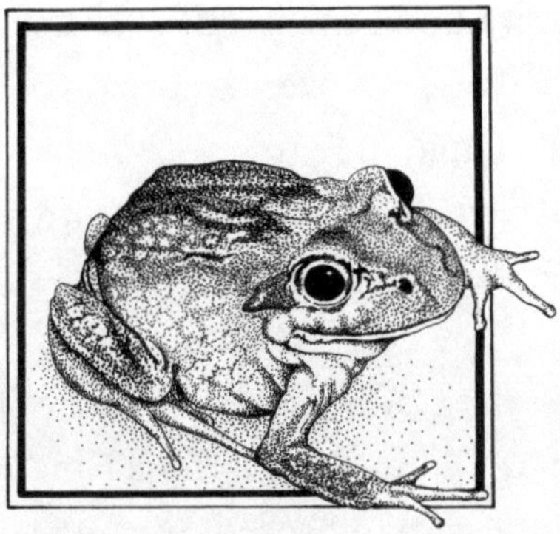

Spring springs forth

Spring really gathers momentum in October. It's the month when you suddenly realise that the sun can burn, and the daily ritual of popping vitamin D capsules must now be swapped for the slip, slap, slopping of sunscreen. It's the month when a few days without rain spells trouble for vegetable seedlings; and when, one day, the fruit trees are bare; and, the next, they're adorned with blossom and emerald-green leaves. Most of the summer-visiting birds arrive this month, if they haven't done so already; and insect life stirs all around, at least on warm days. In farm dams and ditches, pobblebonk frogs twang their far-carrying mating calls, like a discordant orchestra of banjos. The longer days lift the mood, and there's a tangible air of anticipation for the encounters with nature that lie ahead. On walks, I feel increasingly compelled to carry my camera and insect net.

Our Taroona garden bursts into life. Responding to the longer days, bolting silverbeet and spinach plants launch flower shoots towards the source of newfound light and warmth. The flowers of broad beans and snow peas, which began to appear on the plants in late winter, are now attended by a constant stream of bumblebee and honeybee pollinators, and begin to set seed, rather than withering away unfertilised. From the middle of the month, the chance of overnight frost rapidly tapers away and it's time to get tomato plants into the ground (but not before Hobart Show Day—obviously, for frost does appear still, when you least expect it). Before long they can be joined by other summer staples: bush beans, climbing beans and zucchini. Fruit trees burst into flower in waves of white, cream and pink, satiating the nectar and pollen needs of bees, flies, and honeyeaters. Weeds also run amok as soon as you turn your back, and the lawn demands frequent trimming.

October usually brings the first substantive northerlies to Tasmania, as high-pressure systems build over the Australian interior and drift south over the Bight. There they get caught up in the westerly airflows, which have a free rein for thousands of kilometres over the oceans at our latitudes, and are carried over Tasmania, dragging warm, dry air southwards along the way. The air behaves and smells differently during a northerly. On the north coast it can bring chilly conditions and hill fog, having picked up moisture as it crossed Bass Strait; but as it traverses the state it rises, first, over the Central Plateau, and then descends, becoming what's known as a föhn wind (after a similar phenomenon occurring north of the Alps). Its wind speed, warmth and evaporative potential all rise as it punches southwards, vaporising aromatic oils from the gumtrees and sclerophyllous shrubs it passes along the way. By the time it reaches our southern shores it takes on a discernible and characteristic fragrance, and the dry branches overhead rustle in its presence. October northerlies don't usually persist for more than a day or so though. As a high-pressure system slips away to the east, it's liable to flip us suddenly into the power of a southerly airstream, pushing in vigorously from the cold Southern Ocean. The scent of eucalyptus oils is swapped for an altogether saltier, more maritime aroma, redolent of kelp, fish, and plankton. In the ensuing days, the winds will usually settle into a typical westerly pattern, which brings weather characterised by sunshine and showers (more of the former in the east, and more of the latter in the west). The north-south running estuary of the Derwent often seems to sit at that zone of transition: Taroona gets showers while, just across the estuary, South Arm basks in glorious spring sunshine.

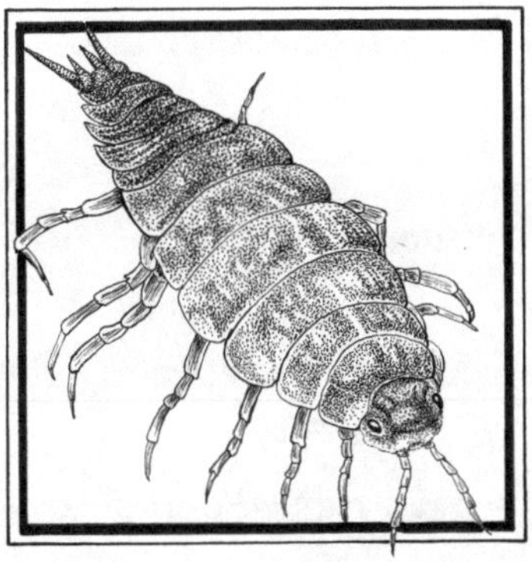

Brine shrimps and salt-lake slaters

In the mainlanders' popular imagination, Tasmania is a thoroughly cold and wet island. Perhaps they (perhaps *you*) haven't heard of the saline lagoons of Tasmania's Midlands? In this region, thousands of years of low rainfall and desiccating winds have caused a build-up of salt close to the soil's surface, much as has occurred in many parts of inland Australia. Salt pans and lagoons have formed where rains flush the salt into low-lying parts of the landscape, aided during dry periods by dust storms. The origins of this wide, open landscape go back to glacial times, when the Midlands' climate was harsher still, experiencing cold, dry winds that carried dust and salt from far afield, infilling earlier creeks and lakes and sculpting lunettes—crescent-shaped sand dunes—downwind. The open landscape proved attractive to Lutruwita's First Peoples, who managed the vegetation with fire to keep it productive of game, yam daisies, and other species useful to them.

Despite dryness, nearly all the Midlands is now farmed, and many of its periglacial features have been turned into featureless non-features. The rolling native grasslands are mostly long gone, and irrigation schemes are increasingly turning rough pasture into bright green paddocks of ryegrass, or cropland. Amid this landscape of intensive human use, there remains the odd patch that somehow escaped. Cemeteries and golf courses often retain echoes of their native grassland past, in both their rare flora and invertebrate fauna. One patch, at Tunbridge, comprises a few hectares of native grassland, known as Township Common. In spring it is a riot of colour as various native daisies and pea-flowers bloom, taking advantage of soil moisture before summer dryness turns most herbaceous vegetation to a crisp. To the side of the Common sits one of the last inland saline lagoons in Tasmania. For a while, it looked as though Township Lagoon (as it is known) might suffer the same fate as others: becoming infilled, drained and turned over to cultivation; but it was in municipal hands, so it suffered a different, though still inglorious, ignominy. In the days before councils applied stringent rules around landfill and rubbish disposal, the Common and Lagoon served as the local dump.

I have taken to stopping in at Township Common and Lagoon whenever I'm travelling the Midland Highway and have an hour or so to spare. A walk or jog around the lagoon is a pleasant alternative to pavement-pounding in Campbell Town or Ross; and, while it may mean that I miss out on a coffee and pastry, I get to see red-capped dotterels and shelducks, and sometimes even a double-banded plover, visiting all the way from New Zealand. Township Lagoon remains a gem, despite its past indignities; but, looking at it, you might not realise that. For much of the summer its retreating shoreline exposes thick slicks of sticky and very smelly black mud, and a tideline of decaying algae. The lagoon's waters are shallow, yet murky; and there are neither fish nor frogs to offer a sense of life to the place. But looks and smells can be deceptive. Beneath those murky, briny waters, and under stones and rubble around the shore, lurk crustaceans both strange and rare.

To appreciate the aquatic life of Township Lagoon, you really need a pair of waders, or at least gumboots. It also helps to come prepared with a dipnet and a white plastic tray into which you can tip the net's contents. Based on first-hand experience, I might add that a crustacean hunt there isn't for the fainthearted; furthermore, it requires a permit. That being the case, I politely suggest that you might be better off remaining an armchair limnologist and making do with my account of what can be found there.

Many years ago, and long before my invertebrate-enthusiast friend Sarah (whom we will meet again, in December), moved to New Zealand, she carried out a survey of Tasmania's inland saline lagoons, looking to update knowledge on which of the special crustaceans discovered there, many decades earlier, were still extant, and if so, where. Sarah's survey coincided with a drought; therefore, she found Township Lagoon a morass of cracking mud with no standing water at all. Needless to say, no special crustaceans were found, and Sarah even suggested that one of them, the threatened salt-lake slater, might be locally extinct. This was thought likely because it has no obvious adaptations for surviving desiccation, unlike some of the other crustaceans. Fast-forward a couple of decades, though, to an October that followed a winter with plentiful rain. I was delighted to be invited to join another hunt for the slater, with Karen—a colleague from the government unit charged with keeping an eye on our threatened species. Karen has worked on invertebrate conservation matters for decades and is a fount of knowledge. My only claim to any prior knowledge of salt-lake slaters was that, a year or two previously, some had turned up in samples from coastal saline lagoons at Musselroe, during one of our annual collecting expeditions (you'll find more on these in February). Even then, I wasn't the discoverer, just the person who had notified Karen of our discovery. Still, I wasn't about to turn down a trip to Tunbridge with Karen on such technicalities, since it promised to be both fun, and an excellent learning opportunity.

And so it proved to be. While I donned the waders and dipnet, and swished and fished as deep as I dared, Karen stuck to gumboots, which, in turn, stuck to the muddy shoreline as she worked the shallows, turning over stones and bits of broken pottery, concrete, and clay piping. Each step I took liberated swirls of pale silt, encouraging me to keep walking to stay ahead of the cloud. Yet the mud sucked at my feet with every step, keeping me in a constant state of near-toppling. It was a fine line to follow. Drawing a blank out deep, I moved inshore with some relief, to where some hardy sedges fringed the lagoon. I jiggled the net along the bottom, and then, tipping its contents into my tray, I was delighted to spot a candidate slater walking across the floor of the tray, before disappearing among some filamentous algae. I exited the lagoon and called Karen over expectantly, but she, in turn, called me over. She was peering intently at a turned-over stone that was lying in a couple of centimetres of water. As I reached her, still holding my tray, she lifted the stone out of the water and revealed several slaters. They looked to me very like rather puny, anaemic versions of the garden variety which some may know as 'woodlice', others as 'sowbugs'. They scuttled over the surface, back towards the underside of the stone, alarmed by the sudden lack of water. My single slater hardly registered in comparison to Karen's discovery. Both, though, were confirmation that the species had somehow weathered the drought years and was alive and well. Our mission was accomplished! Additionally, we learnt that they favour the shallows, and that there isn't much to be gained by searching any deeper (we found none further out). I messaged a photo to Sarah, knowing that her earlier efforts had brought us to this point, and that she, more than anyone, would want to share our delight at the slaters' rediscovery. Her instantaneous response told me that we had hit the mark and reconnected her with an old friend. It made my day.

Salt-lake slaters are exactly what their name suggests. Their ancestry is among the terrestrial slaters rather than the marine versions that are common on our seashores. All are isopods, but the terrestrial versions gave up living in the sea a long, long time ago, while retaining rudiments of the gills that marine species use for aquatic breathing. Salt-lake slaters are an offshoot of the terrestrial lineage, one that more recently decided there was a living to be made from going back into salty waterbodies inaccessible to the marine slaters. The species found in Tasmania is widespread in inland saline country on the Australian mainland, but it has always been a rare beast here, in keeping with the rarity of saline lagoons. The slaters, or maybe their eggs, probably get carried from one lagoon to the next on the feet of birds, such as the red-capped dotterels that feed around the shores of lagoons. However, the salt-lake slater's eggs aren't resistant to desiccation, so how they survive the journey frequently enough to establish populations in such out-of-the-way places remains a mystery. It's why Sarah had concluded that the species was possibly extinct, when the lagoon dried up. Have they recolonised since then, perhaps from one of the few other remaining lagoons in the region? Or, perhaps, beneath the lagoon is a briny aquifer where the slaters can take refuge in times of drought. At this point we simply don't know.

While our focus was on the slaters, the contents of my dipnet told of an abundance of other crustaceans: seed shrimps mostly, like those we encountered in Bill's billabong, back in September. Some of these were monsters—several millimetres across; and all were of species different from those we had found in the billabong, since they were adapted to living in brine rather than freshwater. I collected some for the museum and, not long after they were registered into the collections, I was pleased to receive a request from a mainland researcher interested in borrowing them for a taxonomic revision. There's a fair chance, it seems, that the lagoon's seed shrimps will turn out to include undescribed species. Had I not collected some, we might have remained none the wiser.

Seed shrimps belong to a group of crustaceans known as ostracods, a term that is derived from the Ancient Greek word for shells or shards of pottery; in scientific terminology, it is often used to refer to shell-like structures, such as the hinged, clamshell-like plates that enclose a seed shrimp. Perhaps we would have been better off referring to seed shrimps as clam shrimps; except that this term applies to another group of strange crustaceans of brackish lagoons, which also enclose themselves in clamshell-like plates. Whereas you can find seed shrimps almost wherever there is standing water, I have only found clam shrimps twice: once in a small lagoon at Stony Head, and once in similar circumstances on Flinders Island. The latter occasion was also with Karen.

To digress a little further, if you are ostracised, there's the same Greek-origin word there. In ancient times, the names of candidate citizens in an election whom you *didn't* want to get elected were written on shards of pottery; the more shards bearing a given name that were tossed into the ring, the stronger the case to banish the citizen in question. I rather like that idea.

Though the slaters and seed shrimps were fun, one further beast had me scratching my head, when I first set eyes on it. It was swimming in slow, lazy circles near the surface of the lagoon, in water that was almost as deep as my waders. To my untrained eye it looked like a fish; indeed, its general demeanour and pallid colouration reminded me of a pregnant guppy—an aquarium fish that I had taken great pleasure in keeping, and breeding, as a child (I had named the first of these fecund females Geraldine Guppy). Looking around, I could discern several more. I scooped one directly into the tray—no need for the net—and retreated to dry land so that Karen and I could see it more clearly without risk of losing our footing. Up close, it was obvious that this was yet another crustacean, and that it was swimming on its back—which is fine, because the word 'swims' works just as well on its back. Multiple pairs of legs pulsed in unison, propelling it round and round in the dish. This, it transpired, was a fully-grown brine shrimp.

Brine shrimps hadn't been on our radar at all, but it made sense that they were here. The species concerned is related to the one whose dried eggs are sold as aquarium fish food, or for kids to hatch into so-called sea monkeys (a blatant, taxonomically ludicrous misnomer if ever there was one). Like the salt-lake slater, it is found more widely in the inland of southern Australia, with Tunbridge being its southernmost outpost. Unlike the slater, its eggs are desiccation-resistant, and so it is more understandable that they might regularly get carried on birds' feet from one lagoon to the next. The eggs can also rest at the bottom of a lagoon for years, awaiting just the right conditions of salinity and warmth to trigger hatching. They go through a succession of larval stages (sea monkey, sea chimp, sea gorilla, perhaps?), before reaching adulthood; although, for every one that makes it all the way, tens of thousands fall by the wayside. Those predatory seed shrimps probably eat most of them.

I have tried my luck looking for these strange crustaceans at Tunbridge on many occasions since that visit with Karen. The salt-lake slaters are always there, lurking under ostracised shards of pottery along the shore; the seed shrimps are sometimes there, bumbling around a little further out; but the guppy-like brine shrimps haven't reappeared yet. It seems that our first visit must have been just the right length of time after some perfect set of conditions that had triggered their eggs to hatch. Perhaps those weather conditions—not too wet, not too dry—happen every year or two, or perhaps only every decade or two. In the meantime, it's good to know that the shrimps' eggs will be biding their time at the bottom of the lagoon, safe amongst the carcases of seed shrimps and bicycles.

On pastel wings

This month's butterfly is the Hobart brown. Its brazen emergence at the first hint of warmth, betting against the odds for a favourable flight season, is well aligned with the hopes and wishes of this naturalist. My first encounter with this perky little battler was on Knocklofty, on Hobart's wooded western fringe. With its gentle terrain and a network of paths meandering through the woodland, offering occasional glimpses of the mountain, Knocklofty was an enticing prospect for a spring walk with our young boys. The weather was fine that morning, although the south-facing side of the hill we started on was still in shade. The boys' shoes and trousers quickly became damp as they ploughed through dewy grasses, sedges and rushes along the firebreak, showing considerably less restraint and more *joie de vivre* than their parents. No matter though: we soon rounded the hill and picked our way along its northern slopes, with the sun now streaming through the woodland canopy. There's something magical in the way light plays on eucalypt foliage. It bounces hither and thither among the shiny leaves, with more of it filtering through to the woodland floor than you might expect on the basis of canopy cover—in a eucalypt woodland, shade at ground level is relative, not absolute. The ground on this side of the hill was, therefore, warm to the touch, and the fallen leaves and twigs crunched and crackled under our steps. It was a reminder that aspect is all-important in determining ground-level microclimate, especially this early in the season. So it was here that I saw my first signs of insect life when the nymph of a gum-leaf grasshopper jumped out of the way of my footfall. I hadn't spotted it basking on the path, which wasn't that surprising, as it closely resembled a dead gum leaf, both in colour and in outline. With a few more weeks of basking and feeding on low-lying herbage, and a couple more moults, it would reach adulthood. Its resemblance to a gum leaf would have become even more convincing by then; but, should its camouflage fail it, the grasshopper would have an additional means of escape—through the power of flight.

The path ahead was dappled with a constantly shifting melange of fawn and tan tones, reflecting the lively overhead conversation between sun, eucalypt foliage and breeze. It was against this backdrop that I nearly stood on my first Hobart brown, because it, too, was basking; and it, too, had opted for camouflage. Its wings, in pastel shades of tan, fawn, orange-brown and cream, were a perfect match for the sun-speckled, dead leaves on which it sat, and its camouflage extended to movement as it gently raised and lowered its wings to the rhythm of the woodland's conversation. At my approach, however, it snapped its wings shut, exposing their leaf-brown undersides, and the butterfly instantly melted into the leaf litter. If I hadn't seen it before this vanishing act, I might never have spotted it at all; in fact, I only succeeded in pointing it out to Chris and the boys by moving my shadow over it, at which point, feeling sprung, it opted for flight. Just for an instant, though, it rested there, with wings open; just long enough for us to catch a glimpse of the dark eyespots on its hind wings. This behaviour, too, was calculated to improve its chances of escape. Had I been a bird, I might have pecked at the eyespots and got a beakful of dry, scaly wing, rather than the butterfly's juicy body.

The Hobart brown isn't only found around Nipaluna-Hobart, but it is endemic to Tasmania, comprising three geographically separated races that differ slightly in wing colouration and patterning. All share the same temperament, but those living in mountains, and in the cloudier and cooler west, do more basking and have darker wings—more tan and less orange—which better absorb heat. Their caterpillars feed on grasses, principally the native tussocks, though they will also eat European grasses. That said, the adults are discerning enough to avoid the lawns of suburbia, so if you want to see a Hobart brown you must venture beyond the urban fringe. They aren't always as placid as the first one I identified on Knocklofty, though. Once the day warms up, their flight paths gradually rise, so that by noon they are likely to be zigzagging at head height, and by mid-afternoon they will be at canopy height (if the good weather lasts that long). At Knocklofty, we saw several more as we continued our walk, even as we returned to the southern slopes, but at that point they were all up high, among the sunlit branches.

It has often struck me how the muted and pastel colours and patterns on the wings of most Tasmanian butterflies, Hobart browns included, require a more intimate encounter to appreciate their subtle beauty, compared to the dashing, in-your-face gaudiness of many tropical species, such as the electric-blue Ulysses butterflies, or the vivid green-and-yellow birdwings that were common in our Cairns garden, before our move to Tasmania. This pattern extends to much of nature—think birds of paradise and orchids. I think it ultimately relates to how much of the sun's energy penetrates the atmosphere. In the tropics, the light and warmth of the overhead sun supercharges nature. Plants photosynthesise more and grow more rapidly (as long as there is enough water); which, in turn, provides food and energy for more animals.

In nature, diversity begets diversity. The more individuals are competing for resources, the more their species evolve into specialised forms, developing niche lifestyles as a means of dividing up smaller and smaller slices of the energy or food pie. So, in the tropics, species devote more resources to competing with each other than to basic needs, such as keeping warm. Furthermore, since the elements are less hostile in the tropics, competitors can afford to invest in flamboyance as a means of standing out from the crowd. The trouble is that others are playing the same game, so it becomes an arms race, albeit one of beauty and brashness rather than military might. Perhaps, then, it might be worth celebrating the typically sombre tones of Tasmanian butterflies, for they suggest that their lives are just a bit more civilised and rather less dog-eat-dog than those of brash northerners.

Twitchathons

Birds go berserk at this time of year, and birdos (or birders, if you prefer—the latter is more the international standard) are prone to behaving likewise. For several years, I joined a few birding mates in an annual national twitchathon event, in which teams compete to see how many species of bird can be seen or heard over a set twenty-four-hour (these days twelve-hour) period, raising money for charity in the process. A *twitcher* is a particular type of birdo, one for whom the compilation of lists is so important that the mere prospect of ticking off an additional species is said to send them twitchy. That's the folklore, anyway. Non-twitcher birdos generally resent being called twitchers, since it trivialises the enormous enjoyment to be had from simply observing, identifying and being in the presence of birds; but they may make an exception around twitchathon time. October is the ideal month for this, since birds are at their most vocal and visible when they are establishing breeding partnerships and nesting territories in anticipation of a summer of chick-rearing. In other Australian states and territories, the twitchathon event is taken very seriously and is competitive and highly organised, with strict start-stop times and agreed starting and finishing locations. But in Tasmania it's been a much lower-key affair, perhaps just a slight step up from a chance for a few casual birdwatchers to dust off their binoculars (or 'bins') and put their identification skills to the test in the company of others of a similar disposition; albeit with a bit of goodhearted rivalry among teams. In the years when I participated, we would choose our own route and our own start time, and the final species tallies assumed that all participants were impeccably honest (which, of course, we were). A spirit of camaraderie prevailed within each team, and it would grow stronger as the absurdity of the desperate measures taken to lengthen the team's bird list became increasingly apparent. To add to the fun, I also enjoyed preparing a dummy checklist in advance, padding out the real bird species with made-up ones, such as satin puddlebird, Captain Mandolini's corella and the solitary cockawon. One year our team comprised two older and two much younger men. We called ourselves 'The Has Bins, featuring the Forty-Spotted Teenagers'. In another, we were the 'Pheasant Pluckers'.

I can't now recall any of our final tallies, but I do know that our team won in at least two of the years during which I participated. So, in anticipation of your desire to learn from our example, here's a rundown of a more-or-less typical twitchathon (well, a kind of composite incorporating bits from various years). We would generally start at dawn, at the lighthouse on the southern tip of Lunawanna-Alonna-Bruny Island (having found somewhere to stay on the island the previous night, because dawn comes very early in late October). The lighthouse area regularly offers sightings of tawny-crowned honeyeaters, one of Tasmania's rarer species, which has a particular liking for coastal heathland. Generally, spotting one would involve a fair bit of standing around in puffer jackets and beanies, doing little jigs to keep warm, and peering through binoculars at every silhouetted small bird that popped up from the heathland to perch on a prominent shrub. Usually, the bird was either unidentifiable in the dim light of dawn or would be identified as a crescent or New Holland honeyeater. These are common birds, but still tickable at this stage in the proceedings.

All sorts of other birds might just happen to turn up at the lighthouse.

All four cuckoo species were likely to be at least heard if not seen, along with pipits, shrike-thrushes, dusky robins, and white-fronted chats. One year we spotted a covey of brown quail along one of the fire trails; another, we claimed a peregrine falcon perched on a distant sea stack, silhouetted against the sea. That same year we glimpsed a shy albatross, way out to sea, skimming the crests of the waves and seemingly untroubled by the westerly gale that was making our eyes stream, our noses drip and our binoculars shake. The bays either side of the lighthouse were also great places to look for gannets, cormorants, oystercatchers, hooded plovers and white-bellied sea-eagles.

Next stop was usually a roadside layby conveniently overlooking Cloudy Lagoon and backed by damp eucalypt woodland. The lagoon featured various waterbirds and, if we were lucky, a scattering of migrant waders from the Northern Hemisphere. I recall a curlew sandpiper one year, and bar-tailed godwits—birds whose numbers are, sadly, now greatly diminished, almost to the point of no return. At the same time as our eyes were focused on the waterbirds, we would be listening out for woodland birds at our backs. Dusky woodswallows, tree martins, grey fantails, spotted and striated pardalotes, scarlet robins and yellow-throated, black-headed, and strong-billed honeyeaters were regulars. In years when the blue gums were flowering on South Bruny, we stood a good chance of hearing the convivial chattering of swift parrots. With time being of the essence, we wouldn't linger long, though, but we'd head north, towards Lunawanna, and then east, towards Mount Mangana. Driving along with the windows wound down, we'd be keeping an eye and ear out for additional bird species. Collared sparrowhawks, brown goshawks and wedge-tailed eagles were all possible overhead, while native hens and white-faced herons were usually spotted in the wetter paddocks and kookaburras, rosellas, and butcherbirds on roadside telegraph wires. A short diversion south, towards Cloudy Bay, offered the possibility of twitching the elusive forty-spotted pardalote among the planted white gums lining the driveway of the Inala private bird reserve; but we soon learnt that freeloaders, not surprisingly, weren't welcome, should we have ventured any distance up the driveway.

Approaching the higher reaches of Mount Mangana, we would periodically stop and listen for birds of the wetter forests, such as black currawongs, scrubtits and Bassian thrushes. If we hadn't ticked them off by the time we got to the crest, we would park and march hurriedly up the soggy summit track, eyes peeled for any glimpse of movement in the dense rainforest undergrowth that could signal one of these hard-to-get birds. Consolation prizes could include beautiful firetails, Tasmanian scrubwrens and Tasmanian thornbills. If Mount Mangana failed us, the Mavista Falls walk, on the Adventure Bay side of the mountain, was a further area of wet forest that might offer another chance. It usually delivered on at least one of these target species. On one occasion our Bassian thrush, with beautiful marbling on its chest, was nonchalantly scratching up leaf litter on the edge of the picnic area, while a resplendent male pink robin emerged from the gloom of the forest to check himself out in our wing mirror.

Adventure Bay held the prospect of more waterfowl, on Captain Cook Creek, plus plovers, oystercatchers and lapwings on the beach, and swift parrots in the blue gums at the back of the beach. Then we would make a dash for the Kettering ferry at Roberts Point, devouring a breakfast of gourmet pies and coffee from the kiosk, if we arrived before the ferry. The crossing offered the chance of little penguins and more cormorants, as well as crested terns. Back on the mainland, we would take a diversion down the track towards Boronia Beach to see if the tawny frogmouths were nesting this year among the old pines there. They're hard to spot by day as they sit completely motionless, resembling a branch stub; but their presence could be betrayed by whitewash (a euphemism for spattered droppings) on the ground below. We would stop, briefly, beside the highway near Huntingfield to try our luck with the dusky moorhens that skulked around the fringes of an adjacent pond. One year we called in at the Peter Murrell Reserve to look for forty-spotted pardalotes and got a bonus striated fieldwren in the adjacent paddock.

Next we would head straight for suburban Kingston Beach. This might seem an odd destination, but the edge of the golf course was a good spot to tick off some of the feral species—mallards, spotted doves, rainbow lorikeets and corellas—giving our list another boost. The site usually turned up a few additional natives, also, such as fairywrens, welcome swallows, grey currawongs and cuckoo-shrikes.

If we still needed little penguins, we would stop in Lower Sandy

Bay and peer into the nest boxes that have been erected for them on the foreshore near Red Chapel Beach, in the hopes of finding one in residence and visible in the gloom of the box's interior. We wouldn't stop in Hobart, but would keep an eye out for house sparrows and feral pigeons. Our next stop would be Gould's Lagoon, at Austin's Ferry, where there were opportunities for spotting various ducks and other waterbirds, including purple swamphens, coots, egrets, and the elusive, cryptically coloured Latham's snipe. On my first visit, the reedbeds were alive with the calls of little grassbirds; but the reeds had contracted in subsequent years and the grassbirds were nowhere to be seen or heard. If we hadn't spotted yellow-rumped thornbills in the adjacent grassland, then we might next head up to the nearby Poimena Reserve to try our luck there. That was also a good place for cuckoos and musk lorikeets.

Getting more birds became harder as the day wore on; not only because the birds were less active but also because we would, by then, have spotted all the easier ones. One destination that we took a bit of a risk visiting, because it was so out of the way, was the Shene property near Brighton. This was the closest place to Hobart where we might see, or hear, banded lapwings, in the irrigated paddocks there. I think we only ever heard them; but that was enough evidence for a tick. Also, out that way, we once called in at ZooDoo, where a group of feral peafowls were considered just about tickable. Then to Richmond for reed warblers, hanging around near the bridge over the reed-lined river just long enough to catch one of the birds' intermittent and unpredictable outbursts of rich, fluty notes. On to Orielton Lagoon, where we would trudge out across the dry, cracking saltmarsh, and scan the mudflats for waders—typically Pacific golden plovers, red knots, and bar-tailed godwits, but there was always the possibility of other species. We would also check the small island in the middle of the Lagoon (visible from the causeway) for royal spoonbills, a species that, in Tasmania at least, I've only ever seen in this one spot. Musk ducks and great crested and hoary-headed grebes could also be spotted out on the water, and Caspian terns overhead.

With the hours ticking by we would make a break for the east coast, heading through Sorell (more pies and coffee) and over Bust-Me-Gall and Break-Me-Neck hills to the sandspit at the mouth of the Prosser

River, at Orford. This spit held the prospect of fairy terns, a diminutive and rather rare seabird whose choice of nesting location leaves them vulnerable to disturbance by dogs, people (not twitchers, of course) and high tides. If time was on our side, we would continue north to check out one of the dams beyond Triabunna for more waterfowl, and to see if the white-bellied sea-eagle nest on the other side of the road was occupied. On another occasion, we headed south instead, to the Sandspit Reserve in Wielangta Forest (another chance for Bassian thrushes, pink robins and scrubtits), and then to the salt flats at Marion Bay (red-capped dotterels and white-fronted chats), with bonus cattle egrets in the nearby irrigated paddocks, and finally towards the Forestier Peninsula, calling in at a secret spot where nankeen night herons regularly roosted.

Then it would be time to head back towards Hobart, maybe grabbing yet more eats in Sorell on the way through. Sometimes, by the time we got back to Hobart, we would be so exhausted that we would call it quits. But the twitchathon would traditionally run for twenty-four hours (though that has since changed), so we still had a chance for some night birds. Boobook owls and frogmouths were relatively easy to locate, if only by their calls; but masked owls proved more of a challenge. One year we spent several hours unsuccessfully spotlighting and listening out for them at the Waterworks Reserve. When I got back home and dropped into bed, there was one calling from the garden—the first time I had ever heard one there—but I couldn't count it because the rules require fifty percent or more of team members to corroborate a tick. We never did get a masked owl, nor an owlet-nightjar, on our list.

The twitchathon is one of the few occasions when I carry binoculars for birdwatching. During other times, I prefer to let the birds' calls and characteristic movements ('jizz', in the birding vernacular) provide the necessary clues as to their likely identities, allowing me to absorb an ornithological sense of a place almost subliminally, without overly focusing my attention on them to the exclusion of the rest of nature.

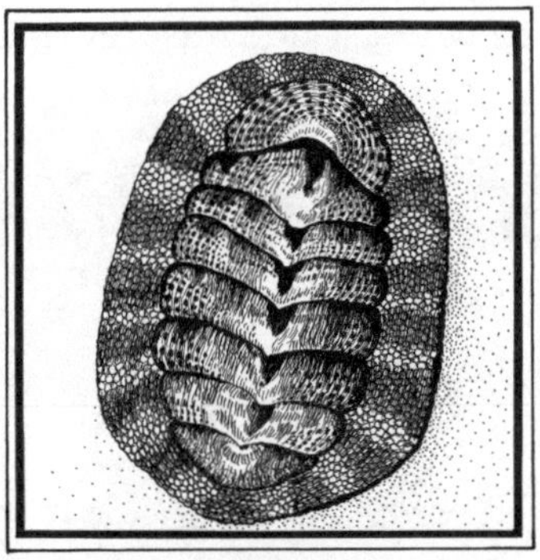

Of cowries, chitons and conchological collectibles

Perhaps because, during my first decade in Tasmania, I was spending many of my working days in the forests, during my leisure time, I increasingly found the seashores more alluring than the bush. Well before we arrived in Australia, and once it became clear that this was our next move, I had a recurring dream in which I was soaring over the ocean, with a deserted, sandy beach ahead of me. I would swoop down over the beach and see hundreds upon hundreds of beautiful seashells strewn across the sand. The beach was my dream come true—except, of course, it *was* just a dream. Still, even before we moved into our house in Taroona, we had acquainted ourselves with the local shoreline, and I had been delighted to find my first cowrie shell on Taroona Beach. I took it as a good omen, a sign that we had chosen the right place to put down roots and raise our kids. And so it proved to be; for all sorts of reasons, not only those connected to the beach. But, for me, there's nothing like an accessible, varied-yet-intimate piece of coastline to engender a strong sense of place. Before long, I was taking daily walks along the Taroona foreshore, soaking up the sights, sounds and smells, in weather fair and foul. Taroona Beach faces south, towards the Southern Ocean and the Antarctic, and cops the full might of a southerly gale. On fair summer days, afternoon sea breezes heading inland, and up the estuary of the Derwent, frequently whip up whitecaps offshore. They can rapidly put paid to notions of passing a balmy evening around a barbecue in the adjacent park. Northerlies, on the other hand, whizz overhead and out to sea, flattening incoming waves and becalming the shoreline. On such a day, the water takes on an azure hue and the only sound from the sea is that of the gentle rise and fall of the swell sloshing around the shoreline rocks.

Taroona rekindled my childhood interest in seashells. To be fair, the interest had never really gone away; but, in Tasmania, I dived in again, deeply. Every walk on the beach would reveal newly cast-up treasures that I would try and identify. Meanwhile, exploring the jumble of intertidal dolerite boulders between the beaches revealed plenty of live molluscs.

I'm surprised that some people don't know where seashells come from, so let's clear that up. They're the permanent exoskeletons—effectively the homes—of living molluscs. The mollusc animal lives within and secretes the shell, adding to its outer edge as it grows and requires more space. It's not like a hermit crab, which has to find a new, preloved, empty (or emptiable!) mollusc shell to inhabit as it grows. Gastropod molluscs (other than the sluggy sort) live within single shells, which may be conical (limpets) or coiled (snails). Bivalve molluscs (clams) live wedged between two shells or 'valves'. Chitons are strange molluscs that huddle beneath their eight articulated shell plates, usually attached to a rock in the manner of a limpet, but with the flexibility of a slater. I later learnt that the name of our suburb was said to derive from the local Aboriginal word for 'chiton', which confirmed my feeling that we had been meant to settle in Taroona.

It's interesting that we call seashells molluscs, because the word 'mollusc' translates from Latin as, broadly speaking, 'soft little thing'. It was adopted in 1758, by Carl Linnaeus, the father of taxonomy (the science of naming plants and animals, traditionally using Ancient Greek or Latin word elements). He originally applied it to all sorts of lowly lifeforms, but signally *excluded* those that made shells. We can't really blame Linnaeus for this; his classification was never meant to reflect evolution—the concept had barely begun to be imagined at the time. In any case, Linnaeus's higher levels of classification were largely borrowed, with translation into Latin, from Aristotle's *History of Animals*, written in 350 BCE. Under Aristotle's schema, snails and oysters weren't molluscs at all, but ostracoderms ('shell-skins'). For a seashell, 'ostracoderm' has a nice ring to it; Linnaeus's latinised version, 'testacean', rather less so. As to why science reverted to calling them molluscs (while tossing various soft-bodied animals out of the category)—that's down to late eighteenth-century French anatomist Georges Cuvier, who recognised that the animals inside those wondrous seashells bore a strong resemblance to those of the soft-bodied slugs and squid, determining that they should be brought into the molluscan fold.

By the way, Linnaeus' coining of the term *Insecta* was also a translation from Aristotle's Greek *entomon* (meaning 'notched into sections'). Aristotle had conspicuously excluded crabs from his *entomon* category. He put them in another, called *malacostraca* ('soft-shelled'), but Linnaeus boldly departed from Aristotle and lumped them in with his *Insecta*, along with most other arthropods, and the odd polychaete worm. Aristotle was vindicated, though: today crabs are once again considered malacostracan crustaceans, rather than insects (though in an evolutionary sense, all insects are crustaceans, in the same way that all birds are dinosaurs).

Back to the molluscs: those who study molluscs, including the soft animals inside, are malacologists; from that Ancient Greek word meaning 'soft'. The term was popularised, if not invented, by none other than the early nineteenth century American writer Edgar Allan Poe—he had a side interest in molluscs and wasn't averse to plagiarising ('editing and compiling') the works of the great scientists of his era. Those whose particular focus is the shells of molluscs are conchologists, from the Ancient Greek word for a cockle (which, as you would know, looks nothing like a conch, but that's lexical evolution for you). So, does that make me a maladjusted malacologist or a cantankerous conchologist, or just a lover of alliteration? You can decide.

Within a year or so of settling in Tasmania, I had bought second-hand copies of Margaret Richmond's self-published (late 1980s and early 1990s), and lavishly illustrated, two-volume set on Tasmanian seashells, and took immense pleasure in matching up my finds and learning more about them. It wasn't long, though, before I was discovering shells on the local beach that she hadn't illustrated. For these, I turned to an older Tasmanian government publication—the 1955 update of a 1923 original, which is more of an illustrated checklist than an identification guide. But I often found that, between the two sources, the names didn't correspond. In frustration, I turned for help to Liz Turner, whose position at the museum I would later fill (on her retirement). Before long I had become a regular visitor, allowed to compare my finds with museum specimens and tap into Liz's deep knowledge. Many of my rarer finds were of deeper-water species, which rarely beached. These were usually very worn by the time they did so, making them all the harder to identify. Still, the Taroona list grew longer and longer, but the rate of increase steadily slowed. Would I soon have found everything there was to be found? The thought dismayed me. That's when Liz warned me that on no account should I delve into the world of micromolluscs—I might never emerge sane! Well, how could I not?

It soon turned out that the real diversity of shells at Taroona lay in this miniature realm of species that grow to a maximum of only about 10 millimetres (there is no hard-and-fast definition; many micromolluscs scarcely reach a fifth, or even a tenth, of this size). On certain days, drifts of micromolluscs would appear on the strandline at the western end of Taroona Beach. Others could be sifted, in a kitchen sieve, from grit collected elsewhere along the Taroona shoreline. With the aid of a borrowed microscope, by the year's end, I had identified two hundred species from Taroona. Within a couple more years, the number had climbed to about 350—all to be found within a few hundred metres of my doorstep. This was indeed paradise!

Shelling became my happy time, my relaxation, my down time. But before long this shell quest, this mollusc madness, started to become addictive. I would get edgy if I couldn't get my fix. On weekends I started venturing further afield—south to Kingston Beach and Blackmans Bay, east to Seven Mile Beach and Marion Bay. Each place has its own characteristic assemblages, contingent on tide and weather—such that one may always anticipate the unexpected, and sometimes the disappointment of no-shows. Rough seas generally wash beach-cast shells away, but high tides in the aftermath of bad weather usually redeposit them, and often offer up a bonanza of fresh material from deeper water. I started plotting: seeding ideas for family holidays to far-flung corners of the state, where child-friendly beaches and enticing shelling opportunities might intersect: Lime Bay, Maria Island, Bicheno, St Helens, Musselroe, Bridport, Narawntapu, Low Head, Rocky Cape, Stanley. By and large, these holidays were a success on both fronts, and mention of these places now awakens cherished memories of simple pleasures and happy summer days by the sea. I hope it does the same for the rest of the family.

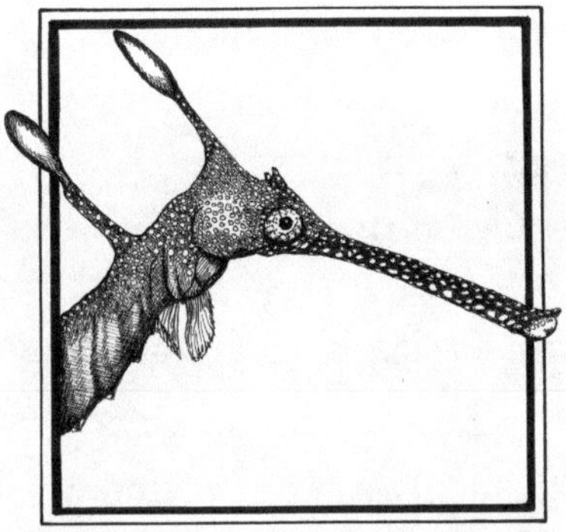

Gulches and dragons

I love the word 'gulch'. Elsewhere in the world it may refer to a rocky ravine but, in Tasmania, it describes a narrow channel or fissure in a rocky shoreline, in which seawater sloshes and gurgles with the rise and fall of every wave, and with the ebb and flow of every tide. It sounds, and probably is, onomatopoeic. Etymologically it's related to 'gush' and 'gulp'. I can't think of the word without hearing those rocky seashore sounds; and when I'm pottering along in gulch country, I can't hear those sounds without peering into the depths of the gulch and pondering the lives of the creatures living there, or picking through the shells cast ashore where gulch meets beach. A good gulch will have steep sides, plunging to depths of a few metres. Gulch geometry seems to condense or concertina the usual vertical zonation of seashore life, meaning species which normally inhabit deeper waters can sometimes live quite happily at shallower depths, within a gulch.

A gulch in spring is a place of beauty, mystery and delight. At this time of year, the upper reaches of a good gulch are typically draped in deliciously glutinous fronds of olive-brown seaweed—crayweed and strapweed in sheltered areas, and bull kelp in more exposed spots. These sway, slither and seethe in unison. They may look like plants, but technically they're brown algae and only very distantly related to land plants. Their presence dampens the waves' energy and softens sound and light. Seaweed holdfasts—the bits that stick them to the rock—offer sheltered niches for all manner of molluscs: mussels, kelp snails, limpets, topshells, turbans, nerites, coniwinks, and their predators, such as lineated whelks, rock snails, and cart-rut snails. Gaps between the holdfasts are where the anemones thrive—bright scarlet waratah anemones higher on the rock face and white-striped anemones lower down. The intertidal zone is difficult to inhabit, a realm of extremes of temperature, salinity and, of course, wetness. It's a wonder that so many creatures make a go of it. Yet kelps and other seaweeds grow rapidly as the days lengthen, presenting an opportunity for herbivorous molluscs. If the weather's not too hot, those are likely to be out and about: some grazing the seaweeds, others guarding their egg clutches.

The best way to get the measure of a deep gulch is from below, by immersion. I find a wetsuit with a hood necessary virtually year-round; and a mask and snorkel are essential for exploring the depths. If you can tolerate the cold (because the warming of the sea lags a good month or two behind that of the land), October is a great month to jump in and take a closer look. The typical algal blooms of later winter and early spring are waning, so the water is likely to have become clearer; and animals that spend the winter further offshore, to avoid the worst of the storms, are moving inshore to breed. Many people dive in search of abalone to prise from the rocks, and gulches are great spots for them. I like to look for abalone, too; but they're such wonderful creatures that I can't bring myself to detach them just to eat them, delicious though they undoubtedly are. Abalones are sea-snails that haven't quite coiled up as tightly as a normal snail; instead, they look and behave rather more like limpets. They feed mostly on bits of loose weed that have been dislodged by storms and currents, but they also browse algae growing directly on the rocks they have clamped themselves onto. In southern Tasmania there's a single species, the black-lip; while along the north coast, the black-lip largely gives way to green-lips and Emma's abalones. The latter is a smaller species that never reaches a legally harvestable size. All three possess shells whose inner surface is coated with iridescent nacre. The shells are so much more suited to the making of jewellery than ashtrays, which has, all too often, been their sad fate.

Talking of jewels, jewel anemones form colonies that can blanket the deeper recesses of gulches in spectacular, luminescent carpets of pink and orange. Their translucent tentacles extend into the currents to snatch passing particles of food. They are really a sort of coral polyp, but one that doesn't grow its own calcareous skeleton. I haven't yet dived the gulches and sea caves along the east coast of the Tasman Peninsula, but I am told that the jewel anemones there transform these into places of great beauty and magic. What I do know is that there is a rare species of sundial snail that feeds on those jewel anemones, and that it can sometimes be spotted *in situ*. My only encounter with Ponder's sundial comes from having found a single shell washed up at the head of a deeply sloping gulch near Bicheno, along with many of its more common, yellow sundial, relatives, which also feed on anemones. Sundial shells are exquisite creations—broad-based, low-spired, shiny yellow or orange shells on the upper side; and, if you turn them over, the hollowed-out spire resembles a spiral staircase complete with miniature steps—which is why they are also known as staircase shells. Local species range from five to fifteen millimetres across, whereas some tropical ones are thrice that size.

Native eleven-armed seastars frequent gulches during the warmer months, where they prey on abalones, prising them from the rocks through the sustained contraction of hundreds of hydraulically operated sucker feet. Much as I am fascinated by the seastars, for me their presence suggests that it would be well worth keeping an eye out for their molluscan predator, known locally as the reddish rock whelk. This is an impressive animal, with an orange body and a shell that may be 25 centimetres long. It occurs all up the east coast of Tasmania, but it takes a particularly low tide or unseasonal storm for them to be visible on the shore without getting wet. Unusually, for a mollusc, the same species is found widely around the world, at temperate latitudes, including in New Zealand, South Africa, the Mediterranean, and the North Atlantic. That's because it has an 'extended planktotrophic larval stage', which essentially means that its larvae live and feed for months, in the open ocean, before looking for a suitable bit of seafloor to settle on. In that time, ocean currents could have transported them thousands of kilometres from where they hatched.

The animal that most entices me into the water at this time of year is neither mollusc, nor seastar, but dragon. A close relative of seahorses, the weedy sea-dragon is one of the world's more bizarre-looking fish species. It is only found in the cool coastal waters of southern Australia, including Tasmania. Unlike seahorses, which have prehensile tails and pootle along vertically, sea-dragons have stiff tails and swim horizontally, snout first. They can grow to 45 centimetres in length, and sport orange or reddish bodies with blue highlights. These colours appear rather more muted in real life (i.e., below the waves), because light towards the red end of the visible spectrum is the first to be absorbed by water. The strangest parts of sea-dragons are their fins, which end in outgrowths that closely resemble the leaflets of crayweed. In spring, sea-dragons move inshore to breed, taking up residence in shallow water where crayweed abounds. They're docile, ponderous creatures, and not hard to net once spotted, making them vulnerable to those who would collect them for aquaria. Unlike rock whelks, sea-dragon larvae aren't ocean-going; instead they pass their childhood adhered to the underside of daddy dragon's tail and pop out, a few weeks later, as tiny versions of their parents. Those which survive to adulthood will spend the rest of their lives within a few tens or hundreds of metres of where they were born, making them poor re-colonisers if locally extirpated. That's why I'll recount my first meeting with a sea-dragon without mentioning where it took place. Suffice to say that it was a memorable encounter and one that I would have liked to repeat, but on return visits to the same area I have found no sign of sea-dragons.

Even though I wore a wetsuit and hood, on first immersion, the icy water took my breath away and elicited involuntary and audible gasps. I cleared the mist from my mask and summoned the strength of will to submerge my head. The dull pain of first contact between forehead and ocean was instant and excruciating, but I knew, from experience, that it would subside. I would then have maybe fifteen minutes in which I would feel either comfortably numb or curiously toasty. After that, my fingers, hands and feet would stop working, my mind would fog, and I would risk serious hypothermia if I didn't get out and warm up. With these sobering thoughts in mind, I swam off at a brisk pace, eager to leave behind the sandy beach, where I had hesitantly waded in, and reach the nearby, gulch-indented shoreline. The sea was calm but, even so, there was enormous energy in the low swell. It buffeted my ungainly landlubber's body from the seaward side, and again, from the landward side, on the rebound. The crayweed, which had looked so serene from above, took on a life of its own, performing wild, sinuous flagellations in response to the invisible forces of moving water. Clouds of tiny mysid shrimps darted in and out of the crayweed, with brightly coloured wrasse in hot pursuit. Here and there, crevices concealed huddled groups of abalone, and a pair of long, white antennae protruding from between two weed-encrusted rocks told of a rock lobster in residence.

I pushed on into deeper water, and the entrance to the gulch that I had peered wonderingly into, from above, earlier in the day. The steep rock faces pressed in on both sides and left much of the gulch in deep, claustrophobic shade, adding to my sense of unease about intruding into this alien world. Crayweed still slithered from side to side, but the water moved my body in unison with it and, for a moment, it seemed as though the rock beneath the weed—or perhaps some half-hidden octopus—was moving instead. I had to take a few deeper breaths at the surface to calm my nerves, before plucking up the courage to duck-dive to the seafloor some four metres below me. There I found myself staring at what I initially thought was just another piece of crayweed swaying in the water, a couple of metres before me. It took a few seconds to recompute that, no, this wasn't weed, it was a sea-dragon, suspended quietly in the water, feeding nonchalantly on shrimps. Out of breath, I headed for the surface, but I was soon down again, keen for another look. The dragon was still there, consistently about two metres from me. If I tried to approach, it would magically drift away, seemingly without any bodily wriggling at all. Up for another breath, back down again, this time holding onto a nook in the rock face to keep my buoyant form from bobbing up to the surface again. I clung, motionlessly, to the rock for as long as I was able, taking in the sea-dragon and its strange, watery domain. This time it seemed to take exception to my continued presence, and with no more than a slight flick of its tail it took a ninety-degree turn and exited the gulch. Much as I would have liked to have followed it after another breath of air, I knew that it would easily outswim me. In any case, my fifteen minutes were nearly up. All things must pass, and these encounters are meant to be fleeting, I decided, as I swam hurriedly back to the beach.

My legs were cramping as I stumbled ashore, and my numb fingers fumbled ineffectually with my mask and hood. My first coherent thought, after a warming coffee from a flask back at the car, was one of renewed respect for Lutruwita's First Peoples, for whom fishing in the frigid shallows must often have involved full-body immersion, all without the luxury of neoprene and masks.

Proud as a peacock spider

You don't always need to travel far for wildlife encounters: sometimes nature comes to you. Such was my experience at work, one warm October morning. Our museum zoological collections are kept in a separate building from my office, accessed via a concrete connecting pathway. I had been about to step through the sliding doors, to enter the collections building, when, looking at my feet, an unexpected movement caught my eye. A dark dot was jumping hither and thither, apparently to avoid being trodden on. I could see, now, that it was a tiny jumping spider, no more than three millimetres across. Curious, I let the doors slide closed again and crouched down, all the better to watch the antics of this funny little bouncy creature. These days I keep my close-up glasses at hand, on a cord around my neck, so I put them on. Being able to see the spider more clearly, I felt a sense of delight ripple through me as I twigged that it wasn't just any old jumping spider, it was a male peacock spider. And it was my first-ever encounter with one of these little gems!

Male peacock spiders (can I just call them 'peacocks', and the females 'peahens'?) have become darlings of the camera-wielding arachnological fraternity; and with good reason. They look gorgeous, and they know it just as well as any proud avian peacock does. The plucky individual in front of me was one known as the common peacock spider. It's the species that seems most at home in suburbia, both in Tasmania and across the southern Australian mainland. Presumably, he was taking advantage of the extra warmth radiating from the concrete, to help jump-start his tiny body for the rigours of the day ahead. He turned towards me, staring at me rather myopically through four rounded, alien, forward-facing eyes—two extra-large ones flanked by two slightly smaller ones. I got the sense (as I often do with jumping spiders) that he could not only see me quite clearly through his multiplicity of eyes, but also understand that I was no threat; indeed, that I might be just as curious about him as he seemed, about me. If so, how right he was. He jiggled his mitten-like, white-haired pedipalps as if sizing me up for a fight, but then turned his back on me; and what a back it was! It is the part of the spider that earns it the 'peacock' moniker. Emblazoned across its reflective, blue-and-green dorsal surface was an intense, fiery red, symmetrical motif. It is hard to describe, but it looks a bit like part of an artsy archery target; or perhaps a letter in some alien alphabet.

I was so engrossed with admiration that I didn't stop to think why the spider had turned his back on me. Before I worked it out, he had taken another leap, away from me, and so gone beyond my field of view. By the time I had him back in sight, he had hopped at least 30 centimetres, and now I could see why: he had spotted the object of his desire, his peahen. Whether he had seen her even while he was sizing me up, I can't say; but his singular travel towards her suggested that this was the case. Regardless, the two were now about five centimetres apart, and the peacock seemed reluctant to advance any closer; and with good reason—if he didn't play his cards right, he could end up as her next meal. I watched and waited. Both spiders shuffled on the spot, but neither seemed prepared to make the next move. What I was hoping for was, of course, that the male would begin the dance routine with which he would attempt to woo the female; but, for whatever reason, he decided against doing so. Instead, he leapt sideways, into the adjacent flowerbed. There I lost him amongst the twigs and branches.

The peahen, in the meantime, stood her ground long enough for me to appreciate her more-camouflaged, mottled brown, orange and white body pattern—a trait shared with the females of many jumping spiders. Then she, too, took a leap into the flowerbed, leaving me to ponder my good fortune of having been in the right place at the right time to witness the whole encounter. I have had more encounters since then—two of them in the same spot; and all of them were in October or November. Perhaps later in the year the spiders no longer feel the need to warm up out in the open and, instead, opt to remain amongst the twigs and branches, where they are much less readily spotted. Or perhaps their lives end after mating and egg-laying in spring, leaving their offspring the rest of the summer and autumn to mature on their own.

Peacock spiders—an exclusively Australian lineage of jumping spiders—went largely unrecorded and unremarked upon until recently. That changed with the advent of affordable digital macrophotography and videography, as well as social media. The peacocks' varied and species-specific dance routines are a remarkable sight up close. Flanges on either side of his abdomen may fan out so that the whole lot pops upright, very much in the manner of an avian peacock's tail and wing coverts. Additionally, he will strut and shudder, while simultaneously waving his contrasting black-and-white arms (okay—legs) in the air. The whole performance is quite mesmerising for a human, and I imagine for the peahen too. And when put to disco music (just search online for 'YMCA peacock spider' if you don't know what I'm talking about) —is it any wonder that the footage quickly went viral, catapulting peacock spiders to pop stardom? (In certain circles, at least.) Even among those of us unmoved by such digital enhancements, peacock spiders now have lots of friends. But, ladies, if you're amused at the males' puffed-up showmanship and extravagant routines, remember this: as with the avian peacock, it's all down to sexual selection on the part of the females. It's *their* genes that give rise to their behavioural preferences, insisting on mating only with the fanciest-looking and best-performing males. In an evolutionary sense, the hapless males have no choice but to deliver what the ladies want, even if it costs them dearly in energy, as well as predation risk.

Maybe now is the time for me to confess that, when I first began my museum job, I was a dyed-in-the-wool arachnophobe. It was a debilitating condition that I managed to hide in my interview for the position. Peacock spiders were my gateway towards what is now almost a spider habit. I'm quite proud of that—proud as a peacock, even. I still avoid huntsmen in the house though; and I occasionally jump out of bed from spider-themed hallucinatory night terrors—but otherwise my phobia is largely cured. That is probably just as well because, with their cult online status soaring, suddenly peacocks, in all their splendid finery, have been popping up all over the place; and many of them have turned out to be scientifically undescribed. In the last decade or so, keen arachnologists have competed for naming rights, which has had the effect, at least, of getting the job done rather more quickly than is usually the case in arachnology. New species are still turning up, but most are now formally described and named. The list currently runs to well over eighty species. More recently, the taxonomic concept of what constitutes a peacock spider has expanded, so it now includes quite a few rather nondescript jumping spiders. Distinct parts of

Australia host different species, and the most flamboyant ones are in warm, dry regions. Tasmania, by contrast, has rather few species but, nevertheless, still plenty to get excited about. At least four can be found within a few kilometres of the centre of Nipaluna-Hobart.

My favourite local peacock spider species is Harris's peacock spider. It was first described in 2011, from the high country of the Australian Capital Territory. It subsequently turned up in similar environments in Tasmania, first on Turapina-Ben Lomond, and then on Kunanyi-Mount Wellington. That is where I would go to search for it, on hot summer days rather than in October, among the dolerite boulders on the summit plateau. Sometimes I even met with success. The diminutive males have deep, powder-blue abdomens, fading to white at the front end, and bearing oblique, ink-black pen strokes and a couple of white-and-peach teardrop-shaped spots. Again, the pattern looks somehow alien, or maybe tribal. As far as I know they don't have the expandable abdominal side-flanges, but they are still very dapper up close. That was the sum of my knowledge of the species until quite recently, when I happened upon a male bouncing along in the middle of a sandy path in lowland eucalypt woodland at Bluff River Gorge, near Buckland. Comparing notes with other fans of peacock spiders, it soon became apparent that they, too, no longer considered it a montane-only species. Perhaps the common thread among its various favoured habitats is simply an abundance (at the micro-scale) of open ground. But why did it take us so long to find this out?

We do have one rather more flash-looking peacock spider in Tasmania, and it's called—wait for it—the Tasmanian peacock spider. It's not one that you'll encounter around Hobart, but it is widespread along the north coast, especially in the lee of coastal dunes. It also occurs in Victoria and South Australia. I have come across it a couple of times, both in the far northeast. Again, it's a tiny spider, and on first glance lacks any bright coloration. But, on close inspection, the peacock's abdomen is adorned with tiny scales in a range of pastel hues—blues, reds, orange, white—which are arranged in a most pleasingly symmetrical pattern of longitudinal and lateral bands. His head, too (more strictly his cephalothorax—the body part, in spiders, that combines both head and thorax) bears contrasting patterning in black, chestnut, grey and white. He not only dances, but also has somewhat expandable side-flanges to bolster his routine, earning him my vote for the loveliest spider in Tasmania. From a recovering arachnophobe, that's praise indeed.

© K. DAVIS 2022

November

See-saw weather

If October is spring's warm-up month, November is the real deal, and the rapid march of the season inevitably catches me unawares. Not that spring's progress is linear; it's always a matter of fits and starts. Indeed, the only thing predictable about spring weather, in Tasmania, is its unpredictability. Some weeks will see temperatures soar, and those indicator arrows for forest fire danger advanced ominously from green to red. In other weeks, things seem to go backwards—southerlies prevail, insect life will hunker down, and growth in the veggie garden will all-but grind to a halt. In one recent, cold, wet, eminently forgettable November, during which snow descended to sea level and Hobart recorded its coldest daily maximum for the *winter*, I felt moved to declare that spring had been cancelled altogether. It certainly pays to be prepared for all weather possibilities if you're heading out for the day—sunscreen, puffer jacket, sunhat, beanie, shorts, and raincoat. But Tasmanian nature has had an awfully long time to get used to this state of affairs. It generally copes much better with flip-flop, see-saw, stop-start spring weather than I do.

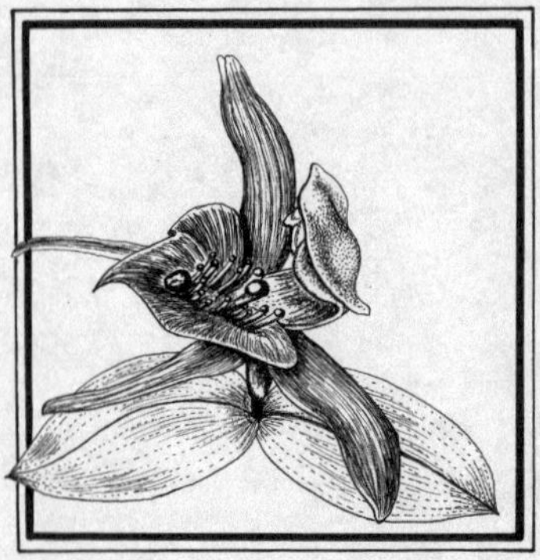

The orchid and the flower wasp

Often as not, the fleeting flowering season for most of the spring orchids comes and goes while my weekends are preoccupied with planting and weeding, which is a shame, because some of our orchids are very beautiful, or at least remarkably interesting. In culture and, some would say, appearance, orchids seem to exude sensuality and sexuality. Their very name comes from a Greek word for testicle, referring to the shape of the tuber of some European species. The flowers, on the other hand, often have more than a hint of, shall we say, femininity about them. The internet is replete—so I'm told—with images of orchid flowers that have been manipulated to appear, well, even more sexualised. But I digress, again.

My favourite local orchids are the flying duck, donkey, spider, and bird orchids. Their names speak of the refreshingly non-sexual associations that the flowers of these species conjure in our minds; yet, if we were thinking more as pollinators, rather than fanciful aesthetes, we might well resort to sexual innuendo. In fact, by rights we would be calling most of these plants 'female flower wasp orchids'. That's because both their visual appearance, and their aroma, are designed to hoodwink male flower wasps into 'mating' with them, pollinating them in the process. The deception relies on the fact that only male flower wasps can fly, while the females typically climb up grass stems or clamber onto flowers to await a passing male.

How does the male find his mate? Initially it's by scent—she emits pheromones that males can detect from a considerable distance away. Though they are unobtrusive, if you get your eye in you can often see dozens of male flower wasps zipping along silently, low to the ground, in a frenzied search for mates. Their transparent, cellophane-like wings often reflect the light in ways that make spotting these appendages easier than looking for the wasp's usually dark-coloured bodies. Once close by, the male will then visually confirm that he's on the right track. The female is usually much smaller than the male and quite distinctively shaped, rather like a lumpy, tubby, purplish-brown ant. She also strikes a characteristic pose, usually vertical, with jaws agape. Details vary by species (and there are a lot of species); but typically, the male will hover within range of those jaws and the female will clamp on, grasping him under the chin. The male may have collected nectar on his earlier foraging trips, in which case he may now feed her some by regurgitation, since she is unequipped for such trips herself. They will then mate in flight, the male often nonchalantly stopping off, periodically, for further nectar meals, with the female still coupled to his rear end. When mating is over, the female detaches and drops to the ground; with any luck, in an area suitable for her to burrow down into the soil. There she lays her eggs on, or close to, the root-feeding larvae of scarab beetles, on which her own larvae will feed, as external parasitoids (half-way between parasites and predators).

Back, though, to the orchids. Tasmania alone has well over two hundred species, and they can be found in almost all natural terrestrial habitats—sometimes in gardens and parks, too. But the best places to look for them are heath, scrubland, or open woodland, particularly along tracksides or other areas subject to occasional disturbance. To find them, you need to think small. Most are only a few centimetres tall; but, for what they lack in stature, Tasmanian orchids more than make up with appearance and charisma. Most sport extravagantly individualistic, bilaterally symmetrical blooms, which may be fleshy, crisp and compact, or wispy and ethereal. These may be carried either as solitary blooms or in bounteous inflorescences.

Most local orchid species only attract males of a single species of flower wasp, initially by emitting an odour (a 'kairomone') similar to the pheromones of the corresponding female wasp. When the male arrives, he is already in the frame of mind for mating; and his own confirmation bias leads him to conclude that the lumpy structures on the orchid's labellum ('little lip'), known as calli, are part of his intended. It's worth getting down on hands and knees and viewing orchids from a wasp's-eye perspective; you may even be taken in by the deception yourself, since the resemblance can be quite striking if you get the angle of view right. Bird orchids are particularly convincing. Some orchids even possess 'love handles' resembling the gaping jaws of the female. The male will grasp these while he attempts to mate with the other end of the labellum. Doing so triggers the flower to dab its two pollinia (sticky packets of pollen) onto the back of the male. Realising his mistake, the disgruntled male takes to the air, carrying the orchid's pollinia with him. His memory is so short, or his sexual drive so strong, that he is quite likely to be duped again by the next orchid (once bitten, twice smitten?). What he finds, however, is that, rather than his own reproductive parts lining up as he had hoped, those of the receiving orchid line up with the pollinia on his back; thus orchid cross-pollination is effected.

Not all flower wasps pollinate orchids. We have a couple of much larger and more fancy-looking flower wasps in Tasmania, in which the abdomens of both the males, and the flightless females, sport bright yellow or orange bands or lunules on a black background. It seems, though, that no orchid has evolved to mimic the females; and the males are more usually seen foraging at the flowers of bottlebrush, tea tree, grasstree and paperbark bushes, especially in sandy coastal heathland. Though they look a little like European wasps, they're completely harmless. A wasp's sting comes from its modified ovipositor (egg-laying apparatus), and no male wasp, ant or bee has one of those.

Furthermore, not all orchids are pollinated by flower wasps. Gnat orchids are pollinated by frit-flies, which are presumably also chemically duped into providing this service. It is not uncommon to see butterflies, bees and beetles with pollinia attached; although this isn't to say that they are all duped frequently enough to act as reliable pollinators.

The tail of the silky hairstreak, and other stories of butterfly deception

Butterfly-wise, there's more going on in November than in the previous months; therefore, there are several species to choose from for my butterfly of the month. My vote goes to the beguilingly named silky hairstreak. November is towards the tail end of their main flight season, since they may first emerge as early as September; but over the years our paths have intersected more often in November—perhaps because I'm likely to be wandering further afield, come November, than earlier in the spring. The hairstreaks comprise a worldwide group of butterflies named for the hairline of contrasting colour that is usually present as a streak across the undersides of both wings. Our silky hairstreak possesses quite a thick, dotted, black line in this position, which, on the hind wing, is embedded in a continuous orange streak, contrasting with the off-white of the rest of the underside. A subordinate, parallel row of black dots may also be present, closer to the base of the hind wing. The upper sides are more boldly marked, with patches of orange and creamy yellow on a background of chestnut-brown.

Silky hairstreaks form colonies, but these may extend to only the crown of a single, favoured tree. They are easy to miss. You could well walk past a tree they have colonised without realising that they were there. At least this habit means that, if you see one, you're highly likely to see more; but you do need to know where to look in the first place. To this end, the first thing to note is that silky hairstreak caterpillars feed only on the buds and feathery foliage of silver wattles. The bigger and older the wattle, the better. The adults tend to hang out on the same trees. Then, it helps to know that silky hairstreak larvae rely on ants to protect them from the depredations of parasitoid wasps (as is the case for most members of the butterfly family they belong to, along with the blues and coppers). They keep their protectors keen by supplying them with droplets of sweet honeydew, which is ejected on demand from their rear ends. As if this isn't enough, the silky hairstreak caterpillar stretches the friendship further, by exuding chemicals (it is thought) that encourage the ants to chaperone a fully grown larva as it descends the wattle tree to find somewhere to pupate. Apparently, the larva usually pupates under the loose bark of a nearby eucalypt, either near or within the ants' own nest. Thus, the larval protection offered by the ants extends to the pupal stage, boosting its chances of surviving until the following spring. At that stage, the adult butterfly will emerge, running the gauntlet of corridors busy with ant traffic on its way out of the nest.

Silky hairstreaks can be seen quite close to the centre of Nipaluna-Hobart, but my favourite spot for them is further east, at Lime Bay. The woodland there is recovering from past disturbance—the area may once have supplied timber for the coal mines that operated just a few kilometres to the south, as well as wood or charcoal for lime kilns (I will have more to say on that subject later). Wattles are often the first trees to colonise after disturbance, partly because their seeds can survive for years in the soil, but also because they can grow in denuded ground since their roots harbour nodules of nitrogen-fixing bacteria. At Lime Bay, silver wattles must have regenerated copiously over quite an extended period, perhaps as a result of occasional bushfires or deliberate, ecological burns. Many are now getting towards the ends of their lives, but in the meantime, they provide excellent habitat for silky hairstreaks.

Hairstreaks could be described as the birdwatchers' butterfly, because the best way to spot them—and the way I first encountered them, one fine, calm November evening at Lime Bay—is through binoculars. That's another thing about hairstreaks—they are late risers, and most active in the late afternoon or early evening, so long as the weather is cooperative. On this evening, I had been following a male satin flycatcher as he flitted through the treetops, chatting to himself all the while, as is their wont. These sleek, blue-black birds with contrasting white bellies are one of the last of our summer visitors to arrive; so this one was probably a recent arrival, hungry for a meal. I watched him perform an aerobatic sortie, as he chased what I presumed was some sort of moth flushed from the crown of a silver wattle, until I realised that it wasn't a moth but a male silky hairstreak. At this point I lost interest in the flycatcher, which had failed to snatch the butterfly and moved on in search of easier prey. The hairstreak, meanwhile, had run into more difficulties. It had, apparently, encroached on the territories of a succession of other males of its kind, who must all have been perched in the crown of the wattle, waiting expectantly for the excuse for an aerial dogfight. (Perhaps the first hairstreak had initially mistaken the flycatcher for a rival, himself.) Three, four, then five hairstreaks joined the fracas. They spiralled upwards for ten or fifteen metres before shooting off in different directions; then each made its way back to what I surmised was its original perch.

One of the combatants was now posing nicely on a high sprig of wattle foliage, with wings cocked half-open and poised for action (they seldom open them fully when at rest). From my angle of view, down below, and through binoculars, its wings were translucent in the evening sunlight, so that, although I was viewing the hairstreak's pale underside, the pastel colours of its upper side were also plainly visible. Perhaps this translucence is the origin of their 'silky' moniker. In any case, the overall effect was most pleasing to the eye. I enjoyed the moment while it lasted, but it was cut short by the butterfly darting away on another combative foray. From the aerial activity, I estimated that there must have been a good dozen or more butterflies staking out the one wattle tree that evening.

My opportunity to get up close to one of the Lime Bay silky hairstreaks came the following morning. We were camping there, along with several other Taroona families, as part of a Scouts weekend. With clear, starry skies overnight, the temperature had dropped, and dewdrops on the grass and bracken at the edge of the campsite glinted in the sunlight. A slight movement on a bracken frond caught my attention, so I approached for a better look. The movement turned out to be a rather bedraggled, dew-spangled silky hairstreak who was inching her way along the frond towards a speck of sunlight. One of her hind wings was deformed, which, presumably, had prevented her from ascending to the canopy on emergence. The best I could do for this unfortunate individual was put my forefinger in her path, so that she climbed aboard. Her weight was almost imperceptible to me, as I gently transferred her to the low-hanging branch of a nearby silver wattle, now in full sun. I had wondered where hairstreaks got the energy for their frenetic evening jaunts; and now I found out, as the butterfly unfurled its tongue and probed the extrafloral nectaries—little dispensers of sugary drink—that wattles often provide at the base of their leaflets. For the wattle, the benefit lies in the protective, herbivore-eating ants that the nectaries attract; but for this butterfly it was a lifesaver, for now at least.

As the butterfly fed, she alternately wiggled each of her hind wings up and down in a scissor-like movement. Such behaviour is common to many in the blue-copper-hairstreak family. In this one, it accentuated the movement of the wispy, tail-like extensions on her hind wing. When one also observes the black and orange markings on the underside of the hind wing, just behind the 'tail', it is easy to see why a bird might mistake the wiggling tails for antennae and the hind wing for the main meal, a ploy that potentially allows the butterfly to escape with just a beak-shaped nick punched from its wings. All the same, I didn't rate highly this one's chances, in the event of such an encounter. She seemed content enough for now, though, so I let her be while I returned to camp for a cuppa.

Silky hairstreaks come in six slightly different races, three of which are exclusive to Tasmania, while the others are found only on the southeast Australian mainland. The differences are subtle, and I do wonder at the motivations of the lepidopterists who first named them, mostly a century or more ago. Were they driven by dispassionate science, a desire to better catalogue creation, or by a hankering to have their own names immortalised as the authors of the subspecific epithets in question? Put it this way: if we accept that distinct parts of Tasmania host distinct races of hairstreaks and other butterflies, then it seems likely that many other insects, too, might also be divisible into races, if only we could discern their differences. Modern molecular techniques quite often reveal what we call genetic structure among geographically distant populations of the same species of insect; but, these days, few of the scientists involved feel that it's worth formally assigning new Linnaean names to these, unless the differences are deep enough to warrant the recognition of new species.

For better or worse, it's only in butterflies and moths, with their showy and intricately marked wings, that we find such a tempting opportunity to discern subspecific differences without resort to DNA. Then, once a race or subspecies is formally named, it can't be rescinded. That said, its name can be ignored, as I have chosen to do, here, with the races of the silky hairstreak. That, too, is easier said than done, though, as only one of the Tasmanian races—the one found at Lime Bay—has found its way onto the threatened species list.

The wiggling tail of the hairstreak is a deception that we can spot quite easily, and can readily imagine a bird falling for. Perhaps surprisingly, it's far from unique in the butterfly world; it's just that most other deceptions aren't intended for our eyes. While it's nice to imagine that mother nature made the patterns on the wings of butterflies purely for our delectation, she usually had other motives, and not all of them were entirely honest. According to Philip Howse, author of the 2014 book *Seeing Butterflies*, they are often better interpreted in the context of the butterflies' imperfect mimicry of a range of inedible or downright dangerous animals, including millipedes, salamanders, frogs, snakes, falcons, spiders, hornets, bats, caterpillars, wolves and owls. The mimicry may rely on viewing the upper side, or the underside, or from the front or an oblique angle, rather than the overhead view in which butterflies are usually portrayed for us in books or collections. At times, the deception plays out in the part of the light spectrum visible to us; but it may involve ultraviolet light. Often, it might only work for animals with vastly different visual processing capabilities, compared to our own. Even the pleasing, patchwork patterns adorning the wings of our cheerful painted lady butterflies are there for deception, Howse contends: essentially, they provide a visually confusing backdrop, helping to cement an illusion involving the iridescent scales that adorn a butterfly's body. These form a more coherent outline against the backdrop, and, in combination with the streaks of similar scales radiating out from the wing bases, evoke a passing resemblance to the body and legs of a big, hairy spider.

Imperfect mimicry may sound like a waste of time and effort, so why does evolution bother coming up with such deceptions if a predator isn't convinced by the con trick? The answer is that mimicry doesn't have to be perfect for it to give an individual mimic the edge. Just a moment's hesitation by the would-be predator may be enough for the mimic to make good its escape.

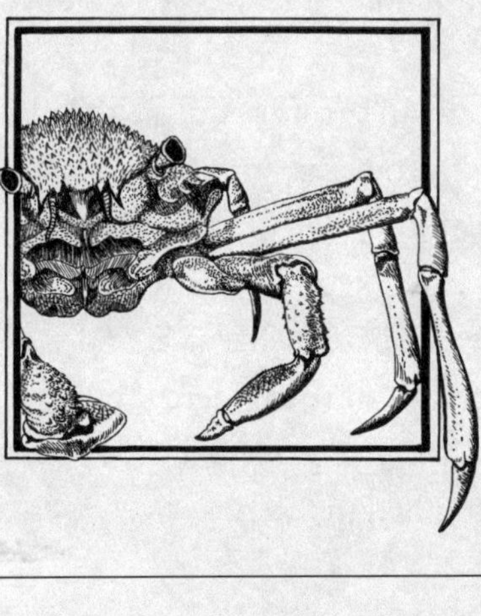

The long march of the spider crabs

Our Scouts camping trip at Lime Bay that November is full of fond memories. Besides my butterfly encounters, there were plenty of scouts-oriented activities, including campfires, walks to the dunes, snorkelling forays and kayaking trips. The broad, shallow bay—part of the western shore of Norfolk Bay—is sheltered from westerlies and a wonderful place for the youngsters to hone their water skills in relative safety. Chris and I enjoy kayaking, too, so, one morning, before the breeze got up, we rounded up a few like-minded scouts, donned lifejackets and carted our craft the few metres from the campsite to the beach. The sea was cold on our feet, our ankles, and then our calves and thighs, but not icily so, as we launched our kayaks and clambered in. Once everyone had progressed past the initial wobbly-kayak stage, we decided to first head northwards, to low sandstone cliffs; then traverse the deeper waters of the outer bay south-eastwards, to the headland marking its eastern end; and then back across the shallows to our starting point.

For much of the outward paddle we were escorted by an overhead party of greater crested terns, which seemed to be banking on us stirring up fish for their breakfast. I found their antics fascinating. Every now and then their elegant, buoyant flight (mariners of old called them sea swallows, and with good reason) would be interrupted as one or more birds peeled off from the group and plunged vertically into the water, folding wings at the last possible moment for maximum streamlining. While we couldn't see what they got up to beneath the waves, we enjoyed watching them bob to the surface, fish in bill, and then, in an instant, find the stamina to lift themselves back into the air again, their plumage seemingly perfectly dry. Generally, the tern would then fly off with its catch rather than re-join its mates, perhaps for fear that one might steal its breakfast. There were also plenty of beady-eyed silver gulls watching and waiting onshore, ready to harry any tern rendered less agile by the weight of its fish.

You get a quite different perspective of the land from a kayak. What seemed like low, stepped sandstone cliffs from the landward side now towered above us as we reached the northernmost edge of the bay, appearing more like the ramparts of a fort rather than an inviting stairway linking the realms of land and water. Ancient she-oaks and yellow dogwoods protruded at unlikely angles from the undercut cliff edge, and cascades of bower spinach spilled over the stone ledges. These same eroded slabs, I had read, had once entombed giant, dog-sized labyrinthodont amphibians which, way back in the Triassic period (before the rise of the dinosaurs) had made their living in the extensive deltas and outwashes of the northern shores of a vast, largely barren continent. Their fossils had been discovered here, just a few years previously. The labyrinthodont's period in Earth's history lay in the aftermath of a runaway greenhouse effect, thought to have been caused by a sudden and massive spike in volcanic activity, which abruptly marked the end of the Permian period. The heat had burnt much of the planet's land surface to a crisp, stripping it of its protective covering of plants and leading to a massive pulse of erosion. Tasmania's convict-era sandstone buildings comprise one modern-day legacy of that distant time, which also gave rise to the honey-brown sandstones that are emblematic of locations as far apart as Maria Island's Painted Cliffs, Sydney's waterfront, and the Blue Mountains.

Ahead of us, a pale shape on a bare she-oak branch, almost overhanging the cliff, gradually resolved into a magnificent, white-bellied sea-eagle. It was sitting proud and aloof, as though sensing it had nothing to fear from our little flotilla, so long as we remained out at sea. Most likely it had a nest somewhere nearby, and was acting as sentry while its mate was incubating their egg, or eggs. It was rather less enamoured with an agitated pair of kelp gulls (yelp gulls might be a better name for these birds) as they taunted the eagle from a safe distance, constantly calling to alert all to the eagle's presence. Kelp gulls are relatively recent colonists of southern Australian seashores. Over the past century they have expanded their range from southern South America, first taking South Africa, then Australia and New Zealand. Why they did so is unclear, but being opportunists, they probably benefited in some way or other from our own species' expansion along the fringes of the Southern Ocean. Eventually the eagle tired of the commotion and launched itself into the air, sailing off around the corner on powerful sweeps of broad, stiff wings. Those stiff wings can sometimes make sea-eagles look rather unstable in flight, as the birds wobble from side to side—not unlike our kayaks. The wobbling makes them look as though they are unable to cope with being buffeted by the wind; but this individual demonstrated that they pack ample power when the need arises.

Our little party, having enjoyed the spectacle, now took off too, powering out across the bay and slicing through glinting wavelets born of a strengthening sea breeze and bright November sun. With each stroke of our paddles, sea spray wafted characteristically maritime aromas—salty and slightly fishy—out onto the breeze and into our nostrils and lungs. I felt invigorated, my senses heightened by the combination of exertion, lungsful of sea air and the feeling of cold seawater on my hands and arms. I could see that our scouts did too, realising with just a touch of envy that their youthful minds and bodies were much more adept at living in the moment than we older folk, with our ever-present concerns about safety and sunburn. We regained our breath in the more sheltered waters close to the headland, before striking out again for base.

This part of our voyage took us back into the shallower reaches of the bay, the sea turning from deep blue to milky turquoise as the sandy seafloor rose into view. Peering down over the side of my kayak, I thought that I could discern an irregular scattering of brick-orange shapes on the seafloor, but swirling reflections made it hard to judge their size, shape and animacy. I called the other kayakers over and we formed a raft of kayaks that somewhat tamed the waves in our lee, enabling us to see the seafloor more clearly. Now the orange shapes sprouted legs! Each was a brick-sized spider crab, and there were several within our field of view. As our kayaks drifted in the breeze, we passed over more and more crabs. There were hundreds, perhaps even thousands of them! Unfortunately, despite our best efforts, we couldn't really make them out clearly from the kayaks, so attentions began to wander, and we decided it was time to head back to shore. After dragging our kayaks up the beach, however, Chris and I immediately hurried back to camp, changed into our wetsuits, grabbed our masks and snorkels, and headed back out into the bay.

We met the first crabs only about fifty metres from shore, in water shallow enough to stand in, had we wanted to do so. Much better, we decided, was to drift passively on the sea surface, face down, watching the spectacle below. Each crab bore a bumpy, swollen carapace; rounded at the back, tapered towards the front, and with a series of blunt spines around the rim. Though their background colour was orange, wispy tufts of colonial hydrozoans, in shades of beige and olive, sprouted from the bumps on the carapace of some of the crabs. Four pairs of pale, gangly legs raised the crab's body off the seafloor, while an additional, more robust, pincer-wielding pair at the front reached out to probe the substrate ahead. A pair of articulated eyestalks protruded from near the front of the carapace, each seemingly capable of independent movement, such that the crab could peer in different directions at the same time. As the crabs advanced—forwards, not sideways like other crabs—their tippy-toe march left tell-tale tracks in the sand. It was clear that the entire seabed hereabouts had been traversed by one or more of these ungainly crustaceans. A little further from shore, in slightly deeper water, the crabs formed dense aggregations, in some places piling on top of one another but disentangling and dispersing in all directions as we glided overhead.

We would have liked to have spent more time with these strange creatures and their sublime, watery world, but coldness eventually got the better of us, forcing our retreat to the beach and the welcome, delicious warmth of the late morning sun. I did, however, return the following morning and enjoyed the encounter all over again.

I have subsequently investigated the natural history of these crabs, commonly known as great, or giant, spider crabs. It seems they usually dwell in deeper waters, down to about eight hundred metres, where they live a life of scavenging, picking over the remains of dead or sickly molluscs, seastars, fish and crustaceans. But every so often they feel compelled to head into the shallows. The long march across kilometres of seafloor may take weeks, and somehow they synchronise their migrations so that they all arrive in certain shallow bays at the same time. In Victoria and northern Tasmania this is usually in winter, but further south—such as at Lime Bay—they may delay their arrival until spring.

It was long assumed that these aggregations were all about mating, but this may not be the primary motivation. Indeed, it transpires that the main purpose is to give them safety in numbers while they moult, which they need to do periodically to grow. During moult, they shed their entire exoskeletons, rupturing the rim of the carapace, then struggling out of their old skin and casting the whole lot aside in one piece, like an all-encompassing onesie—from the membranes over their stalked eyes right down to the tips of their toes. It's not the moult itself that is especially dangerous, though it can't be a lot of fun; it's the hours afterwards, while the new exoskeleton is still soft, that leave them exposed and vulnerable. Aggregations of moulting spider crabs attract the attention of predators such as stingrays, eagle rays, Port Jackson sharks and octopus; but none appears to make serious inroads because there are just so many crabs to eat and so little time in which to do so.

Why don't the crabs congregate in deeper water instead, and save themselves the bother of a long march to the treacherous shallows? It's probably because there are no universally agreed meeting points in the deeps. Each crab might decide that its home turf was the best spot to meet, meaning there would be no aggregation. By all heading shoreward and funnelling into bays and inlets, aggregations occur by default. As for the idea that these aggregations are about mating, well, there may be some truth in that, too. Many crab species engage in soft-shell mating, which is where coupling is only feasible shortly after the female has moulted and before her new exoskeleton has hardened. In such species, a male crab is often observed guarding 'his' female, to ensure that no other males have access to her at this critical time. This strategy works well for crabs that continue to moult periodically as adults, since each female will have multiple mating opportunities. However, spider crabs don't moult again after reaching adulthood; and they don't have the necessary mating organs until after that final moult. The newly matured males that moult first, before the females, are in with a chance at soft-shell mating, and sightings of such encounters may be behind the older notion that this was the purpose of these aggregations. But if they had to rely exclusively on soft-shell mating, the females, at least, would only have one chance in life to do so. Therefore, spider crabs are also endowed with the option of hard-shell mating, and those liaisons could take place many months or even years after their final moult—well out of sight of the prying eyes of shore-bound human observers.

As for all those surplus exoskeletons, many will end up cast ashore by the tides, where their lifelike appearance may fool beachgoers into concluding that there has been some sinister mass stranding or pollution event. Over time, the beach fleas and other shoreline scavengers will strip them of their proteinaceous membranes; then the wind, sun and waves will disarticulate, dissipate, and eventually dissolve what's left.

After learning about Lime Bay's spider crabs, I looked into the origins of the place's name. I had been under the misapprehension that it alluded to a bay back in the 'Old Country', in the county of Dorset, centred on the town of Lyme Regis. That, unambiguously, is the origin of the identical name applied to the settlement at the mouth of the Little Musselroe River, on Tasmania's own Dorset coast, in the far northeast of the state. But our southern Lime Bay reflects a more utilitarian derivation: from having been a place where lime was produced, in convict times (for use in cement, plaster and fertiliser). Why did the colonial authorities source their raw materials here? There are a couple of reasons. The first stage in the manufacture of lime involves cooking crushed limestone or shells in a kiln, which converts the calcium carbonate into oxide (quicklime), so that, with the addition of water it becomes calcium hydroxide, which we otherwise know as slaked lime. Limestone is all well and good as a primary material for this process, but it takes a lot of effort to quarry, transport and crush it. It is far easier to mine ancient deposits of oysters and other seashells, which are just as rich in the all-important calcium carbonate. Even better is if those deposits are matched by a ready supply of wood for the kilns and within striking distance of where the lime is to be used or exported. Lime Bay met all those conditions, lying just a few kilometres north of the Saltwater River coalmines where, in the first few decades of the nineteenth century, convicts were put to work mining coal for the penal settlements and for the growing colony of Van Diemen's Land.

Even today the shoreline at Lime Bay is periodically turned white with the crushed fragments of millions of clamshells—mostly wedge shells and Venus clams ('cockles'), attesting to the productivity of the shallow bay in which the molluscs thrive, embedded in soft sediment and filtering the seawater to extract microplankton and detritus. Scattered amongst these smaller shells are the larger shells of our native mud oyster, or flat oyster. Many of them have been worn smooth by decades or centuries of tides and weather. You wouldn't spot it today, but these oysters once formed a discontinuous shallow-water reef along Tasmania's south-eastern coast, from Norfolk Bay to the southern end of the D'Entrecasteaux Channel. The oysters of those rich reefs sustained Lutruwita's First Peoples for thousands of years. The evidence is—or was—all around us, in the form of massive middens of discarded oyster shells that indicate former living places. Maybe the First Peoples of the region also enjoyed a meal of spider crab; but if they did so, the more fragile exoskeletal remains of these have long since vanished. When European colonisers murdered or forcibly displaced the First Peoples, or subjected them to European diseases, the middens remained—for the mining.

Meanwhile, the wooded hinterland would have been plundered for firewood for the kilns (hence the preponderance of short-lived silver wattles today, with several generations having grown up in the aftermath of this disturbance). The oyster reefs, too, were mined, both for their shellfish (exported in barrels of brine) and for the shells—they were all but extirpated. That is not to say that native oysters have been extinguished, but they no longer form reefs. It's one of those chicken-and-egg situations, because oyster spat seek out existing oyster shells as points of attachment, when they are ready to settle out from their planktonic stage and start secreting their own shells. In the absence of oyster reefs, they may try their luck on other shells such as whelks and scallops; but even if they manage to cement themselves to these shells, their efforts are unlikely to result in the establishment of a new reef. These days, plans are afoot to seed the establishment of new reefs by dumping masses of waste king scallop, abalone, or oyster shells in strategic locations, in hope that oyster spat will settle out and cement the whole lot together, triggering the formation of new reefs which will then attract more spat. It's an enticing prospect and may mimic the origins of at least some of the old reefs, which could have been founded on shell middens submerged by rising sea levels over the past ten thousand years. The benefits of restoring oyster reefs are manifold. They include protecting shorelines from erosion, providing habitat for fish and other reef-dwelling animals, and acting as highly efficient biofiltration systems for pollutants.

Time spent at Lime Bay can reveal our relationships with fellow human beings and nature, if we choose to look. We can reflect on the shifting sands and deltas of the distant Triassic; on rising sea levels; on the long and continuous history of habitation by Lutruwita's First

Peoples, and their relatively harmonious relationship with nature; on the brutality and utilitarian outlook of the colonial project; on persistence and drive in the natural world; on ecological healing; and, of course, on the privileges of recreation and nature immersion that we enjoy today. The same could be said about much of the rest of Tasmania and beyond, but at Lime Bay those layers of history and prehistory, of use and abuse, of healing and learning, of working and playing, seem more real, more poignant, more organic, and more alive.

As the gull flies, Lime Bay is not very far from Hobart—you can hear the planes descending towards the airport—but driving there involves heading east, to Dunalley, then south across the Forestier Peninsula, then back west, along the northern shores of the Tasman Peninsula; and, finally, north again, through the reserve to Lime Bay. All up, it takes a couple of hours. By the same token, it takes two hours to get home again; so, after our Scouts weekend, we knew that much of our drive would be in the dark. Nevertheless, the sun was still just above the horizon as we exited along the potholed and corrugated dirt road through the wooded reserve, and then out, onto the bitumen at Saltwater River. I always relish the first few kilometres of bitumen here. In part it's because the drive seems so smooth and quiet after travelling on the dirt; but it's mostly because of the views over the series of shallow bays that form the southern flanks of Norfolk Bay. The bays' profiles are even more gently shelving than at Lime Bay, which means that an ebbing tide reveals wide expanses of wet sand, aesthetically arranged into meandering ripples and channels that catch the light. They were looking especially appealing this evening, with the water surface flushed apricot, and the sand a warm, fawn brown. The sands were also pockmarked with shallow depressions, where stingrays had been feeding when the tide was in; and I could just about discern bands of darkness parallel to the water, which I knew were armies of feeding soldier crabs. A fly-by of pied oystercatchers and silver gulls completed the picture. As I drove home, all passengers now asleep, I reflected on our good fortune, once again, at having such easy access to this scene of serenity, and countless others across Tasmania, and to the nature that so perfectly stitches it all together.

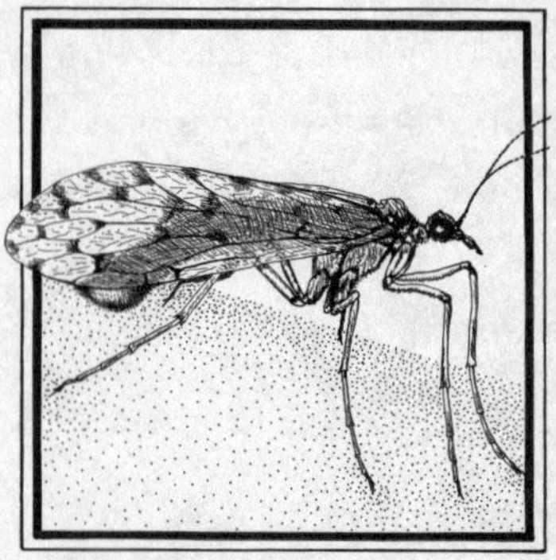

Hardy's hardy nannochoristids

Exactly a hundred years before I became a curator at the Tasmanian Museum, another Englishman, with the grand and delightfully Pythonesque name of George Hudleston Hurlestone Hardy, arrived in Hobart. Not long afterwards, he took up the position of Assistant Curator and Secretary to the Trustees of this same august institution. He brought with him an abiding interest in entomology, and soon set about building an unrivalled collection of Tasmanian flies and other sweep-nettable insects (he seems to have had rather less interest in fossicking for beetles and bugs). Before long he became the *de facto* national fly expert and a prolific publisher of taxonomic papers describing scores of species new to science. He no doubt picked up his dipterological interests from his father in England, also G. Hurlstone Hardy, who in 1915 had authored *The Book of the Fly*, in which he introduced a no-doubt enthralled global readership to the life and times of houseflies and stable flies. Despite such productivity and scientific accomplishment, the junior Hardy's lack of formal museology qualifications apparently precluded his elevation to the level of Curator, which he must have found a bit galling. Perhaps this is why, a decade after he arrived at the Tasmanian Museum, Hardy moved to the University of Queensland, where his entomological career prospered. The bulk of his Tasmanian collection was sent to the Australian Museum in Sydney, leaving just enough in Hobart to give future curators—myself included—a tantalising glimpse into the breadth and depth of his Tasmanian entomological achievements. In one of those satisfying, circle-closing twists, the trusty brass microscope, through which he would have first viewed and studied his Tasmanian insect collection, was recently donated to the Tasmanian Museum by descendants now living in Canada.

Flies, of course, owe their name to the fact that they can fly; but not only flies fly. Four-winged insects first took to the wing hundreds of millions of years before true, two-winged flies came along. Members of one major group of insects that beat the true flies go by the name of 'scorpionflies'—and therein lies a tale involving our George Hudleston Hurlestone Hardy. It's a tale that enthralled me when I first heard it, and set me on my own path of entomological enlightenment.

Our story begins in November 1916, when Hardy mounted a collecting trip, with a colleague, to the accessible lower reaches of Mount Wellington. I like to imagine them chugging up the road in one of those new-fangled motorcars, sporting goggles and scarves. More likely, given Hardy's modest station in life, they took public transport, or simply walked. Strolling along Strickland Avenue, they came to a '*little watercourse flowing from the leakage of a portion of the Hobart Waterworks scheme*'. There, Hardy swept several little lacewing-like insects that were the target of this trip. The previous month, when he had first encountered these insects nearby, he had sent the specimens to Robin Tillyard, a respected entomologist at the University of Sydney. Confirming Hardy's suspicions that he had found something new and unusual, Tillyard—who had just been sent specimens of a clearly related insect from the cool uplands of the New England region of northern New South Wales—'*at once wrote and urged him to obtain more*'. In his paper, published by the Linnean Society of New South Wales, Tillyard describes how Hardy '*became fully seized with the importance of his discovery*', hence this follow-up trip.

Hardy relayed his successes on this second trip to Tillyard, who was sufficiently moved to travel down from Sydney, later that summer, to see these strange insects and their habitat for himself. This would have been quite an undertaking for Tillyard, since he suffered from crippling, chronic pain arising from a train crash a few years previously. Alas, though Hardy took him to the exact same spot, Tillyard concluded that their season must have been over as not one was found. Nevertheless, they had more success higher up the mountain, at The Springs. Later that month, presumably as part of Tillyard's same Tasmanian visit, they proceeded to the then barely explored locality known as Cradle Mountain (perhaps staying at the Weindorfer's newly opened Waldheim lodge), where they collected two further species by working the shores of Dove Lake and Lake Lilla. Tillyard now had enough specimens to formally describe these insects, and to announce to the world that Hardy's discovery constituted a whole new family of insects, the nannochoristids. Though lacewing-like, they fitted within an entirely different group of insects—the scorpionflies. Scorpionflies, it goes almost without saying, are neither scorpions nor flies. The name alludes to the somewhat scorpion-like telescopic 'tail' of one lineage common in temperate parts of the Northern Hemisphere.

Tillyard was clearly very taken with these discoveries. Finding nannochoristids in these cooler, wetter parts of Australia led him to suggest to colleagues in New Zealand that it would be worth keeping an eye out for similar insects there. Lo and behold, specimens emerged in a collection from the South Island, dating from 1913. Tillyard moved in rarefied academic circles and was well versed in theories about how plants and animals came to inhabit distinct parts of the world—a field that we now call biogeography. At the time, people were starting to find similarities in the fauna and flora of temperate Australia, New Zealand, and southern South America; and some had postulated that Antarctica might have been their ancestral home, from which they had since dispersed to the other southern continents. Bear in mind that this was many years before the theory of continental drift formalised by Alfred Wegener, in 1912, had gained broad acceptance; and even earlier compared with today's understanding of plate tectonics.

One of the proponents of this 'Antarctic Theory' was Charles Hedley, curator of molluscs at the Australian Museum in Sydney (we will meet Hedley later, in connection with dredging for deep-sea molluscs). Tillyard wrote, of nannochoristids, that '*the bearing of this discovery upon the Antarctic theory as advocated by Mr Hedley is obvious … if another form in this family were to be found in South Chili or Patagonia, the evidence would be complete.*' He added, somewhat ruefully, that, '*it seems almost hopeless to expect this region to be well-searched for such out-of-the-way insects, for a very long time to come.*'

Science—and the European colonial project—advanced more rapidly than Tillyard anticipated, however; and two nannochoristid species were described from southern South America just twelve years later. We now know that the larvae of these hardy southern survivors live in cold, clear streams and shallow, peaty pools, which is why the adults are usually found close to such features. Today we would label the distribution of nannochoristids as Gondwanan, but that isn't the full story, since fossils found in South Africa extend their lineage way back to the Permian era, before Gondwana came into existence. Furthermore, younger fossils, of Jurassic age, have been found in Mongolia, which was never part of Gondwana.

Like Hardy, Tillyard went on to achieve great things in entomology, first in Australia and then in New Zealand; but he continued to suffer poor health. Work-related pressure led to a nervous breakdown in 1934, and he died three years later, following another vehicle accident, this time involving a car.

Exactly a hundred years after Tillyard's publication announcing Hardy's discovery, I decided to mount my own commemorative expedition in their footsteps. I dug out my hobnail boots and gave them a good smear of fresh dubbin, donned my tweed breeches and greatcoat, and packed my canvas knapsack with essentials—a killing jar primed with crushed laurel leaves, a foldable silk butterfly net, my trusty wooden pipe, a tin of tobacco, and, of course, a hip flask of whisky. Duly prepared, I strode off along the rutted track into the foothills. In my imagination, at least. In reality, I had the modern convenience of comfortable apparel, while a fossil-fuelled car swiftly and effortlessly conveyed me along paved roads to within metres of Hardy's original collecting locality. In well-frequented public settings such as this, I feel a bit bashful about swishing a net around, but I told myself that this was an important historical re-enactment and constituted entirely reasonable behaviour, given the circumstances. I advanced towards the gushing rivulet, at a point a little above where it dives into a culvert under the road, and began energetically sweeping the overhanging ferns, dogwoods, and mountain currant bushes, to the bemusement of a party of onlookers. Bemusement turned to concern as they witnessed the next step in my search, which involved inserting my head into the net to peer intently at the tiny insects within. And concern may well have turned to alarm when I retracted my head and shook the net out, muttering darkly to myself, only to move upslope a few metres and repeat the entire process.

Disappointingly, like Tillyard, I failed to find any nannochoristids at Strickland Falls, but I wasn't going to give up that easily. I hazarded that perhaps they had retreated upslope a little, for I thought, or at least hoped, that the impact of the 1967 bushfires on the rivulet's riparian habitat might have been less severe there, and that somewhat cooler temperatures might still prevail, too. Accordingly, I headed up the riverside track, intent on trying my luck at O'Grady's Falls. Again, I was met with strange looks as I scrambled down the steep bank beside a little bridge below the falls. Only a few metres downstream, the rivulet disappeared beneath a tangle of fallen logs, so my effective search area was rather limited; but closer to the falls the bed of the rivulet was wider and strewn with stones and fallen branches, providing convenient, if precarious, mid-stream perches for my swishing, and enabling me to cross from one shady bank to the other, and back again, without getting my feet wet. Walls of glistening, dripping ferns closer to the falls looked enticing, but would have soaked my net, so I restricted my sweeping to the drier overhanging shrubs. And, yes! The very first sweep of a small sassafras overhanging the rivulet produced a fine specimen of Hardy's nannochoristid. About half the size of a green lacewing, it had a narrow body with downward-facing mouthparts at the front end and a downwardly arched abdomen at the rear. It held its strongly net-veined, but otherwise transparent wings tucked along the line of its body yet vertically orientated. It was a very flighty insect despite the cool, damp environment, constantly engaging in short flutters from one part of the net to the other, yet tucking its wings closed on each landing. Elated, I invited a walker, who had been warily watching me from the bridge, to come down and take a closer look, but she wisely declined.

That was the only nannochoristid I found that day, but I wasn't done with them yet. Later that summer, I made a pilgrimage to my favourite happy spot in the lee of Kunanyi's lofty, windswept plateau. There, among the peaty pools and mossy hummocks, I netted my second species of nannochoristid—the spot-winged nannochoristid, one of those that Hardy and Tillyard had first found at Cradle Mountain. I like to think that they would have been as delighted with my discovery as I was.

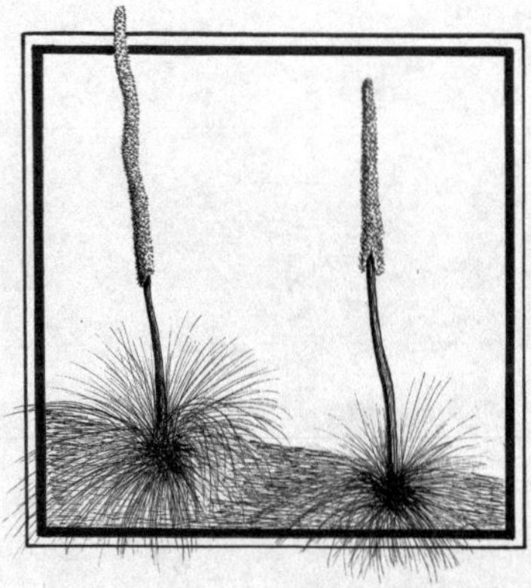

Grasstrees

The early British colonists who fanned out across what is now Australia were a motley bunch: free settlers and other opportunists, convicts and ex-convicts, and upstanding and less-than upstanding representatives of the Crown. They brought with them not only their crops and livestock but also their values, prejudices, worldviews (both utilitarian and religious), laws, customs and ways of speaking. The idea prevailed that there was little to learn from the natives, who had occupied what they regarded as '*terra nullius*' for tens of thousands of years; and the invaders' approach was, virtually universally it seems, one of cultural annihilation—and *vox nullius*— for the locals, rather than reverence, deference, or even a willingness to hear what the First Peoples might have wished them to know.

From these unsteady—and, I suggest, deeply unsavoury—foundations was the modern country forged. And the colonists' legacy is still with us in so many ways. Much has been written on such matters, but I want to highlight an aspect peculiar to plants: the vernacular names that became attached to some of Australia's iconic trees and other flora.

It's not often that a culture gets the chance to invent a swag of new plant names, but that was the prospect that seemed to present itself to the colonists. I have to say that, for the most part, they weren't particularly creative. Rather than inventing new words, they tended to prefer to reuse old words, applying them in new ways. To be fair, the etymological origins of British plant names mostly go back hundreds or even thousands of years. Nobody suddenly invented the word 'oak', for example. So perhaps we shouldn't be surprised that the colonists weren't in the business of inventing brand-new terms for the plants that they encountered in Australia, although some imaginative exceptions come to mind—for example 'running postman' and 'pigface'. The trouble is that the English botanical lexicon was rather limited and not up to the job at hand. No matter, though—I suggest that the narrative went something like this:

> *That tree over there, with the stocky trunk and gnarly bark and upright foliage? It must be an oak! So must that one with the droopy foliage. Let's call one a bull-oak and the other a she-oak … even though their leaves look like pine needles … Speaking of which, that one bearing cones and with strongly scented, resinous sap—it must be a pine. No matter that its leaves are more like a cypress. We'll call it a cypress pine. Those tall trees with aromatic foliage? No idea what to call those, but they seem prone to exuding gum at the slightest provocation, so why not call them gumtrees? Mind you, some of those gumtrees produce timber reminiscent of an ash tree from the Old Country. I reckon we should call them ashes. Then there's that jungle full of useless trees that needs clearing before we can make use of the land—we'll call it scrub, and the big patch of jungle—it can be The Big Scrub.*
>
> *But back to individuals: there's that understorey tree with soft, oval leaves. My best guess is it's a dogwood, but on the other hand we might sometimes call it a native pear, just because why not? The seeds on the shrub next to it look a bit like hops, so that must be a native hop, even though it's not a twiner; wonder what it's like in beer? Those berries look a bit like cherries, so that tree must be a native cherry, even though its leaves (does it even have any?) bear no resemblance whatsoever to those of a cherry; come to think of it, nor do its berries …*

So on it went, across the country, and through every botanical branch of the tree of life. Eventually, however, the colonists happened upon a plant that defied all classification. Its fine, grass-like leaves emanated from dense tussocks. So far, so 'tussock grass'. But from the centre of the tussock a tall, woody flower spike several times the tussock's height might protrude; and the tussocks themselves often sat atop sturdy, frequently bifurcating woody trunks that were entirely cloaked in strange, gnarly tessellations akin to reptile skin. '*Well, I guess we'd better call these grasstrees!*' someone might have said, while another might have pointed out that '*after a bushfire, when their trunks are charred to blackness, how can we not refer to them as blackboys? It will do us good to keep the memory of our genocidal conquest of the great southern land alive, after all, don't you think*?'

Grasstrees number some thirty species, and they are an iconic feature of Australia's nutrient-poor sandplains, heathlands, and heathy woodlands. They occur from west to east, and from the Top End all the way to Tasmania. Individual plants may live for decades, or even centuries. They thrive in terrain where the soil is poor and fires are frequent. Fire stimulates the plants to send up their flower spikes, the top section of which is a fragrant, insect-attracting inflorescence packed with thousands of tiny, creamy-yellow florets. A fire may destroy the foliage and char the trunks, but these soon re-sprout. Grasstrees' scientific name translates as 'yellow flow', alluding to the resin—usually more orangey red than yellow—which oozes from damaged plants. The resin was long used by Aboriginal peoples across Australia as an all-purpose glue, particularly for attaching spearheads and axeheads to their shafts. Settlers soon got in on the act, using it in varnish and incense, as an alternative to shellac and—of course—in explosives. The Aussie term for strenuous physical work, '*hard yakka*', incorporates a South Australian Aboriginal word for grasstrees, and reflects the great effort (of Aboriginal labourers, no doubt) that was required to extract resin from the plants on the industrial scales demanded by the settlers.

All of this is a rather longwinded way of saying that grasstrees come with a lot of baggage, but at the end of the day, they are wonderful plants—perhaps Australia's finest. Who could gaze at a heathland recovering from a bushfire and not be awestruck by the sight of thousands of flowering grasstrees emerging defiantly from the ashes? Alas, quite a few people, it seems, since grasstree country has long been viewed as wasteland, ripe for taming: ripping, ploughing, fertilising, and re-seeding to provide succour for livestock; or simply built over. If we feel a certain unease around the treatment of forests in Australia, we should feel utter disgust for how our heathlands have been abused. Those that remain are acutely vulnerable to phytophthora root-rot, an invasive, water-borne, soil-dwelling fungus-like microorganism that is often unwittingly carried from one place of infestation to the next on the tyres of vehicles or the boots of bushwalkers. It has a particular liking for grasstrees. Infected plants turn straw-yellow, then brown, and eventually shed all their leaves. Their bare, blackened trunks then stand for decades as deathly-silent witnesses to our collective indifference.

One sunny November day a couple of years back, while carrying out a reconnaissance for a forthcoming collecting expedition in the far northeast of Tasmania, I found myself approaching a sandy, heathy ridgeline near the sea. The heathland had been burnt the year before, apparently by some local arsonist with a grudge. Or perhaps the fire-starter had a deep understanding of fire ecology and was intent on doing the heathland a favour, since the consequences of their actions were now a sight for sore eyes (and, in fact, a smell for sore noses). Approaching on foot (wearing boots scrubbed in disinfectant to reduce the risk of introducing root-rot), I could see that the ridge's outline was broken up by the blackened stems of innumerable grasstrees, now topped with fresh new tufts of deep-green foliage; and from every one of them soared a vertical, cream-coloured pole of flowers. Luck was on my side, it seemed, and I had arrived at the peak of the grasstrees' post-fire flowering! I could hardly contain my excitement as I knew that there were also special insects that timed their emergence into adulthood to coincide with these flowering events.

The smell intensified the closer I got to the ridgeline. I wouldn't call the aroma of grasstree blossom especially pleasant—hints of suburban cherry-laurel and privet, but with sweeter overtones—yet it certainly evoked a strong sense of place; and I knew that it would be bringing in the insects. I arrived at the first grasstree. It was a relatively small and stocky individual with a trunk about a metre tall and a flower spike adding a further metre to its height. That put the flowers at a perfect height for bug-watching. Before long I had spotted the first of what proved to be a long succession of special insect species that afternoon. This one was a fly, but it looked more like a beetle because the rear halves of its smoky-black wings were folded abruptly downwards along a hinge-line, rendering the fly's dorsal outline strongly angled rather than lying on a single plane, as would be the norm. I have no idea why a fly would evolve this strange appearance, but it must be a useful thing to do since the flies continue to exist, and evolution tends to weed out maladaptations. The fly's body was also matt-black, though the dark, compound eyes bore horizontal bands of slightly paler facets. Two spindle-shaped antennae projected obliquely from the front of its head. It was my first ever sighting of this particular soldierfly species, whose scientific name translates as the brown-winged soldierfly. I was very pleased with myself. The species was described by one Mary Fuller, back in 1934. Her paper notes that she reared these flies from larvae found in the grasstrees. Regrettably, I'm none the wiser as to whether she meant the trunks, leaf bases, inflorescences, or leaf litter, but I suspect she meant the crumbly, woody, and resinous material that often accumulates around the base of a grasstree. Mary must have been an interesting character. In the same year, she also published a whole book on *The Insect Inhabitants of Carrion*. A woman after my own heart.

I left the beetley soldierfly to its feeding and moved on to the next grasstree, mere metres away from the first. It was a much larger specimen, and its flower spike towered over me. Right at the very top, two males of our largest species of flower wasp, gaudily striped in black and yellow and with cellophane-like, slightly reflective wings, competed for what must have been a particularly special patch of flowers (though they all looked the same to me). Further down were numerous blowflies and tiny pin-tailed beetles. These were all nice to see but they weren't grasstree-specific insects, so I moved on. Now I was into the main stand of grasstrees, surrounded by them in all directions. I found myself entering that state of heightened, rapturous mental arousal during which I can feel almost giddy with elation. Wherever I looked there were insects zipping back and forth from spike to spike: native bees, native wasps, stinkbugs and, especially, flies of all shapes and sizes. I stood there, transfixed by all the happenings, for perhaps ten or fifteen minutes. I don't know now whether I was unable to move, or whether I chose not to because I was in the thick of it where I was. Either way, finally the fly that I had been really hoping to see put in a brief appearance. It's a grasstree-specific hoverfly, whose name translates as the greasy straight-face.

Believe me, it's a lot more appealing than this name would suggest, though perhaps only to fly afficionados (and to others of its kind). It's a stocky, dark grey fly with a couple of triangular wedges of creamy white on the sides of the abdomen. Its grey thorax is densely haired, other than two bald, black eyespots either side of the midline. More striking is the fly's bright lemon-yellow face sporting short, rounded antennae in apricot-orange. All of this I got to see momentarily, because, although the hoverfly settled on the flower spike right in front of me, my presence must have spooked it and it buzzed off almost immediately. I had seen this fly before, but only ever on the flower spikes of grasstrees. As with the soldierfly, the scientific literature informs me that its larvae have been reared 'from grasstrees', but again, I don't know which part thereof. The species occurs around Australia's eastern, southern, and southwestern seaboard, but in Tasmania it seems to be confined to the heathlands of the northeast.

To find the last of the flies that I was looking out for, I needed to change tactic. Instead of focusing on the flowers, I turned my attention to the foliage; and, within minutes, I had found it: small and rather slender orange-brown flies with long wings, sparse but long black bristles, reddish eyes, and short, dark-grey antennae. These were lauxaniid flies, but not any old lauxaniid (a fly family with members found in all sorts of habitats), since they belonged to a genus associated exclusively with grasstrees—though, again, in what capacity I remain ignorant. The genus was only recently taxonomically reviewed, at which time the number of known species was increased from one to nine. Those that I was looking at now belonged to one of the new species, the 'long lauxaniid'. I stood and watched them for a few minutes. Each fly seemed to have staked out its own leaf blade and sat on the outer portion, where the blade curved downwards, with head facing down. Occasionally one would launch into the air and pounce on one of its neighbours, perhaps hoping for a mating opportunity or, more likely, to intimidate a potential rival. It soon became clear that all these flies were male, because a rather more rotund individual then flew into the melee, prompting those stationed nearby to take flight in pursuit of her. That was the last I saw of them, as the whole contingent promptly dashed off out of sight.

I was meant to be checking out plenty more of the potential expedition site, besides this fabulous ridgeline; so I reluctantly left the grasstrees and continued on my way, to spend the rest of the day exploring woodlands, wetlands, and coastal cliffs. The site certainly got my thumbs-up; but when, eventually, we did return for our main expedition (later in the summer), the grasstrees had finished flowering and their spikes were brown, dry, and brittle. The lauxaniids still busied themselves on the leaf blades, but, for the time being, the magic had left the place. The sweet memory, however, lingers on.

© K.DAVIS 2022

December

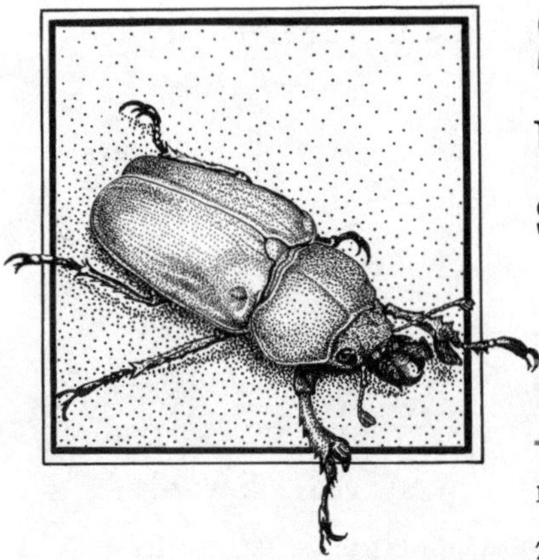

Summer's promise unfolds ... sometimes

During the long, dark weeks of winter, I yearn for December—that far-off month when the sun reaches its southern zenith; when the days are blissfully long and everywhere is bursting with life. Then December arrives and takes me by surprise, again. Where did spring go? For weeks I had been enjoying the lengthening days, and the extra possibilities for after-work gardening or after-dinner walks; but, *'surely it's not summer already?'* In truth, spring didn't go anywhere, because spring-like weather usually lingers well into December. I'm rather grateful for this, because I relish the freshness of late spring, both in the natural world and in the garden. But there's no guarantee of any stable weather this month, and every chance that there will be days and nights that feel as though the weather gods had capriciously plucked them out of September or October. Still, when the sun does shine, the intensity of its glare and its readiness to burn the skin does justify comparisons with Mediterranean Europe, even if the air or sea temperatures may not. Like the sun, insect diversity also reaches its zenith in December, although peak abundance comes a bit later in summer.

Gentle rain in December is great for boosting soil moisture levels and topping up the groundwater, postponing the day when dams dry out and creeks run dry. Creeks and stormwater drains fed from wooded catchments can fill with what looks like soap suds, which isn't far from what they are. Many trees protect their foliage from insects and microorganisms with bitter-tasting, soap-like chemicals called saponins; and rain after a dry spell can rinse these into the frothy water that trickles down their trunks and into creeks. In Taroona, where wooded gullies meet the sea, the suds are often washed up on the beach, sometimes in great billowing masses, when they can be mistaken for some hideous industrial pollutant.

Summer rain can perk up the local frogs enough to extend their breeding season. In such weather, our local creek (more a weed-choked ditch really) echoes to the upwards-inflected *shree-shree-shree* calls of brown treefrogs and the high-pitched crackles and croaks of common froglets. It's not only frogs that enjoy warm summer rain either. If I'm in the right frame of mind, I will also go out of my way to get wet and feel the pleasure of raindrops gently caressing my face and scalp. Rainwater trickling down one's face tastes good, too. I love to breathe in the distinctive, earthy smell of freshly dampened ground. It has a name: petrichor, which was coined in 1964, from the ancient Greek words for 'rock' and 'the blood of the Gods'. A year later, its chief constituent, a microorganism metabolite, was named geosmin, meaning 'earth smell'. Geosmin appreciation is all about context: in a glass of drinking water, the same chemical reminds me of ditch water.

For many of us, for better or worse, December is almost synonymous with Christmas. What better way can there be, then, to bring out a little festive cheer than with a beetle? This is one, of course, whose appearance at this time of year has earned it the name of Christmas beetle. It's no secret that Tasmania likes to be a little different from the rest of the nation, and this is true of Christmas beetles, too. What we call Christmas beetles are a species of stag beetle, also known as the golden stag beetle (whose generic name is *Lamprima*), whereas, on the mainland, the moniker more usually refers to any of several related species of scarab beetle in the genus *Anoplognathus*. What these have in common is bright, metallic colours that may remind some of Christmas baubles, and a tendency to take to the wing in warm, sunny weather, especially during the days leading up to Christmas. Indeed, any such beetle or bug flying about around Christmas time is likely to be called a Christmas beetle. Our own golden stag beetle occurs not only in Tasmania, but also along the eastern seaboard, all the way north to the Queensland tropics, as well as in South Australia and southwestern Western Australia. Most individuals are a gorgeous golden-yellow colour, but others may reflect orange or apple-green. Some individuals carry a purple sheen, which may be applied so richly that the whole animal looks almost black. Such forms seem to be particularly prevalent near the coast, for reasons unknown.

Golden stag beetle larvae feed in rotting wood, particularly in the decaying roots and stumps of long-dead trees. They may take a couple of years to mature and pupate. Strangely, the adults may emerge from their pupae in autumn, but if they do so, they will remain within the rotting wood throughout the winter. It's only when temperatures climb, in late spring or early summer, that they claw their way to the surface and take to the skies. Although perhaps 'skies' is pushing it a bit: when flying over open ground, they generally bumble along at a cruising altitude of a mere metre or so, which puts them at a perfect height for being admired by humans—especially children.

Since room was made for our suburbs by clearing bushland and remnant trees, they start out with a helpful supply of rotting tree roots—which are hidden from our view beneath lawns and open spaces. That's probably why golden stag beetles seem so at home in suburbia. When people report that there aren't as many as there used to be, it may be that they're not so 'in your face' as they were during childhood; or perhaps it's just that, in that particular suburb, the supply of rotting tree roots has been exhausted and the beetles have needed to move to more recently established suburbs to find the habitat their larvae need.

I would prefer to celebrate the summer solstice, in lieu of the Christmas festivities that follow just a few days later. A dawn or dusk solstice swim in Taroona is likely to be a bracing affair, but you've got to mark the passing of the seasons somehow. While the solstice signifies a gradual return to shorter days, it also anticipates more summery weather ahead, since the last week of the month (days after the summer solstice) is often when summer finally settles upon us. It's not uncommon for temperatures on Christmas Day to struggle into the low teens, and fresh falls of snow to settle on the mountain, so that festivities are carried out huddled round a fire; but by New Year's Day they may well have soared into the high thirties.

The swallow's tale and the swallowtail

Tasmania's Overland Track is one of the state's, the country's, and indeed the world's, most iconic long-distance walks. Its rather bland name commemorates its original 1920s working title and its function at the time, which was to provide a direct 'overland' route to connect the scenic reserves of Cradle Mountain and Lake St Clair. Today, the entire area is a National Park and is, itself, encompassed by the vastness of the Tasmanian Wilderness World Heritage Area. Walking the Overland Track is almost a rite of passage for outdoorsy Tasmanians, but numbers on the track are strictly limited. Accordingly, during the high season, one needs to book in advance and then hope for good weather—or at least for weather that makes the walking feasible. Picking a date early in the season is likely to mean less competition for a pass, but the chances of foul weather are higher. A later departure improves the odds of better weather, but you have to book well in advance to secure a pass. We compromised, settling on a mid-December departure for our Overland initiation some years back, joining with some other local families we knew on the six-day, 83-km walk.

Much has been written about the walk, and I'm not going to repeat here what others have already said about the experience and landscapes. Suffice to say that, for a naturalist, the walk is hard to beat. It passes through remote and spectacular country and a variety of habitats, including wild rivers, tarns, buttongrass plains, rainforests, windswept boulderfields, ancient conifer groves and extensive tracts of old-growth eucalypt woodland and forest. The only drawbacks are that one must carry all one's supplies and keep moving so as to arrive at successive designated overnighting spots each evening. Collecting specimens is also out of the question! Consequently, for the most part, one completes the walk having gained an overall impression of the nature and weather experienced along the way, rather than notching up deep and meaningful encounters with nature. It hardly needs saying that weather experiences are all-pervading, however, and often win out over nature experiences. Even in high summer, all eventualities must be prepared for—we experienced snow, hail, freezing rain, wind, fog … and intense sunshine and heat.

A long-distance walk can induce a trance-like state, particularly towards the end of the day, when energy levels are low and you've done so much walking that putting one foot in front of another becomes automatic. Boardwalked sections of track, designed to avoid erosion and ease progress, only add to the risk. I was in just such a state on our fifth day of walking, as we wearily trudged southwards from Narcissus Hut towards Echo Point, on the western shore of Lake St Clair. Passing through callidendrous rainforest and some of the largest myrtles that I have seen anywhere, it was a stunningly beautiful section of the track, sadly bypassed by most walkers, who opt to take the ferry from Narcissus Bay to the walk's official end at Cynthia Bay; but I was near the end of my tether by this stage, and almost oblivious to my surroundings.

It took a butterfly to snap me out of my trance. I glimpsed a movement out of the corner of my eye and instinctively turned to identify its source. Along this part of Lake St Clair, the shoreline and adjacent forest is thick with Christmas bush—a lanky species of mint-bush whose showy, lobe-petalled, purple-flecked white flowers are at their peak around Christmas time. Hovering in front of a sprig of these flowers was a large butterfly whose wings were mint-green inside and out, overlaid with chocolate-brown panelling. The hind wings bore long 'tails'. The late afternoon sun accentuated the butterfly's colours as it adeptly inserted its watch-spring proboscis in a succession of flowers, hovering all the while. This was Macleay's swallowtail, one of our finest and most charismatic butterflies. Perhaps 'hovering' is the wrong word to use, since the butterfly's feet periodically made contact with the flowers and foliage, as though the insect were in a permanent state of just-about-but-not-quite landing. This is characteristic behaviour among swallowtail butterflies, most of which live in tropical forests. Macleay's swallowtail is tropical, too, but its range extends all the way south along Australia's eastern seaboard, from the Iron Range in Far North Queensland to Tasmania, where it is our only swallowtail species.

Once I had been reawakened, I couldn't help but spot more and more of these enchanting butterflies. Every Christmas bush hosted at least one; and over the following hour I must have notched up at least a hundred sightings. You don't have to travel to the wilds of the World Heritage Area to spot one, since they can be found wherever their larval foodplants, chiefly sassafras, occur. Sassafras is a rainforest tree, but one that seems able to survive in tiny pockets of rainforest, even in otherwise dry country. Furthermore, the butterflies travel widely in search of new patches of habitat and can turn up almost anywhere, particularly on hilltops. Butterflies start to emerge as early as October and can be seen into February, but December is when numbers peak. It's for this reason that I have selected Macleay's swallowtail as butterfly of the month for December.

Newly enlivened by the swallowtails, I rather belatedly took in my surroundings. Mighty myrtles towered overhead, their trunks sprouting luminous orange bracket fungi; while the forest floor was an emerald carpet of moss and lichen with an intermittent overstorey of low ferns. We also passed through groves of some of the most massive treeferns that I have seen—and smelt. The forest exuded a pungent aroma: earthiness and ferniness, for want of better words, punctuated with wafts of mintiness from the leaves of the Christmas bushes. I recall few bird sounds, other than the near-constant jingle of pink robins and the occasional squeaky squawk of green rosellas.

At long last we arrived at Echo Point, our destination for that night. We dropped our packs near the hut and felt as though we might levitate, such was the relief at the sudden lack of weight on our backs and shoulders. Steadying myself, I grabbed my water bottle and wandered over to the small wooden jetty, following it to its end and basking in the views across the lake and the sense of accomplishment at a walk well done. Sitting, now, with my feet dangling over the edge of the jetty, I glanced back towards shore and was delighted to see a family of welcome swallows,

Mum, Dad and four fledglings, all lined up on the wooden handrail. Every now and then one or other of the parents would ruffle their feathers and the fledglings would follow in unison, dislodging the last of their nestling down. Caught in the slightest of evening breezes, the tiny feathers drifted over the lake, only to be plucked from below by brown trout, once they brushed the water's surface, mayfly-like. Presumably, the fledglings had been reared in a nest of daubed mud bracketed somewhere beneath the eaves of the walkers' hut at the start of the jetty.

Welcome swallows are common summer migrants to Tasmania and are also found right across Australia. They apparently owe their name to sailors of old who, encountering them at sea, welcomed them as a sign of nearby land. Here was I, doing much the same but in reverse: welcoming them as a sign of nearby water, over which they would be hawking for mayflies when not at rest on the handrail. The tail streamers on a welcome swallow are trifling compared to the extravagances worn by some swallow species in other parts of the world, yet they imbue the birds with a certain daintiness nonetheless, pleasingly complemented by the glossy blue-black feathering of the birds' backs and their rich, chestnut-brown facial plumage. Swallows seem to think they can sing, but to me their endearingly feeble attempts, more of a scratchy squeak than a song, hardly qualify. But at this moment, I was content to let their calls wash over me. I was exhausted and had been dreading the prospect of one more day of walking; but in the here and now of the swallows' moment of family intimacy, I couldn't have been happier.

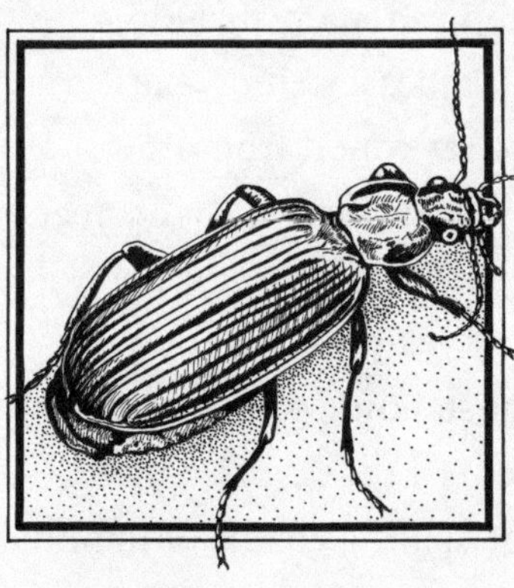

Booming bitterns and other frog-eaters

Watching birds brings immense pleasure to thousands, perhaps millions, of Australians, myself included. But the word 'birdwatcher' feels constraining. It seems to imply that one's sense of vision is all-important; yet one of the characteristics of birds is that they are a noisy bunch, and appreciating them demands that we use our ears as much as our eyes. Thus, I prefer to go *birding* rather than birdwatching. Sometimes, indeed, a bird's calls are all you can hope to experience, because the bird refuses to show itself. Nocturnal birds, such as owls, frogmouths and owlet-nightjars, are a case in point; but many diurnal birds are also secretive and, if not for their calls, would go unnoticed. I rarely feel hard done by with such non-visual encounters, though, finding that natural sounds can be committed to memory as readily as sights and are just as delightful on the recall. So let me relate my first—and so far, only—delightful, Tasmanian encounter with a bittern.

As is often the case with memorable encounters, I wasn't expecting this one. I was revisiting a private grazing property in the Midlands where, earlier in the year, I had participated in a weekend bio-blitz—a survey of the property to identify its natural values, following a few years of nature-friendly management by the owners. During the survey, I was particularly looking for insects and other invertebrates. I had made a few notable finds (more on one of these later), but the cool and showery spring weather wasn't particularly favourable for the work, so I had left with the feeling that the species list I had contributed hadn't done the place justice. I arranged, therefore, to pop in later in the summer, if I happened to be passing during better weather. The opportunity arrived one December afternoon when, after a trip up north, I was ready for a decent leg stretch and net swish.

The property encompasses a series of sheep and cattle paddocks and some fenced-off woodlands, grasslands, and wetlands, including a segment of Diprose Lagoon. This 'lagoon' occupies a shallow but expansive depression in the landscape and receives runoff from the surrounding low hills (not much more than hillocks, really). In dry years, when runoff is minimal, the lagoon is, at best, moist. Often it is bone-dry and encrusted with deeply-cracking clay. In wet years, as was the case when I visited, the middle of the pan may hold standing water, which I suspect is slightly saline; while its fringe sprouts a thick, waist-high carpet of rushes, sedges and grasses. The fringe spans a couple of hundred metres across and is occasionally broken-up by the tracks of wombats and wallabies. Further out, there is more fissured clay, which is partially concealed by creeping, ephemeral herbs and the rosettes of thistles and other weeds.

Spying all of this lush vegetation from the ridgeline farm track, I was pleased that I had decided in advance to home in on the lagoon for my insect hunting. I parked at the nearest point in the adjacent paddock and climbed over the fence separating paddock from lagoon fringe. Skylarks sang continuously, overhead, as though on a tape loop (indeed, Pink Floyd once used such a loop as backing for one of their early album tracks, and I now found it replaying in my head). Both 'joyous' and 'monotonous' are reasonable words for describing the songs of skylarks, but on this day, any monotony was allayed by a jingling party of goldfinches, which I paused to watch as they industriously extracted seeds from the downy heads of a patch of thistles. The brightly coloured adults displayed much competence, while the dowdier youngsters seemed rather inept. A pair of white-fronted chats, a bird I associate more with coastal saltbush country, also flew up in front of me, uttering quiet, nasal alarm calls as they made their undulating flight low over the rushes, before they dropped out of sight. From the amphibian world, across the rushlands, a medley of frog calls reached me. It always amazes me how frogs can reappear *en masse* in a wet year, seemingly out of nowhere. To my shame, I cannot now recall all the species, but I think spotted marsh-frogs predominated. My excuse is that my thoughts were on other things—insects.

Perhaps that's why I only gradually became aware of the significance of a sound that, I now realised, I had been hearing since my arrival. I had subconsciously dismissed it as the distant, persistent lowing of some lovestruck bull; but now that I was paying more attention to it, I could tell that it was coming from the far fringes of the lagoon. Then the penny dropped—it was a booming bittern! 'Booming' is a perfect descriptor for the call of this secretive, frog-eating member of the heron family. It sounds like a series of deep, single notes—forceful hoots, really, more foghorn than bull. Australasian bitterns inhabit reedbeds, irrigated rice fields and other swampy areas in which the vegetation is tall enough to allow these shy but bulky birds to skulk out of sight. Their nests, too, are built close to ground level (or indeed water level), hidden in the reeds or rushes. Streaky brown plumage enables the birds to blend perfectly among the stems; and if disturbed, bitterns typically remain motionless, beak pointing skywards to accentuate the deception that there's nothing to see here—nothing but reeds and rushes. Add to this the bittern's tendency towards crepuscular (dusk and dawn) or even nocturnal activity, and it's easy to understand why one seldom *sees* a bittern. Presumably, other bitterns also find it hard to spot their potential mates or rivals, which may explain their far-carrying, booming, territorial calls, which they emit both day and night.

Australasian bitterns are only ever found in low numbers, but they can be found right across southern Australia and New Zealand. Tasmania has few permanent wetlands capable of supporting them. The reedbeds lining the Derwent between New Norfolk and Granton comprise one such area; as do the low-lying Egg Islands in the Huon estuary, near Franklin. In most years, Diprose Lagoon would be totally unsuitable; yet bitterns are accomplished trackers of suitable patches of habitat and they can home in on isolated lagoons such as this, in wetter years.

Hearing the bittern was a real thrill for me, fulfilling a lifetime ambition. My first notion of bitterns had emerged from music lessons at primary school, in which we were taught a strange little ditty from Wales, the words of which, and tune, I can still remember. It concerned a 'hungry heron' and the rather unlikely scenario of the bird descending from the hills of Brecon and bumping his head on a barrel of apples. The heron's response to this predicament was to 'boom' repeatedly.

Even then, I knew enough about birds to deduce from its transliterated call that the 'hungry heron' would have been a bittern; and I can still recall the beautiful painting of one, in 'freeze' position, in a childhood nature book. It's been a long time since Eurasian bitterns boomed over Brecon, however. I was never lucky enough to see or hear one in Britain; although they are now making a tentative comeback. My first and only sighting of the near-identical Australasian version involved two individuals, which I spotted in a marshy roadside paddock not far from Geelong, in Victoria. My binoculars had been trained on some very dapper looking shelduck in the foreground, so it was only by chance that I spotted the bitterns; they had been standing motionless among the tall rushes beyond the ducks. That was three decades ago. I had longed for another encounter ever since.

Seeing the birds is one thing, of course, and a rare thing at that; but what I had really craved was *hearing* them. Bittern's English name stems, ultimately, from Latin (via Old French), in which it meant 'bull heron'—for then, as now, the bird's call was thought to resemble the lowing of cattle. In Australia, however, bitterns' haunting, nocturnal calls may have been likened to, or even attributed to, an entirely different creature—the bunyip. This mythical, swamp-dwelling spirit-monster appears in the marvellously inventive oral traditions of many First Peoples' societies, although not only in bittern country. Other contenders for 'bunyip' include cassowaries, errant inland seals, and various now-extinct giant marsupials. Perhaps each contributed something to the myths in different parts of the continent, with the elements eventually coalescing into the chimeric bunyip many of us imagine today.

For several minutes I stood still, among the rushes, delighting in the booming bittern. Turning my head slightly, to pinpoint the sounds, it became apparent that there were two birds involved, each calling from a different patch, somewhere on the far side of the lagoon. I hoped that there were silent females as well as booming males out there in the rushlands, and that they would successfully raise offspring on the bountiful supply of frogs at their feet.

I eventually did stir from my bittern-induced trance and find a few interesting insects that afternoon. These included a couple of tiny wetland hoverflies, belonging to an undescribed species. I had been keeping an eye out for these since having been sent a photo of the same species, taken at the same place a few weeks earlier, by Karen, my invertebrate biologist and naturalist friend (whom we met at Township Lagoon, in November). We had both forwarded the photo to a Canadian hoverfly expert (on whom you will hear more, in January), who had asked if I could collect some specimens to help him with a planned description of the species as new to science. Encounters with insects are always unpredictable, so I was gratified to be able to follow up on Karen's photos by providing some actual specimens.

Karen and her partner, Chris, had also participated in the spring bio-blitz, and it was the entomological highlight of that visit, shared by the three of us, that had drawn Karen and Chris back to Diprose Lagoon later that spring. Old records from this part of the Midlands had indicated that the lagoon might provide just the right habitat for one of Tasmania's strangest and rarest ground beetle species, which I have dubbed the green frog-eater (it is otherwise known as the green-lined ground beetle). At three centimetres long it is also one of our largest beetles, so you might imagine that it would be hard to miss; yet the green frog-eater has eluded many a would-be collector over the century or so since its local discovery. Most finds have been serendipitous; for instance, it has been found under logs rolled during searches for other beetles, frogs, or skinks; but there have also been reports of beetles seen emerging from fissures in the clay of dried-out wetlands and nearby grasslands—a habitat that they are known to occupy on the Australian mainland.

Over the previous few years, Karen and Chris had selected the green frog-eater, a threatened species, for special study. Their meticulous observations had revealed much about its natural history, including its frog-eating habits (the beetles will also opportunistically eat crickets and other invertebrates) and the remarkable fact that the adults can live for several years. But even these accomplished naturalists struggled to find more than the odd individual during their daytime forays. Perhaps the beetles only emerged to hunt at night? Neither they, nor I, would normally be in frog-eater country at night to find out; but during the blitz we were. The prospect had motivated me to venture down to Diprose after dark, one evening, armed with a headtorch and camera.

On this cool, spring evening, the lagoon was an eerie place. No birds or frogs called, and my torchlight threw pale clumps of rushes into sharp relief against the blackness of the night. Orb-weaving, long-jawed and wolf spiders had emerged from daytime retreats, and dozens of pairs of spider eyes twinkled at me as the beams of torchlight swept across the rushes and cracked ground. I noticed a particularly deep fissure in the clay, which extended across the ground in front of me, and decided that my best approach was to advance on hands and knees, doing my best to avoid pressing my limbs into the spines of the aptly named prickfoot plants (also known as blue devils). I had only gone a couple of metres when I met with my first ground beetle, but it was a much smaller species than my quarry, so I shuffled on. A metre or so further, I struck gold (or green, I should say). The green frog-eater really is most strikingly coloured, being garbed in what I can only describe as metallic lime-green and black livery. I was taken aback by this one's size, and it was taken aback by the bright light shining in its eyes—so much so that it rapidly reversed into the fissure. I turned the beam off for a while, in hope that the beetle would re-emerge, but it must have been thoroughly spooked as I never saw it again, which left me wondering just how deep the clay crevices could be, for them to be able to comfortably swallow such a large beetle.

I moved further along the fissure, and came across another frog-eater! This one was roaming twenty or thirty centimetres from its retreat. Caught out in the open, it froze when I shone my torch in its direction, which allowed me a closer look, and even a series of photos. The body was predominantly dark but reflected green throughout; and the outer rim of its longitudinally ridged wing cases and of its pronotum (the upper side of its thorax) was unambiguously lime-green. Its head sported a pair of bulging eyes (perfect for night vision) and a pair of formidable-looking mandibles that appeared to be more than capable of piercing frog flesh (and, I suspected, human skin). Eventually the beetle scuttled off into the gloom, before tumbling down another opening into the clay underworld.

I had decided that this was probably my last frog-eater encounter for the evening, so I was heading back towards the distant lights of our base when I spotted another! What's more, this one appeared to be engaged in some sort of tussle. I got down on my hands and knees, and only then realised that the beetle's jaws were embedded in the soft abdomen of a rather large wolf spider. The spider was putting up a good fight in attempting to break free. At times like this it's tempting to intervene, but, instead, I gingerly scooped both combatants into a handy container and brought them back to base to show Karen and Chris. By the following morning, the spider was dead, its abdomen now sadly deflated, and the beetle was clearly victorious, albeit battle-scarred—having lost a leg during the night's mortal combat.

We returned the now well-fed beetle to its lagoon-side home in the evening, and spotted more of its kin in the vicinity. It seems that frog-eater beetles must live much of their lives hiding away in fissures such as those at Diprose Lagoon, venturing out only at night, to hunt and mate. What they did when it rained and the swelling clay closed the fissures … that was a puzzle for another day.

Karen and Chris have documented the lives of many of Tasmania's threatened invertebrates over the years, and had great plans for documenting many more together in the years ahead. Tragically, Chris was diagnosed with cancer a year or so after our visit to Diprose Lagoon, and died soon afterwards. Chris was a self-taught naturalist with acute observation skills and an unbridled curiosity for the natural world. His legacy is immense. He is sorely missed.

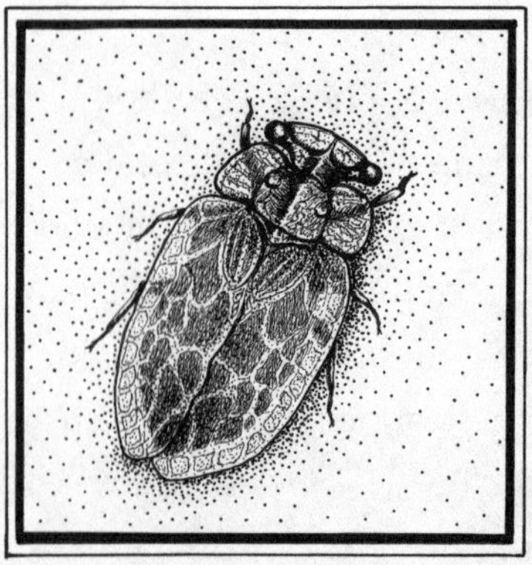

The quest for the moss bug

By now you will have come to appreciate that I take great pleasure in the little things in life—teasing minuscule micromolluscs out of beach grit, say, or being in the company of obscure little insects; particularly if they're captivating to look at up close and have intriguing backstories that add meaning to the encounter. There's one particular bug that ticks many of these boxes for me; it's one that had eluded me for years, from the moment I first heard of its existence. It's the moss bug.

Moss bugs are an ancient lineage of true bugs. That is, they are bugs in the entomologists' taxonomically restricted sense of the word, as well as in the looser sense of being 'little creepy-crawly things'. In evolutionary terms, they are the sister group to all those bugs that have semi-hardened forewings protecting their membranous hindwings when the animal isn't in flight; shieldbugs, leatherbugs and stinkbugs, for example. Moss bugs have been around a long, long time, with fossils dating back to the Permian, some 240 million years ago. These ancient fossils demonstrate that moss bugs once lived worldwide; but they seem to have died out on the northern supercontinent of Laurasia while surviving on Gondwana. In today's world, moss bugs are restricted to the cooler and wetter parts of eastern Australia, New Caledonia, Lord Howe Island, New Zealand, and southern South America (an archetypal Gondwanan distribution). There are only some thirty-two species alive today, and all but one, a Chilean species, are flightless. Two species live in Tasmania and neither of them occurs anywhere else.

I find it humbling to contemplate the millions of generations of moss bugs that have been quietly getting on with their lives on clumps of moss, doing little more than crawling or hopping from one clump to the next, as the land beneath their feet has nudged its way northwards, from once-temperate polar latitudes to where these southern refuges sit today. This passive mode of getting around through deep time is called vicariance. Essentially it means going with the flow of the continental plates. Moss bugs, it would seem, are seasoned practitioners of vicariance. Not for them the alternative method of getting around: dispersal—flying from one continent or island to the next, say, or catching a lift on a floating log or migrating bird.

Another characteristic of moss bugs is that they are almost impossible to spot in the field. Their choice of moss as both a home and a food source is unusual for a herbivorous, juice-sucking insect, because mosses lack a vascular system (not much juice, then!) and are rather poor in nutrients. Perhaps mosses were the tastiest plants on offer back in the Permian, and these bugs simply know what they like and like what they know. On moss's plus side, it provides a very intimate, three-dimensional habitat. It's more like a micro-universe for the moss bugs and other tiny creatures that call moss home, such as springtails, mites, moss *beetles*, litter beetles, cranefly larvae and many more. More inconveniently, for the bug-hunter at least, moss's scale-like leaflets and shoot tips enable moss bugs to all but disappear, for the bugs are only a couple of millimetres long, saucer-shaped, and greenish-yellow, so they are very well camouflaged within the moss. And they occur at very low densities.

All this means that looking for moss bugs can be like looking for a needle in a haystack. They can't be caught in flight intercept traps like many other insects—because they are flightless; however, they do occasionally turn up in pitfall traps set on the mossy forest floor; and they can also be extracted by scooping up copious quantities of moss and leaf litter, spreading this out on a white sheet, and looking for the movement of any creatures disturbed by this upheaval, which may be easier to spot against the white background. Then there is a more sophisticated version of this approach, which involves sifting the litter first, then stuffing all the siftings into a big funnel suspended over a pot of ethanol, and illuminating it with a warm electric light from above. This gradually heats and dries out the contents, driving the moss bugs and other creatures downwards until, eventually, they drop out of the bottom of the funnel and into the ethanol. You will then have a concentrated bug soup, known in the trade as berlesate—after Italian entomologist Antonio Berlese, who devised the technique in 1905. We owe the improvement of using an overhead light as the heat source (instead of Berlese's wraparound hot water jacket) to a Swedish entomologist, Albert Tullgren, who was working on the method a few years after Berlese.

My own rather haphazard attempts to find moss bugs while out walking in likely habitat had always come to nought. Admittedly, they were not helped by my impatience when kneeling on wet moss in the company of mosquitoes and leeches, nor by my worsening longsightedness, which came to mean that I couldn't focus on anything closer than about 40 centimetres (it's now about a metre). It seemed that moss bugs would forever remain my *bête noire*, or maybe my Holy Grail. Then, out of the blue, I opened my museum email to find a message from a group of Russian palaeoentomologists (professionals interested in insect fossils and ancient insect lineages). They were reaching out to us for logistical support for an impending Tasmanian expedition, which they would be conducting on an absolute shoestring, since their trip was self-funded. One of their target creatures was our moss bug.

I replied that we would be happy to facilitate their visit, and could I please come along too? (Of course.)

Alas, leukaemia reared its ugly head—this was not the first time, but you will learn more about this in due course. What it meant, on this occasion, was that I spent most of their visit in hospital. However, I resumed the conversation by email, the following year, when I was sufficiently recovered. I was delighted when Dmitry sent photos of a particular clump of sphagnum (bog moss) on the summit of Kunanyi-Mount Wellington, where I was guaranteed, he said, to find one of our moss bug species (the one whose scientific name translates as the faithful moss bug). '*But don't bother checking any other clumps on the summit*', he advised me—Kiril (the moss bug enthusiast in the group) had spent hours doing just that, and found none at all. They would only be found in this one clump, Dimitry reiterated.

The next bit of good fortune was a rare visit by my friend Sarah to her home state from her adoptive home in New Zealand, where she has forged a successful career in the museum sector. Besides her passion for natural history collections, Sarah is a deeply knowledgeable all-round naturalist with a particular fondness for weird invertebrates. When I suggested we go on a moss-bug hunt Sarah generously made some space in her crammed schedule. It wasn't long before we were driving up Kunanyi's Pinnacle Road together.

It was one of those gorgeous summer days of intense sunshine, with a gentle breeze from the south moderating the temperatures to perfection. As we parked the car, donned our sunhats, and slip-slop-slapped the sunscreen, we were greeted with the buzzing, clicking songs of countless mountain cicadas—they seemed fine-tuned to wind me up for a fine day's entomologising. I had remembered to bring my close-up glasses, as well as my camera, and we had a vague idea which way we should head to locate Kiril's moss clump. With a bit of squinting, we had actually managed to match up Dmitry's emailed photos with satellite maps on our phones, so this gave us a precise location. All we had to do was navigate successfully across several hundred metres of alpine shrubland, peatland and boulderfield.

As a rule, I don't like to leave the formal walking tracks on the plateau because of the damage it can cause to the mountain's fragile vegetation. We made an exception on this occasion, and carefully stepped and skipped our way from one bare slab of dolerite to the next, checking our position on our phones along the way and recalibrating our bearings as necessary. Each footfall seemed to send legions of basking and mating grasshoppers leaping in panic, their hind legs flailing as they crash-landed into adjacent shrubbery. They were mostly Tassie hoppers and D'Albertis's grasshoppers, which are the most frequently encountered of the many grasshopper species that live on the mountain. Both are endemic to Tasmania and are flightless. Losing the ability to fly is common among montane grasshoppers and other montane insects worldwide. There are a few theories as to why this is so, and they are not mutually incompatible. One suggests that, since uplands are windy places, no montane grasshopper wants to get blown off its mountain top as soon as it takes to the air. Another is that growing wings requires extra protein, which can be in short supply in a diet comprising the tiny, tough leaves of alpine shrubs. One more suggests that diverting scarce resources to making wings adds time to the grasshopper nymphs' development towards adulthood, and time is of the essence at higher altitudes because they have shorter and cooler summers than the lowlands. Here on Kunanyi, I could imagine that all three factors might conspire to favour flightless grasshoppers over their winged cousins.

So it was that the soft sounds of scattering grasshoppers gradually replaced the cacophony of car doors slamming and shuttlebuses disgorging parties of chattering sightseers. As we made our way across the summit plateau, we were soon enjoying the feeling that we had almost the entire area of the mountain uplands to ourselves. I savoured the scene before us, shielding my eyes from the bright glare of the sun and enjoying the breeze. The whole western horizon was rimmed with rugged and remote mountain peaks and ranges, which shimmered in the haze. The prominent ones I could name: Adamsons Peak, Hartz Peak, Mounts Picton, Weld, Mueller, and Field; but there were many more that I couldn't. For mountain insects, each was like an island, and I contemplated how many an ancestral, cold-adapted insect species, having enjoyed unimpeded movement across the landscape during glacial times, had found itself isolated on a mountain peak or perhaps a high range, whereupon some came to evolve into distinct species.

Turning the other way, towards the east, all we could see was the broad bulk of Kunanyi's high plateau, stretching beneath a vast sky. Marking where we had come from, the upper half of Tasmania's tallest construction was still clearly visible: a 130 metre-high transmission tower, which had been built just a couple of hundred metres from Kunanyi's summit. But for this beacon of modern civilisation, we could have been in the middle of nowhere. Other than our own breathing, the only audible sounds now were the calls of cicadas and grasshoppers, and a lone striated fieldwren singing from a distant thicket.

After a final bit of triangulation with Dmitry's photos, we found The Clump. It looked much like all the others in the area: a rounded cushion of sphagnum moss, about 40 centimetres high, mint-green towards its damp base but bleached white on top. One side was embedded in the bank of a small, peaty pool, from where the sodden moss wicked moisture and distributed it to the rest of the clump. Meanwhile, from its landward side, some stout sprigs of pink swamp-heath protruded. I figured that they might have provided the initial scaffolding for the cushion's growth, which would suggest that this venerable moss cushion might be several decades old. Notwithstanding such credentials, I was intrigued as to why the moss bugs would have chosen to proliferate on this cushion while apparently scorning all others in the vicinity. Regardless, I was grateful to Kiril and Dmitry for having saved us the trouble of searching them all.

I put my glasses on and we sank onto our hands and knees on the cool, moist turf and gently began to prise apart the tightly-packed crowns of individual stems of moss. We sampled from various parts of the cushion. Up on the cushion's highest point the moss was too dried out, we decided. No luck near its base either: too wet, perhaps. Moving on, Sarah was the first to meet with success. Goldilocks-like, it came from a strand of moss plucked from the damp 'just-right' zone between the other two extremes. Beaming with delight, she extended her cupped hands for me to inspect the sprig. At first, despite my glasses, I couldn't see anything that didn't look like part of the sphagnum's spongy, water-saturated crown. It was only when I picked it up and delicately teased apart some of the tightly packed 'branchlets' within the crown, that I saw that some were broader and flatter than the rest, and they weren't attached to the stem! When one of them moved, I was filled with a sense of relief that my epic quest had at last come to an end! I unpacked my camera and clipped on the close-up lens, and for the next few minutes we both got to admire these strange little bugs—an adult and a couple of nymphs—up close and personal, through the camera's eyepiece.

Despite their diminutive proportions, the moss bugs were the most wondrous little creatures. Although they were perfectly able to walk, their preference seemed to be to stay precisely where they were. Their body outline was expanded on all sides: by translucent extensions to the front and sides of the head, to the pronotum (the upper covering of the thorax) and to the hemelytra (those semi-hardened forewings). All upper parts of the body were criss-crossed by a coarse network of interconnected, raised ribbing, which divided the entire surface into a series of shallow pits. The bulging eyes were on long stalks, yet they didn't protrude beyond the rim of those translucent extensions. I was so engrossed and enthralled that I nearly forgot to take photos, and I barely noticed that, by now, my knees had sunk deep into the sodden peat, soaking my trouser legs.

At length, I handed the moss and its resident bugs back to Sarah so that she could replant it in the cushion as well as she could, while I resumed my own bug hunt. Now every other moss stem in the Goldilocks zone looked like it would be ripe for bug detection. Many times, my hopes were raised, only to be dashed by the bug-like appearance of aberrant branchlets and nascent spore capsules peeping out from among the tight-packed crowns. When I finally met with success myself, it was sweet indeed.

It's hard to put my finger on the feelings that moments like this elicit, but they involve a heady mix of joy, wonder and elation. Peering at the little bug in its mossy microcosm, and sharing the moment, I was in heaven. Above all, I felt an overwhelming sense of gratitude towards life in general and towards the forces of nature that had brought about this special place and these special, never-to-be-repeated moments. Who needs drugs when you can get high on Kunanyi's plateau, amidst the moss hummocks and the cicadas? It very nearly reduced me to tears; as does the recollection of the moment, as I write these words. Granted, it all sounds a bit weird now, long after the event. How could an encounter with a moss bug evoke such intense emotions? All I know is that it did, and it didn't surprise me because it was neither the first nor the last time that nature has woven me into its magic like this. '*In all things of nature there is something of the marvellous*'—Aristotle said this, not me, more than two thousand years ago. He could not have encountered a moss bug, ever, but he surely knew something of the feelings experienced by those who spend quality time in the company of nature's wondrous creations.

Eventually, Sarah and I made our way back over the boulderfield to the car park with an extra spring in our steps. Sarah had another engagement, later that afternoon, but I didn't want the trip to end, so I suggested we briefly try our luck further down the mountain. Kiril had also had success along the path that leads from the Chalet down to the Springs, albeit using the white sheet technique rather than peering at moss *in situ*. Below the canopy of twisted snow gums, we found it harder to find substantial clumps of moss, and we had no photos to guide us this time; but Dmitry had directed us to look for a particular arrangement of moss cloaking the lower trunks of the eucalypts and waratahs, where the moss extends to the ground. We lost no time in getting down to business, despite knowing that our chances weren't particularly high. Shaking a small clump of moss, I was surprised to see several Griffith's moss beetles drop into the palm of my hand, looking for all the world like tiny, green, shiny, Christmas baubles. Though they don't rank quite as highly as moss bugs, for me, they're yet another weird inhabitant of this mossy world and, like the moss bugs, they are endemic to Tasmania.

With the shadows deepening and time running short, we were about to call it quits when I spotted a tiny bug sitting on an erect sprig of moss with spirally arranged leaflets. With little more than a cursory inspection, I blithely declared that I had found what we were looking for and hastily took some photos. We hurried back to the car and were soon back in town saying our goodbyes and reflecting on our good fortune. It was only when I uploaded the photos onto my computer, that evening, that I realised that the last bug I had seen was an impostor. It was not a moss bug at all, but an equally beautiful lacebug. Like the moss bug, its body plan involved expanded margins to both pronotum and hemelytra, but these possessed a much finer network of lace-like ribs superimposed on the coarser network. The sides of the pronotum were, additionally, fringed with four black tooth-like extensions, and the bug bore long, clubbed antennae. Furthermore, it clearly lacked the eyestalks of the moss bug. I chided myself for having made such an elementary misidentification, then tried to figure out what species of lacebug I had found. To my astonishment, it turned out to be one whose scientific name translates as the toothed lacebug. This species was known from only a handful of specimens, it seemed. Nothing is known about its ecology, but it would seem to be one of the few lacebug species that feed on moss. It's a fair bet that this similarity between two very distantly related bugs is the result of convergent evolution. Such excellent camouflage surely offers enhanced chances of survival to bugs such as these, when they can so effectively disappear into their chosen home among the mosses.

Two rare bugs in one trip! I slept soundly that night, immersed in semi-lucid dreams of strange, trundling, lace-like creatures, and a vivid revival of the euphoric feeling of our uplifting day on the mountain.

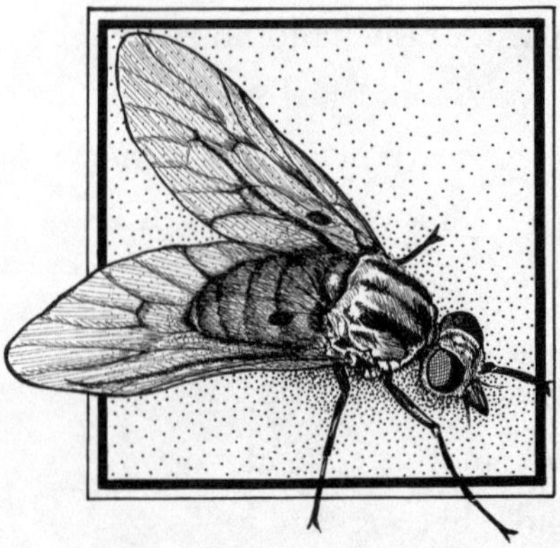

Laud of the flies

In entomological circles I used to declare that I was a lapsed dipterist—for many years, I had abandoned a longstanding interest in flies (Diptera). In my youth, I had participated in annual, week-long, fly-recording gatherings, and had subscribed and occasionally contributed to that well-known journal, *The Dipterists' Digest*. During the intervening decades, however, the world of beetles had drawn me in. It had begun during my doctoral research; then, during my work at Forestry, beetles (comprising the insect order Coleoptera) had proven themselves to be much more cooperative than flies, at least when it came to exploring forest ecology and management issues. Working at the museum rekindled my interest in flies, though, so I am proud to call myself a fully-fledged dipterist once again (as well as a coleopterist).

Diptera is the insect Order comprising the true flies. They were given this name, which means 'two wings', because one of their distinguishing features is the presence of just one pair of wings, as opposed to the usual two that other flying insects have. (In beetles the forewings are hardened into shield-like elytra that protect the hind wings when at rest). In fact, flies do actually have the vestiges of a second, hind pair, but they have turned these into drumstick-like balance organs. This repurposing has enabled flies to become some of the most adept and accomplished flying organisms on the planet.

For many people, 'fly' means 'pesky', but that's probably because the insect they have in mind is the bushfly, blowfly or housefly. Mosquitoes are also flies, and there's little that's positive to say about them, despite (or maybe because of) their pivotal role in human history. Yet these nuisance, and even dangerous, flies are an extreme minority in an Order comprising millions of species. There are many thousands in Tasmania alone and, for me, their sheer diversity of form, behaviour, and lifestyle, is part of their appeal.

Much has been made of the idea that, in our modern lives, we are seldom more than a few metres from a rat; but the same can be said of flies. Indeed, there is a tiny scuttlefly crawling on my computer screen as I type these words. But you haven't heard of scuttleflies? This particular one is a member of one of the most speciose genera in the animal world—there are at least 1,700 described species of *Megaselia*, as well as countless more still out there, awaiting description. Surely that is enough to make anyone interested in naming a new species seriously consider dipterology? Indeed, one taxonomic wit has described this scuttlefly genus as 'open-ended', since the true number of species is currently incalculable. And within the genus, its various members embrace a broad range of lifestyles, counting amongst themselves predators, parasitoids, kleptoparasites ('cuckoos'), and more. Flies, you see, are much more interesting inhabitants of our world (or theirs, really) than rats.

To study flies you really need a net—more than for any other insect group. I seldom leave home without one if I'm going bush during summertime. A standing joke in our household is that I'm having an affair with 'Annette', such is the degree of my attachment to this useful implement. Granted, I may look eccentric when entomologising, net in hand; but most people seem remarkably tolerant, and many seem genuinely interested in what I'm up to. I'm often asked if I'm collecting butterflies—an activity that doesn't seem to attract the same opprobrium in Australia as it would in England, these days. But, no, I'm more interested in flies and beetles, I reply. This flummoxes most people and usually brings the conversation to an end; but sometimes it initiates a bit of a discussion about the wonders of insects and how little we still know about them.

As I've mentioned back in November (on the subject of Hardy's nannochoristids), a standard netting technique that I employ is called sweep-netting. It involves sweeping a net back and forth through, or over, vegetation, then peering inside to find out what's been caught. There's a knack to sweep-netting, which takes advantage of insects' usual tendency to head upwards and towards the light when fleeing. There's also a knack to avoiding getting the net snagged on prickles or abraded through overzealous sweeping. Even for the most competent sweep-netter, each net bag may last only one or two seasons, with regular use; the frames a little longer. The most productive sweeps involve a series of forthright but carefully aimed strokes; preferably as slight downstrokes. Then, to ensure one's catch remains caught in the net, it's best to flip the bag end of the net over at the conclusion of a series of sweeps, and leave it that way until one is ready to take a closer look. Since most of the insects that interest me are small, I can't see them through the white netting. This means I have to insert my whole head *and* one hand, bearing a specimen tube, into the mouth of the net. To do so, I must let go of the handle, so that the whole thing is held in place by resting on the crown of my head. My other hand holds the bag end of the net out and a little up, which encourages the insects within to congregate at that end, rather than making a dash past my ears to the exit. I find it interesting to observe how the behaviour of passers-by changes when they see me in this state. Most wisely give me a wide berth.

There are so many interesting flies and fly encounters that I could write about, but I will limit myself to one memorable December day of dipterology (although I am sure that, like flies themselves, other encounters will emerge). This particular outing was on yet another of those glorious days, perfect for entomology, with clear blue skies and a gentle breeze. Such days have me itching to head for the hills, just me and Annette. On this one, I set my sights on Wombat Moor—beautiful, peaty-pooled and runnelled open country, on the high saddle that the road to the ski slopes of Mount Field passes over, between Lakes Fenton and Dobson. Bright orange poles on either side of the road helpfully indicated that, today, the snow depth was zero. It isn't always the case, even in summer. Instead of snow, a heat haze was generating mountain mirages, which hovered incongruously above the road.

Although striking, the mirages compared poorly with the rugged dolerite grandeur of the actual peaks beyond. From the gravel road ahead, a lone Australasian pipit took flight, a blur of streaky-brown, except for its white outer tail feathers. I parked in the layby for the walk to Lake Belcher. Opening the car door invited in the sounds of the moorland: bleating Tasmanian froglets, zizzing mountain cicadas, chirruping pygmy crickets and buzzing marchflies. I sat there for a few minutes, reacquainting myself and mentally recalibrating—it's something I often do on trips necessitating a long drive. Then I donned my boots and gaiters, took Annette in hand, and ventured forth. Turning my back on the boardwalk, I crossed the road and strode out across the moor instead. The ground was pleasingly soft underfoot, verging on squelchy in places—just as it should be in a place like this.

One of the tributaries of the Tyenna River has its headwaters on Wombat Moor, but the gradient is so slight that much of the moor is an intimate mosaic of incipient shrubland, sphagnum-lined pools, peaty trickles and rivulets. It's perfect fly country. I say that because many flies, especially those of the more ancestral lineages, have aquatic larvae. Even later-evolved lineages that are not strictly aquatic tend to have soft-bodied larvae, which are vulnerable to drying out. Therefore, hydrologically well-endowed areas like Wombat Moor are much favoured, although the moor experiences rather more cold weather, with cycles of freezing and thawing, than is ideal for maximum fly diversity; so the flies there are, by necessity, a hardy lot. (Hardy—there we go again.)

Up ahead, I spied a pool partly overrun with sphagnum moss. I squelched my way towards the shoreline, where some protruding dolerite slabs enabled me to tread close to the water's edge without damaging the fragile habitat or getting my feet wet. A mating pair of cup ringtail damselflies almost immediately caught my attention, and I watched them climb purposefully down the stem of a rush and submerge themselves in the peat-stained waters. Throughout their descent, they remained in the 'wheel' position, by which I mean that the rear-end claspers of the male (the upper of the two, sporting a powder-blue and black abdomen) were holding the female behind her head, while the tip of her blue-blushed (and otherwise fawn and black) abdomen was curled forwards to clasp a special receptacle on the underside of the male's thorax. Earlier, he had transferred his sperm there, so that she could now take it in to fertilise her eggs. It's a strange arrangement, but it must have stood the damselflies and dragonflies in good stead—they've been doing it this way since, at least, the Carboniferous period. Under the water, I next saw the female detach her abdomen from the male's thorax and dab its tip onto, or perhaps into, the stem of the rush. I knew that she would be pushing out a single, tiny egg at the same time. Following this, the mating couple remained underwater for a few more minutes, crawling around among the submerged stems of the rushes, and laying a succession of eggs that would soon hatch into predatory nymphs. The male's determination to continue clasping seemed aimed at guarding the female from other suitors, although the chances of any spotting her here, in the depths of the pool, must have been slim.

Meanwhile, flies of several distinct types were patrolling a few centimetres above the pool's surface. The smaller ones were midges of various descriptions—meniscus midges and other non-biters. The larger ones—slender but stocky flies about a centimetre long—were danceflies, which are so called because the males often form mating swarms that dance around bushes and above pools such as this one. I netted some for a closer look and found that there were several species, and that all were males—which are identifiable by their abruptly swollen terminal segments. All had dark bodies and smoky-dark wings. This is common in flies inhabiting cold areas, where the dark coloration may help them rapidly absorb heat whenever the sun shines.

In one of the species, the basal segment of the front feet was inordinately swollen, a bit like Popeye's biceps after he has downed the contents of a can of spinach. I knew that the purpose of the swelling was to house silk-producing glands within the segment. Who knew that flies could produce silk? Male danceflies do so to bundle up a prey item, such as a smaller fly caught on the wing, which will form a nuptial gift for a potential mate. If he doesn't present the object of his attention with a gift, he's at risk of becoming her next meal, pierced by her downward-pointing, needle-sharp proboscis. With a gift, he can keep her front end occupied while their back ends are conjoined in mating. Mating takes place on the wing, but away from the hordes of bachelors. Some males, termed 'sneaky males', adopt a different strategy, dispensing with the prey item and just presenting the female with an empty ball of silk. If he's lucky, she won't notice until they've finished mating. There are also entire species of dancefly in which that's all the males ever do. In these, apparently the females judge the males' merits by the size of their silken balls.

I became aware of a loud and insistent buzzing coming from the middle of the pool, so I swished my net inside out to release the danceflies. This movement caused the source of the buzzing to shift by a metre or so. Its maker was one of our most enchantingly attractive hoverflies, whose name roughly translates (with me, at least) as the beautiful little southern dronefly. It's a native relative of the very bee-like European dronefly, whose larvae, indelicately known as rat-tailed maggots, sometimes infest septic tanks and crawl around the u-bend to emerge into toilet bowls when it's time for them to seek a dry place to pupate. Our species has much more refined tastes. It lays its eggs in peaty pools and permanent swamps. The one in front of me was about 15 millimetres long, and it was hovering on the spot, about a metre above the water's surface. It was facing me and very obviously watching my every move through bulging, compound eyes. If I advanced, it retreated by the same distance, only to advance again if I retreated.

The hoverfly was an absolute wonder to watch, not only for its remarkable hovering ability but also for its appearance. Its head, thorax and the first part of the abdomen were a bright golden-brown, with a pair of prominent black spots on the thorax. The rest of the abdomen was a very pale, metallic green-blue, also banded and spotted with jet black. This was the first beautiful little southern dronefly (you see why using scientific names is sometimes preferable) that I had ever seen outside of an insect collection. The fact that the eyes met above the antennae told me that it was a male. In females, the eyes are separated throughout. This is a common pattern in hoverflies and some other fly families, and may reflect the males' need for better spatial awareness and split-action decision-making, since they are the ones that do all the hovering, patrolling, and defending of territory, while awaiting the arrival and approval of cruising females.

I cautiously extended my net in the hoverfly's direction; but I think

I already knew that this would be in vain. Sure enough, he took fright instantly, and zoomed off to who-knows-where for a couple of minutes; however, he returned, and resumed his hovering in the exact same place. I sensibly decided to let him be—or hover—and moved on to explore fresh fields.

A little further on, where the terrain began to drop gently, a small runnel emptied into a nascent, meandering creek complete with riffles of fine shingle interspersed with quieter pools. I swished over the shingle in the hope of provoking any flies resting there to launch themselves into the path of the net, then performed my head-in-net examination of its contents. Sitting quietly at the far end of the net, amongst scores of tiny midges, was a much larger fly. At about a centimetre long, it was one that I had forgotten to expect in this country: a water snipefly. Like the hoverfly, it was my first live encounter with one. I was delighted with my find and how the day was panning out. As with the danceflies, this one had darkened wings. Its specific name, one of Hardy's, translates as spotted-winged. It also had a rather droopy abdomen and held its wings in a sort of V-shape, with the tips resting on the net. This made it look rather down-in-the-dumps, an appearance which persisted even after I coaxed it onto my finger and returned it to a streamside stone. It sat there for a while, still looking miserable, before lazily flying off to another stone a metre or so away. There, it seemed to perk up a bit, perhaps inwardly smiling for the camera that I was now pointing its way. In other respects, it seemed unbothered by all the fuss.

I wondered how water snipeflies ever find mates and keep the species going if all they do is sit around on stones all day. But breed they must, since water snipeflies are a worldwide family. They comprise rather few species, and Tasmania hosts four of them, all of which are endemic to the island. The spotted-winged one before me would have to be the toughest of the lot, as it lives only at high altitudes. The larvae of all species are aquatic and predatory, and the adults seldom venture far from water, which would explain why, thus far, I hadn't swept one from the moorland vegetation.

Pushing on, I made my way towards the flowering shrubs fringing eucalypt woodland at the edge of the moor—teatrees, banksias and hakeas, mostly. I gave them a good sweeping. Each peep into the net revealed hordes of tiny, brown marsh beetles and one or two individuals of a much larger marsh beetle species, the Tasmanian painted marsh beetle (for want of a better translation, or translator). These solid, rather hump-backed beetles, about a centimetre long, come in a variety of colour patterns. Most are orange with a broad, black, transverse band and additional black blotches; others are pure yellow or pure black, or are orange overlaid with fine, deep-brown scribblings. They are common inhabitants of moorland and wet heath throughout Tasmania. Their larvae are aquatic, generally occurring in wet peat, while the adults seem to pass their days among flowers, either feeding or seeking romantic encounters.

I moved further along the woodland edge and changed tactic. Instead of sweeping the bushes, I took to standing and staring. The mass of flowers on the woolly teatrees was almost blindingly white in the early afternoon sunshine and it took a while for my eyes to become accustomed to the brightness. When they did, I started to discern movement among the flowers. Mostly the movers were those marsh beetles, but there were also danceflies, bristleflies, blowflies, male marchflies (it's only the females that feed on blood), male mosquitoes (same story) and a few others. And then I spotted the fly that I had been hoping for: a green-sided soldierfly. The term 'soldierfly' covers flies of a variety of shapes, sizes, and colours in the same family (you may recall we met one earlier, on the grasstree's flower spike). The one in front of me—a male, with conjoined eyes—was predominantly black but with a dense coating of brown hairs on the thorax and lime-green sides to its dorsoventrally flattened abdomen. Despite being almost a centimetre long, it was remarkably well hidden among the flowers. I advanced with my camera to my eye, but the soldierfly objected to my intrusion and dashed off, never to be seen again. Since then, however, I have seen what I believe is the same species in the Hartz Mountains, and at The Springs on Kunanyi-Mount Wellington. Many of the black-and-green soldierfly species still lack scientific names, so I couldn't say definitively whether this was one of the described species or not. Some, on the other hand, may have been described more than once, since the males and females tend to have different patterning. The green sides come in various shades from yellowish to almost blue, and can extend, to a greater or lesser degree, towards the centreline of the abdomen, either as broad wedges or narrow bands. All species inhabit wetlands, mostly where the wet vegetation is kept short and lawn-like by wallabies or wombats. This seems to be where they prefer to lay their eggs.

A little further on, an echidna waddled purposefully along the woodland edge, occasionally stopping for an exploratory dig and snuffle. It halted abruptly at my approach and hunkered down with its snout tucked firmly under an impenetrable thicket of stout spines. Tasmanian echidnas are much hairier than their mainland counterparts, and this one was even hairier than most; but its spines still protruded well beyond the hair and looked menacing enough. Every sound from my footfalls caused the animal to shudder, which made its spines bristle, perhaps in expectation of imminent attack. I stood motionless for a while, and eventually the echidna decided that the danger had passed and lumbered off into the woodland.

I found many other dipterological delights that day, high up on Wombat Moor, but I think I've written enough to give you the gist of why I chose the title for this section. If you can leave your anti-bushfly prejudices behind, you, too, may find that flies are worthy of a lot more praise than is their current lot.

Gone dredging

As land-living animals, we place great emphasis on the bits of the world that aren't watery. Most human affairs revolve around the land: we colonise it, we work it, we divide it up and fight over it. Meanwhile the sea is a handy conduit for our trade and transport, a source of seafood and somewhere convenient to dump our waste. We loosely divide the vast blue yonder into a series of named seas and oceans; but on a grand scale, even the continents are mere islands emerging from a single body of seawater that averages 3,700 metres deep and covers close to seventy percent of the globe's surface. Our planet's skin is overwhelmingly wet, salty, and blue.

I recounted earlier how my shell-hunting addiction took me further and further afield, to the far corners of Tasmania. Perhaps it's time to confess that, all up, I have recorded seashells from some four hundred Tasmanian beaches, or distinct sections thereof. This mad endeavour occupied me for a dozen or more years and transported me to a marvellous range of shorelines, from South East Cape anticlockwise around the coast to Strahan, and to King, Flinders and Clarke islands in Bass Strait. I've paced ocean beaches pounded by surf; quiet inlets and lagoons with tannin-stained waters; silty estuaries; secluded, white-sand coves hemmed in by granite boulders with fluorescent caps of orange lichen; steep banks of cobblestone, driftwood, cuttlebones, and storm-tossed sponges; mudflats and sandflats; and more. Probably few people have been to more Tasmanian beaches. I loved almost every one of them. Through my wanderings I came to appreciate how extraordinary Tasmania's coastline is. Certainly, as an island, it is one of its defining features, but I would suggest that it is also one of its crowning glories (if we can allow it more than one). I love how, driving in the southeast, you never quite know on which side of the road the next sea view is going to materialise, or whether it will be a rocky headland or a quiet inlet. Along the east coast, there are stretches where you drive through a salty haze emanating from surf thumping against a shoreline just a few metres way. On the north coast, the vistas change dramatically, depending on the state of the tide, as the tidal range is at least twice that of the south, and sometimes much more. The west coast is harder to access, retaining a wild and untamed air; and often it is only glimpsed, through gaps in the dunes, as a roiling mass of white water. But the west also boasts the magnificent, sheltered inlets of Port Davey and Macquarie Harbour, which remain largely unexplored, by me, at least.

One thing my wanderings and follow-up studies didn't do, though, is throw light on what lives on the seafloor in deeper waters, safe from the tug-and-tow influence of tides and winter swells. It's a paradox that our accessible shorelines comprise the rarer habitats for marine life—a thin ribbon of half-land, half-sea; while the inaccessible deep waters are much more extensive. I had rather comprehensively documented the seashell fauna of Tasmania's shallows; but our knowledge of what lives further offshore hadn't advanced significantly since the early twentieth century. At that time, a Hobart businessman, William May, devoted much of his money and leisure time to shell-collecting. He, too, realised that little was known about what lived in deeper waters. It had only been a few decades since the British *Challenger* expedition had first determined that the ocean depths weren't lifeless, as had once been assumed. That expedition broke so much new ground that the preliminary reports were still being written thirty or more years after its completion. Here, May teamed up with Charles Hedley, a mollusc expert who worked at the Australian Museum in Sydney. The two of them would charter a steamboat to take them out towards the edge of the continental shelf, east of Maria and Schouten islands, and off Cape Pillar, at the south-eastern tip of the Tasman Peninsula. They dredged in up to 100 fathoms of water (one fathom equals six feet, or about two metres, so 100 fathoms is about 200 metres). Their surveys greatly boosted the number of mollusc species recorded from Tasmania's region, and their own taxonomic efforts described dozens of species as new to science. May eventually included the bare bones of their findings in his *Illustrated Checklist of Tasmanian Shells*—I had relied on the 1955 update of this slim volume for many of my own identifications.

One day I received an email from a Port Arthur-based retiree who had become keenly interested in the local seashell fauna and wanted advice on how to identify them. I pointed him to my shell book and website, and we soon developed a friendship based on our shared interests. Before long, Rob had set himself the challenge of finding every species known to occur on Tasmania's beaches; and within a few years he had pretty much found the lot, including some that hadn't been recorded before. Like me, Rob then looked longingly seaward, pondering all the other species that might be found beneath the waves. Rob had a distinct advantage in this regard, because he was the proud owner of *Rambler*, a 10.4 metre motorised wooden yacht, which he kept moored at Port Arthur. Rob's a can-do sort of bloke, and he reasoned that it would be feasible to build a small dredge and take *Rambler* out to give it a try, using the onboard electric craypot winch for feeding out and hauling in the dredge. He also had a mate with a blowtorch and metallurgy skills, who would be happy to build one for him. It wasn't long before Rob sent me photos of the finished product, which comprised a steel frame lined with coarse gauze and measuring about a metre across, the same deep and about a third of that in height. I was working at the museum by this stage, so I was in a good position to ensure there was some scientific legitimacy to Rob's quest. We were granted a research permit to dredge, and soon it was all systems go.

We can generally expect, or at least hope for, a few consecutive days of calm, settled conditions in December. At the first opportunity the weather offered, Rob sailed *Rambler* around to Dunalley and through the Denison Canal, into Boomer Bay. I joined him there, early the next morning. Rob welcomed me aboard with a mug of hot coffee, and we were soon chugging across Blackman Bay, the soft morning light casting a pastel spell on the waters rippling gently in our wake. Pelicans, Pacific gulls and a Caspian tern warily watched our cautious passage through the Narrows from soft sand at the end of Long Spit; and then we were out into the broad expanse of Marion Bay and the bright blue yonder. Since the whole dredging approach was untested, we resolved to give it a first go in the middle of the bay, where we knew the risk of snagging would be low. With the engine idling, we fed the dredge out on a polypropylene rope, until it reached the seafloor some 20 metres below, and the rope slackened. We let a bit more rope over the side and then tentatively put the engine into low gear and hoped for the best. The rope tightened and the need for a little more throttle implied that everything was in order and the dredge was moving, though we had no idea whether it was bumping over or through the sediments or had launched itself off the seafloor entirely. Our permit had specified that tows were to be no longer than ten minutes, so when the time was up, Rob put the engine in reverse, and I started feeding the rope through the winch. Once the line was both taught and near-vertical, Rob cut the engine and we allowed the winch to slowly continue its hauling. Before long, the dark shape of the dredge loomed up from the depths. We manhandled it aboard and were relieved to see that it wasn't empty. Tipping the contents into a big, white, plastic crate, we could tell instantly that our first dredging trial had been successful—a quantity of brightly coloured sponges had come up, some attached to stones or shells, along with a fair bit of loose rubble.

We were impatient to work through our haul and, in our haste, neglected to put on the rubber gloves we had brought for this purpose. It was a mistake that we would only make once, as it soon became clear that squeezing or pulling apart live sponges bare-handed is not a good idea. Not only do they contain a potent mix of bioactive chemicals for deterring would-be predators, but also their internal skeletons are made of a framework of tiny, needle-like spicules of silica—essentially glass. Before long, our fingers had swollen up with itchy, red, puffy welts. But we weren't going to let itchy fingers dampen our spirits. Buried deeply in a sponge were several live gastropods. They were triphorids, which are hard to find onshore. When they are, the shells are usually very worn, so that the beading and microsculpture necessary for species-level identification is obscured. Finding live examples of these sea-snails was a great start to our dredging. Nearly all members of this unusual family live embedded in sponges, feeding on the soft tissue that surrounds them. We would be hard-pushed to say whether their parasitic existence causes much harm to the sponges. So, too, might the sponges, since these simple, colonial, filter-feeding animals—among the earliest multicellular animals to have evolved—lack a nervous system.

Another odd thing about triphorids is alluded to in their common name, the sinistral-creepers. Their shells coil the opposite way to that of most gastropods, including their doppelgängers, the creepers. To coil 'sinistrally' means to coil to the left, rather than dextrally (right-handedly). If you don't quite know what this means, take a snail in your hand, with its main axis running vertically and the tip of its spire pointing up. Now turn it on that axis, so that the shell's opening (its aperture, or mouth) is facing towards you. If your snail's aperture is off to the right of the main axis, it's a dextrally coiling species. If it's off to the left, it's sinistral. Once you have your eye in, spotting shells that coil the 'wrong' way becomes almost automatic. That's why I cringe whenever I see artistic or stylised depictions of shells that make such a fundamental blunder. It happens much more often than chance alone would dictate and seems to be a special instance of Murphy's Law.

I've taken to calling it Grove's Law of Gastropod Reversibility. Check your shower curtains, your bathroom tiling, your beach towels, your Hawaiian shirts, your holiday brochures. You'll soon see Grove's Law at work.

There doesn't seem to be any particular advantage to coiling one way rather than the other, but the fossil record tells us that, throughout most of deep time, species that coil sinistrally have been very much in the minority. However, while most species of snail are reliably dextral, sometimes a fluke of nature will produce an individual that coils sinistrally. These 'sports' (to borrow a gardening term) are very rare though. Sinistral cowries are so rare that they fetch a ludicrous price among collectors. In Tasmania I have only once found a sinistral shell of a normally dextral species: a wine-mouthed rockshell that I spotted on a beach on lungtalanana-Clarke Island, amongst hundreds of dextral examples of the same species. A sinistral false-olive also turned up in one of a series of CSIRO dredge samples from eastern Bass Strait, which ended up at the museum. It was spotted by an observant volunteer, Don, who was working through the samples at the time. That find got me thinking about how sinistrality arises and why the trait doesn't generally spread in a population.

To answer the first question—how does sinistrality arise?—it seems that there are a couple of main ways. One would be when a baby snail starts coiling the wrong way for its species right from the start. If it does this, it's liable to end up with all its internal organs incorrectly aligned relative to the shell it inhabits. Another way would be for the snail to coil the right way, but for successive whorls to be laid down at increasingly *higher* points around the axis, rather than lower. In such cases it's actually the spire that's the wrong way around (upside down), rather than the direction of coiling *per se*, but the result is the same. As to why the sinistrality trait doesn't spread—well, in species that sidle up to each other to mate, their sex organs would be on the 'wrong' side. Our false-olive would be a case in point.

But from such speculation, back to the dredging. All sorts of living gems were hidden amongst the rubble of stones and old shells that we had brought to the surface. We were particularly thrilled to find our first Port Philip topshells—pearly, button-sized snails with stepped spires which, as far as I know, have never been found washed up on a beach. I had also never seen a living May's margin-shell either. Those before us were graceful, glossy, orange and yellow banded shells with a matching orange animal within. A complete surprise was Dunker's topshell. The only part of the state where this deep-purple, beaded species can reliably be picked up off the beach is the far northeast, particularly around Flinders and Cape Barren islands, where it is quite common. But here it was, living happily within a few hundred metres of the shore, a couple of hundred kilometres further south. It was our first inkling that one can get a rather biased impression of the distribution of molluscs if one only considers beach finds.

We undertook several more successful tows that day and, with the support of Rob's depth sounder, gradually pushed out into deeper and less well-charted waters off Visscher Island. Some hauls consisted almost entirely of introduced New Zealand screwshells, but there were always a few more interesting species to be found amongst them: wentletraps, margin-shells, turrids and a range of native screwshell species, too. Occasionally we would hit a snag and have to double back, and sometimes the dredge would be so weighed down with stones that we struggled to haul it aboard. Fortunately, the rope never broke, and we didn't lose the dredge.

Motoring back into Blackman Bay that evening, Rob and I were exhausted but elated. We knew that we would do this again. Over the next couple of years, Rob persevered with his dredging, venturing into waters up to 100 metres deep, towards the edge of the continental shelf east of Cape Pillar—the same area that Hedley and May had dredged a century previously. I joined Rob on a few of his expeditions, most memorably way out beyond Cape Pillar, where the ocean swell dwarfed our little boat. Albatrosses, shearwaters and petrels checked us out in the hope that we might offer up some tasty bycatch, and at one point we found ourselves in the midst of a super-pod of perhaps a thousand dolphins. But, despite these enticements, I soon learnt that I wasn't really cut out for a life on the ocean waves. More often than not, seasickness would force my retreat below deck, and I would spend much of the trip uselessly asleep. Thereafter, my main role became helping Rob sort and identify his finds back on dry land, and keeping track of these as the geographical coverage expanded and the number of tows rose beyond a hundred.

At about the halfway mark, with Rob's rate of species accumulation tapering off, he decided it was time for some further innovation. Using cable ties, he attached a fine-meshed bag, about 15 centimetres wide, to the rear end of the dredge. This meant that some of the fine sediment passing through the coarse mesh of the main dredge would then filter through the fine mesh. Immediately the rate of accumulation rose again, as the fine-meshed bag retained the tiny micromolluscs that had previously sailed straight through the original dredge. It didn't always work as planned, however—sometimes the bag became clogged with silt; but it gave Rob a boost of enthusiasm and kept him going out, whenever the weather permitted.

Finally, after some two hundred tows, Rob decided that he had achieved what he had set out to do and brought *Rambler* home for the last time. We had, by then, recorded and identified some six hundred species of mollusc, including most of the species that Hedley and May had found, plus a host of additional ones. We had extended the known ranges of several species previously only recorded from off the coast of New South Wales; and we had doubled or trebled the number of known specimens of some rarely found species. About one in every twenty species, mostly micromolluscs, looked to be new to science. Paradoxically, the species that we recorded most consistently was Gunn's screwshell, the only one in our entire haul that is officially listed as a threatened species in Tasmania. Whether this suggests that all those other species are even more threatened, or are naturally rare or merely hard to sample, is a moot point.

Rob eventually moved on to bird photography, and very generously donated his entire shell collection to the museum. As for me, despite my lack of sea legs, I was not done with dredging yet. I knew that to stand a chance of finding additional species, we would need to go deeper than Rob's outfit would allow. Around this time, CSIRO's newly commissioned and purpose-built research vessel, the *Investigator*, sailed into its new home port of Hobart. It's a substantially bigger vessel than its predecessor, the *Southern Surveyor*, and it has more capacity and cabin space for researchers. My health had not been good when the ship arrived, but, as soon as it improved, I let it be known that I would be very keen to join any research voyage plying Tasmanian waters, if it also involved dredging. My museum colleague, Kirrily, a deep-sea coral expert, had been offered a spot on an abyssal voyage that year, and I think she helped spread the word. The Tasmanian Museum is too small an entity to feasibly design and run a research voyage of our own, but we're happy to be involved in other researchers' projects.

So, in late November 2018, both Kirrily and I joined lead researchers from CSIRO and Museums Victoria, and others from the Australian Museum, on a voyage revisiting the Tasmanian seamounts. These are undersea peaks lying to the south and east of Tasmania, mostly in Commonwealth waters. They rise from the ocean floor, at a depth of over 2 kilometres, sometimes to within just a few hundred metres of the surface. Many of these mounts were fished heavily for orange roughy during the 1980s and early 1990s, before the fishery was largely shut down. At the time, fishing for orange roughy caused a lot of damage to the seamounts, because the fish formed spawning aggregations near their summits and that's where they could be reliably caught. Rather than risk damaging their nets by snagging them on corals and other outgrowths, the trawlers would deploy heavy chains and drag these across the surface to even out irregularities, enabling nets to pass freely overhead. Every ten or so years since, CSIRO has conducted a survey of the condition of the seamounts to gauge their recovery. Ours was the latest in this series. Most of the survey work was to be conducted by means of submerged video cameras piloted from the ship; but there would also be some bottom-trawling (dredging) on the flatter seabed between seamounts.

The *Investigator* is enormous: ten decks high and 94 metres long, and, the seasoned crew assured us, with extra stabilisers for the sissy scientists. Nevertheless, as we sailed out of Storm Bay and headed south, into the Southern Ocean, it didn't seem all that stable to me. I was glad that I had brought industrial-strength anti-nausea medication with me. For the first couple of days and nights I could do little more than loll around on couches or my bunk, or rug up and stand on one of the open decks, grimly holding onto the railings and staring stoically at the horizon. The crew, of course, had seen it all before, and assured me that I would come good in a day or two. I sort-of did; well enough to function, at least; although I don't think I ever fully gained my sea legs during the three weeks I was on board.

I was fit and well enough to help process the dredge samples as they came aboard, though. The dredge was a couple of orders of magnitude larger than Rob's, but still with a coarse mesh. We were dredging at depths between one and two kilometres, so it took a couple of hours just to winch it down and up again, on several tons of steel cable (never mind the actual dredging time). Operations were conducted at all hours, and we worked in two twelve-hour shifts: 2 am to 2 pm (Kirrily) and 2 pm to 2 am (me). Food was available at any time, but, strangely, cooked meals were served at set times as though we were still on land. I'm no night owl but, after a while, normal daily rhythms seemed completely irrelevant in our self-contained universe, and I would sometimes find myself still working away at 3 in the morning.

As with Rob's dredging, it was very hit-and-miss as to whether the dredge came up empty, full of rubble or full of life. We had to be prepared for all eventualities. Once the catch was landed on deck, we would don rubber gloves and gumboots and work quickly to pick out everything of interest, transferring it to plastic bins of ice-cold seawater. The low temperature was critical for keeping things alive, or at least fresh, as it approximated the temperature of the water at the depths dredged. The bins were then shunted into the wet lab where we would sort all the animals into different trays as best we could, according to their taxonomic groups: corals, gastropods, urchins, seastars, crustaceans, and so on. In an unusual and gratifying twist, the fish were considered bycatch and the invertebrates the main deal. Each group contained wondrous creatures of every shape and size. The corals (primarily soft corals, rather than the stony corals of tropical reefs) were glorious and often garish—purple, yellow, and even golden structures resembling Christmas trees, or perhaps lavatory brushes. Video footage showed that some of the seamount slopes were adorned with masses of these corals, enough to merit their characterisation as deep-sea coral reefs. Who knew we had such wonders around Tasmania?

Once the primary sort was finished, we would attempt to segregate the component species and then count, weigh, if possible identify, and log their details. Then the specimens would be transferred to glass vials or jars of ethanol, ready for sending on to taxonomic experts around the country on our return to port.

I wasn't the only mollusc expert on the voyage, so we agreed that, initially, all the live-caught molluscs would go to the Australian Museum, where the turrids (relatives of the conesnails) would be extracted for ongoing taxonomic studies there that depended on sequencing their DNA. The rest would, ultimately, be returned to the Tasmanian Museum. Strangely, nobody else was interested in all the dead, empty shells coming up in the trawls. These vastly outnumbered the living molluscs, and it transpired that the usual practice was to weigh these and then discard them. I decided that I would try to work through these 'death assemblages', as they are known, as I had a feeling that they might reveal a lot more diversity than the living molluscs. I had my microscope set up in the dry lab along the corridor. It was bolted to the benchtop so that it stayed put as the ship rocked from side to side. Other benchtops bore elasticised nets, like the ones used to stop garden refuse flying out of a trailer on the way to the tip—and for much the same purpose. As I set to work sorting the samples, I didn't care to imagine how rough the seas would need to be for those nets to prove their worth.

Most of the shells in the death assemblages were in poor condition, and I often only kept one or two of each species, just for the record. Occasionally the trawl would come up completely choked with calcareous grit—the long-dead remains of molluscs and other lifeforms. In amongst this coarse material could be found pockets of finer debris that hadn't been washed out of the net on its long ascent from the depths. I took samples as best I could and, to my delight, found that the debris was rich in micromolluscs. None of these would have been retained on previous voyages, so I felt as though I was really breaking new ground by examining them now. I got used to the shells rolling about in the petri dish as I tried to identify them under the microscope. Up close, an incredible diversity of molluscan life was evident. Putting names to them all was another matter. Some of the larger ones, such as the marvellously channel-spired Ken Graham's whelk, had already been documented from the seamounts during earlier voyages. Many of the smaller species resembled the ones that Hedley and May had described from deep water off Cape Pillar, and there was some overlap with what Rob and I had found. Some looked as though they might be related to species that had been described from seamounts much further north, or from New Zealand waters; others still resembled species from the Antarctic, or even the South Atlantic. A few were shallow-water interlopers which, perhaps, after death, had rolled down the shelf slope from the southern Tasmanian coastline. An umbilicated cowrie was, perhaps, the southernmost cowrie on record. Plenty of shells were like nothing I had seen before. I could assign most to familiar families—wentletraps, margin-shells, pyramid-shells, slit-limpets, ribbed-mitres, whelks, to name a few—but with some I couldn't even get that far. It was clear that I was looking at a very poorly known fauna, and that many of the species were likely to be new to science. By the time I had taken a closer look at the samples back in the museum lab, I estimated that I had some two hundred distinct species, with close to fifty percent apparently undescribed. To put this number in context, the live molluscs from the voyage tallied fewer than seventy species.

Then, one day, as the ship was tracking between two seamount ranges, we were warned of approaching severe weather and advised to make sure that everything was strapped, screwed, bolted, or tied down, or stowed safely away. I did as I was told but felt acclimatised enough to continue with my identification work regardless. The ship's movements increased steadily, but those stabilising tanks kept the rolling tolerable. Tolerable, at least, until the ship undertook a routine ninety-degree turn and was hit broadside by a succession of huge waves. I had taken to occasionally looking up from my microscope and peering through the porthole to check the horizon. Often, I could spot albatrosses or other seabirds which, for some reason, reassured me that all was well. But the next time I looked, I found myself staring at an undersea realm, such was the ship's tilt. Apparently, it was 'only' 23 degrees from the horizontal, but it was enough to spook even some of the most hardened scientists, who had endured many a voyage in the Southern Ocean. Now I could appreciate the utility of the elasticised nets. Without them, my trays of sorted molluscs would have gone flying. As soon as the ship had completed its manoeuvre, the rolling subsided; but, deciding that discretion was the better part of valour, I abandoned the lab work for the remainder of that shift.

During rough weather, the worst place to be was down below, in the cabin, as this was beneath the waterline and lacked any portholes. I discovered that I much preferred to be high up, on the observation deck. Even though the rolling was greatest up there, the views were awe-inspiring and turned the experience into a bit of an adventure. The ship was built for precisely this kind of weather, and most work could continue unabated even in swells of ten metres or more. Such weather was also fantastic for observing albatrosses and other seabirds. During calm weather they had no option but to loaf around on the sea surface; but in a gale, they really showed what they were made of, slicing through the air at breakneck speed, and closely hugging the peaks and troughs of enormous waves.

For all the exhilaration of those stormy days, I was glad when the winds and the swell eased and we could move around the ship more safely and comfortably. Within a day or two the clouds dissipated, as a ridge of high pressure descended from the north. The sea flattened until it was as calm as a millpond. The surface wasn't quite as oily-flat or mirror-like as depicted in *The Life of Pi*, but it wasn't far off. In the distance, to the north, we could just make out the hazy blue peaks of southwestern Tasmania poking above the horizon. The most surreal aspect of being becalmed was, however, the sheer number of albatrosses and other seabirds that settled on the sea all around us. They are used to being fed bycatch, but ours was no fishing vessel and there were strict rules that forbade the dumping of food or waste overboard. The birds, of course, didn't know this, but they get top marks for perseverance. Some of the very same birds that appeared alongside the ship when we arrived in the study area were still milling around when we left, three weeks later. For such rare and special birds, the numbers around our vessel were astonishing. At one point, we counted about six hundred albatrosses alone. Shy albatross—a species that only breeds in Tasmanian waters—were the most numerous, but all up there were close to a dozen other albatross species hailing from all parts of the Southern Ocean, plus giant petrels, cape petrels—a striking, black-and-white mottled bird that I had longed to see since my teenage years—and a wide range of shearwaters and prions. As the sun dipped below the western horizon, the air was so still that we could hear the albatrosses chatting to each other as they huddled together in small groups. Indeed, I can still hear them as I write this. Presumably, they all knew each other from way back—albatrosses are very long-lived birds; if they don't drown on longline hooks, that is.

I had little sense of time passing while on that voyage, though I know it lasted but three weeks. The whole experience was fundamentally surreal. I felt immensely privileged to have been invited on board, and, given my recent medical history (of which more will be said), and my proneness to seasickness, more than a little proud of having conducted myself quite well. While the amazing haul of molluscs was, undoubtedly, my personal scientific highlight, the albatrosses stole the show. They will be forever etched in my memory.

© K.DAVIS 2022

January

High summer, high heat

If the week between Christmas and New Year's Day often heralds the arrival of summer heat, then January is when summer cements its presence and that heat intensifies. Whether its appearance is fleeting, or persists for days, depends on how well the high-pressure ridges that drift over Tasmania can resist being carried eastwards, out into the Tasman Sea, to New Zealand and beyond. Optimal conditions for a January heatwave occur when a blocking high-pressure system sits in the Tasman Sea, funnelling hot air southwards from the Australian mainland and deflecting any cold fronts and troughs approaching from the west by forcing them to slip southwards or stall. Ultimately, it's a losing battle, and sometimes the last hurrah as a cold front forces its way in is a rush of hot, dry air as it squeezes up against the high. This sends humidity tumbling and causes the risk of bushfires to soar. The turbulence induced by such an abrupt change from hot to cold air may trigger thunderstorms; often they don't amount to much in the way of rain, but they spark lightning strikes, which are potent ignition sources for bushfires, if the heat lingers long enough for a fire to take hold. A cold, southerly change almost inevitably follows a heatwave; but in January, these are usually short-lived as the next ridge quickly establishes itself. If the cold interlude didn't extinguish a smouldering fire, it may rapidly flare up into a full-blown bushfire with the arrival of the next spell of hot, dry weather.

The nice thing about summer heat in Tasmania is that it's not usually humid, so it lacks the energy-sapping mugginess of the tropics, and even that which the rest of the eastern Australian seaboard experiences in summer. That dryness contributes as much to the threat of bushfires as the heat though, so it is not surprising that January is peak bushfire season. Bushfires can be devastating when they occur in the wrong place and at the wrong time, as we have seen in recent years; but a 'good' bushfire can foster ecological renewal. Lutruwita's First Peoples became experts at using fire to manipulate vegetation to their advantage, whether for nurturing or flushing out game, maintaining trading and travelling routes, or keeping biting insects at bay. Today's forest managers use fire to regenerate forests, too, or to reduce fuel loads; but none of these practices is desirable in January, or even feasible, because the risk of a fire getting out of control is too great. The corollary is that conducting controlled or 'cool' burns outside of peak fire season can reduce the extent of land vulnerable to uncontrollable bushfires during that peak. That said, some fire weather is so severe that no amount of pre-emptive burning will avert the risk. We live in highly combustible country, in increasingly flammable times.

Notwithstanding a pervasive sense of unease around the threat of bushfires, there is much to enjoy in January. Long days of intense sunshine finally nudge coastal water temperatures up a few crucial degrees, making this a great month for snorkelling and taking a peep at our wonderful marine life. You'll often appreciate a puffer jacket come sunset, even in high summer, as temperatures are prone to drop rapidly. Long, balmy evenings outdoors, after a barbecue or picnic, aren't really a Tasmanian thing. Going home for dessert isn't a bad option, though, because, in the garden, all the soft fruit seems to ripen at once (and don't the local lorikeets, blackbirds, starlings, silvereyes and honeyeaters know it). Insect activity is also at its peak, at least in terms of abundance. Sometimes this is good, but it's just a bit annoying if those insects are biting marchflies. Incidentally, there's no agreement on why they're called marchflies, but it is likely to be a corruption of 'marsh' (since their larvae develop in wet soil or, indeed, marshes), rather than having anything to do with the month of March. In fact, most marchflies will have died off by March. Elsewhere in the English-speaking world, marchflies are known variously as horseflies, deerflies, or clegs; but it seems that none of those words seemed quite right to Australia's British colonists.

We hadn't been living in Tasmania long before seasoned bushwalkers advised us of the perils of summer walks in 'buzzie country'. Buzzies are burrs, the globose seedheads of an unassuming, creeping plant that, for the rest of the year, goes largely unnoticed. They are a relative of strawberries, but sadly the similarity does not extend to their fruit, at least not in any way a non-botanist or strawberry-lover is likely to recognise.

Buzzie season usually starts in early January, when the seedheads mature to become the bane of bushwalkers and wandering entomologists alike. The burr consists of dozens of individual seeds, each bearing long, bristly extensions that all-too readily latch on to bootlaces, socks and insect nets, and must be laboriously teased out before they embed themselves further. Gaiters offer a line of defence for bushwalkers' boots and legs; but nobody has yet designed a gaiter for insect nets. Buzzies evolved to allow their seeds to be dispersed via the fur of passing mammals; or, at times, the plumes of large flightless birds—the species is also present in New Zealand (once home to some *very* large, flightless birds), where its local name, bidgee widgee, must be spoken with a Kiwi accent to appreciate its similarity to the Aussie one.

When extra-furry sheep were introduced into buzzie country, the plants had a field day. They are now widespread and abundant. During the early nineteenth century, when northern England's woollen mills were burgeoning, buzzies hitched a ride in bales of Australian wool and are now naturalised there, too.

January is also when many of our grasses set seed. Some of these bear barbs and long awns, which function in much the same way as the buzzies' bristles. I learnt the hard way to give up on sweep-netting for insects in grasslands during early January. It is, therefore, a sad time for entomology, but perhaps a relief for grassland insects—they only need face the lesser threat of my camera.

Cloud watching

January is a good month for cloud watching, because the summer sun's greater influence on air temperature and evaporation means that clouds can form at a multiplicity of altitudes. It can be wonderfully relaxing to lie on your back on the grass or sand and just gaze at the goings-on overhead. It's not uncommon to be able to discern two or even three distinct cloud layers, possibly all moving in different directions. This is an indication that, even during apparently stable weather, the atmosphere high above our heads is restless. Its constant motion speaks of the important role it plays in evening out temperatures across the planet—it is abiding by that universal law of thermodynamics, on maximising entropy. High school physics actually bored me like no other subject, but, staring at a lively sky, I can see its value. We also have entropy to thank for giving time its arrow, so on balance physics is probably quite helpful.

Our highest clouds, hazy cirrostratus, almost always drift eastwards; sometimes rapidly but often very slowly. Some wispy forms are likened to mares' tails, although they appear to me more like fern fronds or feathers. Sometimes high clouds coalesce into gently undulating mountain waves, which form when Tasmania's hilly topography becomes an obstacle to the smooth flow of air streaming in from the west. Their appearance often heralds a change in the weather, preceding further tiers of stratus-type cloud, which come in at progressively lower altitudes. They can arrive from almost any direction, depending on which part of the approaching weather system, with its gyratory airflow, is encroaching into our airspace.

Even in summer, Tasmania doesn't get much by way of fair-weather cumulus cloud; the sort that has almost symmetrical, puffy outlines, like the archetypal cloud symbol on weather maps. Cumulus doesn't develop well near the coast. Such clouds require a combination of humid air, strong convection currents (usually caused by the sun warming the land beneath) and a lack of strong wind—because that would blow the warm air and clouds away from their point of formation. Such conditions don't arise often around here. Tasmania's clouds are usually more about turbulence than convection. They tend to scud across broad expanses of sky, rather than forming *in situ* and ambling along cloud streets (yes, there is such a thing). Consequently, another cloud type that rarely gets a chance to develop over Tasmania is cumulonimbus—the sort of cloud associated with thunderstorms. On the rare (though increasingly frequent) occasions when thunderstorms and torrential downpours happen, they often become a talking point for locals, to the consternation or amusement of storm-seasoned visitors from New South Wales or Queensland. Sometimes little cottonwool cumulus clouds make concerted attempts to push upwards and graduate to the next level, but as often as not their tops are sheared off by high-altitude westerlies, and any showers that they spawn are usually fleeting.

It's also not uncommon for the upper atmosphere to be much more unstable than the zone closer to ground level. This results in thunderstorms that rumble high overhead, while any rain that they produce evaporates during descent, so that little or none reaches the ground. In these conditions we see downdraughts of cold, moist air punching into warmer, drier air below, which may give rise to mammatus clouds. They look a bit like upside-down cumulus clouds, with lumpy (literally 'breasted') bases. Dry lightning strikes associated with storms such as these seem to be on the rise, particularly in the west of Tasmania where, in recent years, they have seeded peat fires, which can smoulder for months and flare up when the next heatwave hits.

Almost like the inverse of a dry lightning storm is another striking phenomenon common in Tasmania. It occurs when the strong winds hundreds of metres above ground-level push rain far from its source cloud. This can cause rain to appear to fall from clear skies. If this occurs when the angle of the sun is low, sunlight may be diffracted by the water droplets, which is partly why Tasmania is a great place for rainbows.

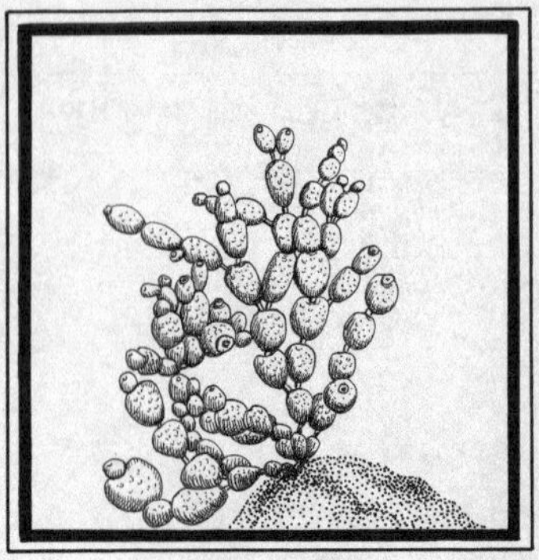

Tide's out

Some of my most cherished childhood memories are of 'poddling' among the rockpools during beachside summer holidays in the west of Britain. 'Poddling' is a Grove family word, but perhaps you've guessed that it means '*bare-footedly exploring rockpools, poking anemones, trying to catch prawns and blennies, and searching for cowries and other sea treasures, before moving on to the next rockpool and repeating.*' That is, of course, how my family passed many a day during the holidays of my formative years. At the time, I thought little about the tides, beyond trying, always, to heed a parental warning not to get cut off from higher ground by a rising tide. In my memories of poddling, the tide was always out; but obviously it wasn't. I suspect discreet parental planning ought to be thanked for engineering visits to castles and countryside when the tides weren't going to cooperate.

Tides are fascinating but rather complicated, so I'll try to offer you a greatly simplified version of what's going on. Essentially, tides are a product of the push and pull of gravitational forces exerted on the world's oceans by both the moon and the sun. Since the Earth rotates on its axis over 24 hours, those gravitational forces vary at any given point on the Earth's surface, causing the tides to rise and fall; often with a bit of a lag, which varies depending on the undersea topography of a specific locality. The daily tidal range fluctuates over the course of a given lunar month; and in some months those fluctuations are more extreme than in others. Most of the tide's amplitude is down to the moon, but the sun's contribution is also significant—it's around half that of the moon. A new moon, which generally arrives once in each calendar month (after all, a month used to be a 'moonth'), signifies that the sun and moon are aligned relative to Earth. That's when their combined influence on the tides peaks as they are both exerting gravitational force from the one side of the earth. It's similar with a full moon, two weeks later, in that the moon and sun are also aligned, except that now they are on opposite sides of Earth. These times are when we have very high 'spring' or 'king' tides. A week after each spring tide is the time of minimum tidal amplitude, known as a 'neap' tide.

But it's more complicated than this. The moon's monthly orbit around Earth is elliptical; so, when it's at its closest to us (the perigee), we can expect greater tidal ranges, on average, than when it's further away. Similarly, Earth's annual orbit around the sun is also elliptical. We're at our closest position to the sun on about the second of January each year (it's called the perihelion), so that's when the sun's individual influence on our tides is at its peak. Furthermore, the Earth's axis is tilted, which affects the angle of the sun relative to fixed positions on the Earth's surface as we orbit the sun; which is why we have seasons. The tide to watch, then, is the first new moon of the year, the one coinciding most closely with the perihelion. It's a particularly good one for low-tide excursions (and poddling), because the tide is out in the middle of the day, while the sun is close to its zenith; and the noon sun is high in the sky because it is not that far past its maximum declination relative to the Earth's equator (which happens a couple of weeks earlier, at the summer solstice).

Slight differences in air pressure can affect the tides, too. This is called the 'inverted barometer effect'. Lower air pressure raises water levels, while higher pressure depresses them. The difference between the two extremes can amount to about 30 centimetres. Low pressure is also usually associated with stronger winds and bigger waves, which can make the tides appear even higher. Together these effects account for storm surges. The corollary is that high pressure is usually associated with light winds, and these allow a low tide to reach its lowest potential. Coincidentally, in Tasmania we often get our first, stable period of summer high-pressure weather around about the beginning of January, which is perfect timing for the perihelion. What's more, it's usually holiday season, so there's really no excuse not to head for the beach for some poddling. But an early January perihelion also means that our high summer is 'higher' (i.e., the sun in closer to us) than at the equivalent time and latitude in the Northern Hemisphere. For once, we may not need to rug up against the cold, but the intense sunlight means that it is especially important to cover up, wear sunnies and a hat, and slap some sunscreen on any exposed skin.

Much as my parents engineered our summer holiday visits to the seaside to coincide with low tides, Chris and I have found ourselves doing the same in Tasmania, with our sons. They're all grown up now, but I hope they will cherish their memories of these occasions as much as I have. One I'd like to share involved a low-tide visit to the rocky shoreline at Trigonia Corner, about ten or twelve years ago. Trigonia Corner is a secluded, sheltered, east-facing spot, just south of McCrae's Isthmus, on Maria Island. Spending a few days on Maria Island over New Year became a family tradition for us, often in the company of other Taroona families. We became adept at squeezing all our camping, bushwalking, and mountain-biking paraphernalia into as few bags as possible for the short crossing from Triabunna to the jetty at Darlington, and at loading up the awaiting cart to the gunnels so that we could drag everything along the track to the campsite in one trip. We would generally pootle around on our bikes for a couple of days, visiting favourite haunts such as the Painted Cliffs and Fossil Cliffs, scattering Cape Barren goslings, turbo-chooks (otherwise known as native hens) and wombats in all directions. After this warm-up, we would tackle the spectacular dolerite peak known as the Bishop and Clerk, and sometimes Mount Maria—with the latter we'd take our bikes part of the way, along rutted, washed-out tracks.

Eventually, we would work ourselves up to tackling the Isthmus. Its sandy trackway usually proved impossible to ride along without falling off and having to push. If the tide was out, cycling on the hard, wet sand of either Shoal Bay (on the west side of the Isthmus) or Riedle Bay (on the east) might be an option. On this occasion we chose Riedle Bay and soon found ourselves at Trigonia Corner. I always liked this particular spot, not least because it was named after one of my favourite bivalves, a species more usually called a brooch-shell—it is a pleasingly shaped, sculpted, pearly pink shell that would, indeed, make a nice brooch.

'*Trigonia*' actually now applies only to ancient relatives of these brooch-shells, which are widespread in the fossil record. For a long time, fossils were all that the scientific establishment knew of their existence. It was François Péron who first found and instantly recognised a modern-day example of these shells. Péron was a French naturalist (officially a 'trainee zoologist') aboard the *Géographe*, on Nicolas Baudin's expedition to New Holland in 1802. He found a *Trigonia* shell while with a party that was exploring what we now call the D'Entrecasteaux Channel, between Bruny Island and the Tasmanian mainland. The *Géographe* visited Maria Island later on the same expedition, anchoring off Riedle Bay (strictly speaking, Riedlé, honouring the expedition's botanist); but, sadly, Péron made no comment about *Trigonia*s there.

When the specimens collected during Baudin's expedition eventually made it back to Paris, the great naturalist Jean-Baptiste Lamarck confirmed Péron's identification, making the find a scientific sensation—here was a real 'living fossil' (even though it was only a dead shell). Others heading to the Antipodes were encouraged to search for live specimens; and two other French naturalists, Quoy and Gaimard, found the first living examples, which they dredged while their ship, the *Astrolabe* (under the command of Jules Dumont d'Urville) was becalmed in Bass Strait, during the late 1820s. Once they had the preserved specimens safely back in France, the word was out and everybody who was anybody in natural science wanted their own fresh *Trigonia*. Among them was an ambitious young English geologist, Samuel Stutchbury, who happened to be voyaging along Australia's east coast a year or two later. The story goes that he went dredging in Sydney Harbour and succeeded in finding three live *Trigonia*s. These were cleaned up (presumably by his man-servant or crew) and put on his seat in the boat. Stutchbury is said to have turned away for an instant, during which time two of the three bivalves extended their muscular feet, gripped and twisted with the same strong muscles, and flipped themselves over the side of the boat, leaping a height of about ten centimetres to do so. As he turned back, Stutchbury witnessed the third and final animal performing the same feat, and then there were none. He was crestfallen, and apoplectic. Apparently, it took him days to find more; but, in the meantime, the story of the molluscs' audacious escape was out and it has been recounted regularly ever since, almost as though their plucky refusal to be gathered into the Englishman's fold might have something to say about the national character of Australians, too.

Back, now, to Trigonia Corner: we left our bikes at the top of the beach, under coastal she-oaks, and picked our way along the rocks, intent on rounding the next headland before stopping for a picnic. Progress was slow, to say the least. The tide was still receding but it had already exposed rocks and weed that wouldn't normally be accessible from the land. Holdfasts of bull kelp and swathes of Neptune's necklace—a brown seaweed composed of interlinked, knobbly segments—had been left high and dry. We explored the nooks and crannies among the kelp's anastomosing 'roots', finding bright orange sponges and sea squirts, tiny crabs, green and orange carpet stars and much else besides. Further out, on the thick, leathery kelp blades, flocks of tiny, mottled green-and-black, air-breathing seaslugs hunkered down against the heat of the overhead sun. There was very little by way of swell and the water was crystal-clear. Peering into the depths, we could see still more seastars and a range of seasnails.

My personal highlights came after we had rounded the headland. 'Rounded' is an apposite word, because there was no one point at which we could agree that we were 'there' (a key consideration for stopping for lunch); but the further we pushed on, the gentler the slope of the rocky shore became and the more we found exposed, for our delectation. Peering between two weed-strewn boulders, my younger son found a large gastropod, and we all gathered round for a closer look. It was enormous—at least 25 centimetres long—and much of its shell was covered in a bright pink encrustation of coralline alga. I turned the shell over to reveal its aperture and the animal within. It was a reddish rock whelk, the same species that I described back in October, and just as exciting to find alive. I had longed to find one of these since childhood, as they were occasionally trawled off England's south-west coast. Seeing this one brought a flood of emotion to me. It didn't seem to be having quite the same effect on the rest of the family, though; and, in any case, out of water, it would soon be suffering from heat stress. So, returning it to its home, we moved on, to other pools amid the rocks. Then I spotted an extraordinary animal—a hand-sized, knobbly, olive-grey blob, like some sort of gargantuan anemone; except that, where an anemone's tentacles should have been, this animal bore a small, leathery-looking, off-white, calcareous cap. The animal was only loosely attached to the rock, and I easily peeled it off so that I could hold it in my hand. Although I had not previously seen one of these, I knew it was an umbrella slug—an unusual, sponge-feeding sort of sea-slug. Umbrella slugs are found widely in the warmer parts of the Pacific and Atlantic Oceans, from the shallows down to several hundred metres; but in Tasmanian waters I was only aware of a few records, and all were from deep water. I had never imagined finding one intertidally, so it made my day; even if I hadn't made its day by removing it from the water to show everyone.

It was unequivocally lunchtime by now, though, and the boys were hungry, which may explain why they scarcely gave the umbrella slug a second glance. They were nonplussed when I suggested that this might be the only one that they would ever see in their entire lives. Duly photographed, I reluctantly returned it whence it came, and we retreated to the shade of the she-oaks for lunch.

The perihelion had come up trumps, but that sun was *really* intense.

Dancing bugs and whistling moths

Long before there was Australia there was Gondwana. And long before Gondwana there was Pangaea. And, before that, there was quite a different arrangement of continents on the face of the Earth. Around a billion years ago, there was a supercontinent known as Rodinia; and, before that, there was another: Nuna. Nuna existed about one-and-a-half billion years ago—three times as long ago as the Cambrian, which is my usual benchmark for 'ancient', since that's when the ancestors of most of today's major animal groups first appeared. Given all that has happened to the Earth over the last billion-plus years, it is no surprise that the evidence for Nuna is sketchy, albeit unequivocal. There is a place in Tasmania where rocks from that period persist. It's Pinmatik-Rocky Cape, in Tasmania's northwest. To be fair, it's not so much the original rocks that persist; rather, it's the siliceous, heavily metamorphosed, tortuously folded strata derived from sediments previously eroded from those rocks. *They* bear the hallmark mineralogical signatures of Nuna. And those same signatures crop up (or 'outcrop') on the other side of the Pacific Ocean, in Colorado—because some 1300 million years ago, as Nuna was breaking up, the regions of seafloor that eventually *became* Tasmania and Colorado were close neighbours. Rocks of this pedigree must have something going for them. Essentially, they are extremely hard and much more resistant to erosion than those which have come and gone around them. Which is why, I suppose, Pinmatik juts so defiantly into Bass Strait.

Chris and I recently took our campervan on a trip to Pinmatik, where we passed our days happily exploring the coastal paths and beaches of this magnificent and underappreciated national park, and snorkelling in its clear, weedy waters. The backbone of the park is a range of hills formed from those most ancient of rocks. Boat Harbour Beach sits at the gateway to the Cape and proved a fine base from which to begin our perambulations. Gardens at the national park end are on the same siliceous bedrock as the Cape, with thin, acid soils. Those at the other end offer better prospects for gardeners, since they are situated on deep, fertile, base-rich basalt that dates from volcanic activity a mere 65 million years ago.

Pinmatik-Rocky Cape, known to some as, simply, 'Rocky', is, well … rocky; very rocky. For the most part, the soils are skeletal and nutrient-poor, as I have described; which means that the vegetation is a mixture of buttongrass moorland, heathland and heathy woodland. These communities comprise plant species equipped to eke out a living under conditions of near-starvation. Paradoxically, the rather special conditions foster a remarkably diverse flora, with a host of endemic and highly localised species. For the casual botanist, the standout among these is, surely, saw banksia, since the Cape is the only place on the Tasmanian mainland where this beautiful tree, with papery, serrated leaves, grows wild. We couldn't stop admiring and photographing these gnarly old trees, their outsized yellow flower spikes and their bristly cones, which sport strangely grinning seedpods.

One warm evening found us on a coastal path among the flowering banksias and teatrees, the grasstrees, and the stunted eucalypts, enjoying views over an uncharacteristically calm and blue Bass Strait. At the furthest point on our walk, with the sun already setting, we stopped for a brief rest on a rocky knoll (there were plenty to choose from) and took in the spectacular coastal scenery as we cooled off before tackling our return. As if by special order, a gentle breeze came out of nowhere, whisking away the heat of the afternoon and ushering in dusk. We would need to get moving, soon.

Eventually we stood up, to begin our return walk. As we did so, the breeze died down again, almost as suddenly as it had arrived. This is often the pattern, towards the end of a hot summer's day. It left an altogether different ambience in its wake. Gone was the background buzz of cicadas, marchflies and hoverflies, and the chatter of crescent honeyeaters. Gone, too, the heady aroma of eucalyptus and teatree oils. Instead, there was silence, deep peace and serenity.

Dusks are long in January, and we continued our walk back towards the campervan knowing that we wouldn't need torchlight for a good hour or so. Over the following minutes we were accompanied by a darkening sea and the sky blushing pastel shades of peach and apricot, while thin, sharply defined ribbons of cloud on the eastern horizon turned inky black. Most of the day-active inhabitants of the heathland had long since settled down for the night by now. As it grew darker, we picked our way carefully over the rocky terrain, eyes firmly fixed on the ground; but, when I stopped for a breather, I allowed my gaze to wander skywards. There, I noticed an erratic, rolling cloud of dancing insects, silhouetted against the fading light. I had seen these performances before and had my suspicions about their identity; but, just to be sure, I netted some for a closer look, with the aid of my headtorch. In the bottom of the net, a dozen Tasmanian gnatbugs were in the process of rapidly folding away their wings as they came to terms with having so unexpectedly made landfall.

Gnatbugs are strange, pin-headed, and rather enigmatic true bugs with an ancient pedigree, not much less so than that of the moss bugs, which we met last month. I say enigmatic because they're hardly ever sighted except when out dancing. Their flattened bodies, about a centimetre long, seem well designed for life under bark, and that is supposed to be their chief habitat; but I have peeled back loose bark from a lot of branches without finding any gnatbugs. I have only witnessed their dancing—which I take to be a mating ritual—on warm summer evenings, within a short distance of the coast. I assume they must live beneath the bark of some coastal shrub or tree, perhaps she-oaks. A much smaller, indeed, gnat-sized species of gnatbug lives in wet forests and does its dancing by day, usually only a few centimetres above the leaf litter. I imagine that is where they spend the rest of their lives. They are very unassuming creatures and their preference for conducting their dances so close to the ground means that they fly beneath the radar (and sweep-net) of most entomologists.

Now that I had my eye in, I could see that there were little parties of gnatbugs dancing silently atop many of the taller shrubs on either side of the path. I wondered how they selected their aerial dance spaces. Perhaps it was a matter of quorum sensing, where the more bugs that congregated above a particular shrub, the more other bugs in the area sensed their presence and felt compelled to join the party, giving rise to a bigger swarm and exerting more pull over bugs from further afield. The term is usually applied to bacteria, but the principle would seem to be the same in some swarming insects. I imagined that most of the dancers were males, with unmated females making occasional dashes into the melee.

I was roused from my gnatbug ruminations by a strange buzzing sound, coincident with a larger, bumblebee-sized insect. It was a moth, zipping across my field of view. I knew it must be a male whistling moth, having learnt about these wonderful Australian insects a couple of years earlier, from my entomologist friend, Sarah. She had sent me video footage of a similar encounter on the Tasman Peninsula. Who knew that moths could whistle? Or at least buzz? In fact, quite a few moths make sounds, although most are so high-pitched that they are inaudible to us. They do so to jam the sonar signals of bats pursuing them through the night skies. Some hawkmoths can squeak like a mouse if they're picked up, doing so to alarm their would-be predator into dropping them off the menu. But male whistling moths are different—their sounds are thought to be songs of courtship. They also emit sounds during the day, as they bask prominently on foliage. These sounds are thought to advertise their presence to rivals, but they are in the ultrasound range and inaudible to us.

There are three species of whistling moth; those in Tasmania are the common whistling moth. All are boldly patterned, with bright red bodies and hind wings and black-and-white striped forewings; not that those colours are at all visible in the dim light of dusk. The sounds coming from this particular moth, as it zigzagged in front of me, would better be described as a continuous, high-pitched buzz or crackle, rather than a whistle; reminiscent of the sound emitted by overhead high-voltage electricity cables; or perhaps someone running their fingers along a plastic comb. The last analogy is quite apt, in fact, because the moth's sound-making apparatus consists of a prominent, crescent-shaped, scaleless area near the leading edge of the forewing, which is ribbed, a bit like a comb. How the apparatus produces sound is still, it seems, unclear. Most likely, the ribbed bits of opposite wings 'clap' together above the moth's body when the wings are beating. For that reason, they are known as castanets. It might be an unusual anatomical feature in insects, but it's common enough to have its own name.

The moth before me was soon joined by several others, all performing vaguely figure-of-eight manoeuvres around some of the same bushes as the gnatbugs. Soon, the still air became quite full of the songs of serenading moths. It was a surreal and haunting experience, which still lingers for me, even though I can no longer recall the exact sounds.

The larvae of whistling moths feed on dodder, a parasitic, leafless, twining plant that scrambles over the heathland's shrubs, sometimes forming dense, knotted mats that remind me of the invasive Martian red weed in H.G. Wells's *War of the Worlds*. Dodder, also known as devil's twine, is also the larval foodplant of the blotched blue butterfly, so I am doubly happy to overlook its parasitic tendencies. Looking at the highest parts of these bushes again, I thought I could make out a few dodder tendrils, which might have been the initial attractant for the moths, but it was getting too dark to be sure. It was time for us to move on. From the path in front of us, a frogmouth took flight, and as the dimly outlined silhouette of this supremely camouflaged, crepuscular bug-hunter passed silently overhead, I pondered how many whistling moths would fall victim to its wide, bristly gape tonight. Or maybe the moths' bright colours in daylight signify distastefulness, and are complemented at night by an acoustic equivalent that bats and frogmouths can recognise.

We could still hear the crackling moths when we reached our van near the beach, as dusk gave way to the black of a moonless but magically starry night.

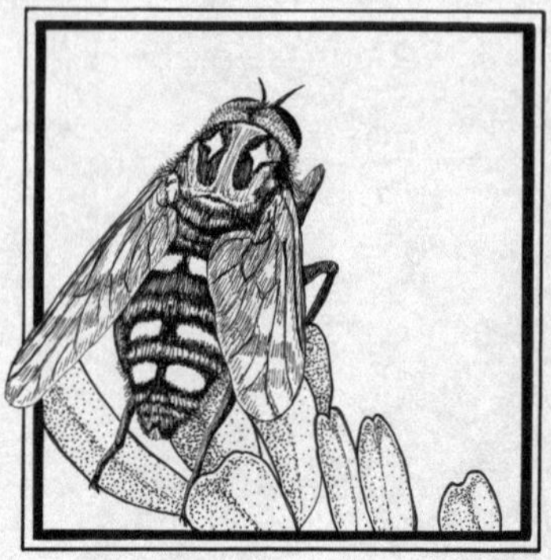

Hilltopping

Our visit to Pinmatik-Rocky Cape reminded me that if there's one thing that Tasmania has in abundance, it's topography. From dunes, hillocks and rocky knolls to imposing mountains and ravines sculpted by wind, water and ice, there are few parts of the state that one could count as flat. Even the plains and plateaus can best be described as 'gently rolling'; which helps explain Tasmania's notoriously winding, albeit picturesque, roads. Another explanation is that they were, for the most part, laid out in an era when it was more practical, and less exhausting for man and beast, to go around a hill or wend their way slowly up, rather than plough straight over.

There are, however, animals that thrive in the hilly terrain and revel in assailing lofty summits. Birds of prey love the updrafts that are a feature of hilly country. With the help of binoculars, it's usually not hard to spot an eagle, falcon, hawk, or harrier soaring, banking, hanging, or hovering above the island's undulating skyline. What is less well appreciated is that some insects also head for the hilltops, engaging in an activity known simply as 'hilltopping'

In the Tasmanian context, I was re-acquainted with hilltopping by Jeff, a visiting entomological colleague from Canada. I had accompanied him on a whistle-stop Tasmanian collecting trip one January, as part of his taxonomic work on the poorly known Australian hoverflies. With a forecast for perfect, clear skies and gentle breezes, we headed out on the Lyell Highway, through dry, rain-shadow, sheep-grazing country. This almost treeless land was one of the first parts of the island that the British colonised. It contrasts dramatically with the wetter, forested country that appears, abruptly, after passing through Ouse—Tasmania's hydroelectricity heartlands. Next come the subalpine eucalypt woodlands, around Derwent Bridge, before the highway descends into the valley of the Franklin River, with its patchwork of rainforests, wet eucalypt forests and buttongrass moorland. It is within this landscape that one of Tasmania's official 'Great Short Walks' takes a gently winding route, from the floor of the valley to the summit of Donaghys Hill. We missed the signpost and car park on our first pass and had to double back a little further down the highway, turning into a convenient roadside apiary site well stocked with hives of bees making leatherwood honey.

The Donaghys Hill walk is like a microcosm of the surrounding country. It ascends through rainforest and wet eucalypt forest, before emerging onto a high, heathy ridge, where the compact flowerheads of buttongrass tussocks wave gently in the updrafts on long, wiry stems. The ridgeline then ascends, abruptly, to a well-defined summit, from which the ground beyond drops steeply away, to plains far below. As we scrambled up the last few metres, towards the lookout, it became obvious why Jeff had wanted to visit this spot. All around us were airborne flies, beetles, bees and wasps. Hoverflies, marchflies, and the very beautiful, dronefly-like, axe-snouted flies (my translation)—all hovered high above our heads, occasionally engaging in what looked like dogfights. Beetles, and the odd Macleay's swallowtail butterfly, drifted about on the updrafts; and bees, wasps and winged male jackjumper ants patrolled invisible pathways in the air. Once I had tuned into the combined buzz of hundreds of pairs of tiny, beating wings, the sound became all-pervasive and impossible to ignore.

A party of tourists arrived at the summit, oblivious to the insects, but gazing in awestruck silence at the 360-degree views. Their presence was a reminder that I, too, needed to allow myself time to appreciate the extraordinary surroundings. The views included Frenchmans Cap, in one direction, and the West Coast ranges in another. But it was hard for me not to feel as though pleasant views were rather trivial when there was so much exciting insect activity in the air. I'll admit that I was soon back to fly-watching.

Much like the tourists, the insects had homed in on Donaghys Hill because of its commanding position in the landscape. Any insect within a radius of several kilometres, that decided to head towards the highest point around, would end up here. But why would they choose to do such a thing? It must take an awful lot of energy. Watching the flies around the summit, the answer was clear. This was *the* place to find a mate. Most of the hoverers and patrollers were males, and, for them, the hilltop was clearly a singles bar. Occasionally a female would appear, and the best-dressed, most aerobatically gifted, or best-positioned males would have the greatest chance of mating. If you're single and live in a big city, you know that if you want to find a mate, it's going to be far more productive—not to mention safer—to head to the singles bar in the city centre, rather than wander aimlessly around suburban streets in the hope of bumping into a potential partner.

So, too, with the insects of the similarly endless buttongrass plains and forests. But this only really applies to species that are sparsely distributed, as they would struggle to find a mate by staying put. Heading to a popular hilltop vastly improves their chances of mating. What's more, they don't even have to agree in advance on where to go. If their instinct is simply to head to the nearest high point, topography will do the rest.

Recognising the Donaghys insects' urge to head upwards, Jeff had an especially long-handled net, which he was soon swishing high above his head. His pursuit was mainly of hoverflies and their diminutive relatives, the big-headed flies (parasitoids of leafhoppers), which were his other area of expertise. Jeff's approach was to work one zone, then move a few tens of metres to work another high point, rotating through several such spots multiple times. I adopted his technique and was soon rewarded with many good sightings. As the afternoon wore on, and the sun descended towards the north-western horizon, the frenzied insect activity, similarly, died down. We, too, decided that it was time to bring our hilltopping adventure to an end. Camping, that night, at a secluded spot a little further along the highway, I ended the day with a river-cooled beer. Meanwhile, Jeff was still pinning his catch by headtorch well into the dusk. A distant owlet-nightjar chuckled, as if to say 'look at the work you've made for yourself'. Across the valley, a boobook boobooked. Or is that 'moreporked', or 'mopoked'? It depends on your preferred degree of taxonomic lumping or splitting, which will determine whether you accept that Tasmania's owls are the same species as those of the Australian mainland, or the same as those of New Zealand, or a Tasmanian endemic.

I have since enjoyed hilltopping encounters all around the state. I highly recommend it as a wholesome and absorbing, if a little eccentric, pastime. All you need is warmth, and calm, sunny weather, and a piece of prominent topography. North of Corinna, on the west coast, I have admired panoramic views in the company of dozens of axe-snouted flies and several painted lady butterflies on the summit of Mount Donaldson; on Wukalina-Mount William, in the far northeast, it was leaf beetles and longhorn beetles; at Echo Sugarloaf, at Randalls Bay, it was bee-flies and butterflies; on Collins Bonnet: bee-flies and hoverflies; on Mount Maria: marchflies and butterflies; and on Mount Wedge, off the Gordon Road, I found a rich diversity of hoverflies and leaf beetles. Mount Wedge is, however, where I first came to appreciate that different hoverfly species buzz or hum at different pitches. I learnt to tell the buzz of a hovering native dronefly from that of its European interloper; and to discern the sound of a common half-band hoverfly from several other native species. I lay on my back on a helipad near the summit, transfixed by the sights and sounds of overhead battles and on-the-wing pairings, while my walking companions napped. Exhausted by all the activity, I may have eventually nodded off, too.

Butterflies on high

Hilltopping is one thing—you fly to the top, hang around in the sun, find a mate, and leave for lower ground before the weather turns. Among the butterflies, painted ladies and Macleay's swallowtails excel at this strategy. But living your whole life high in the hills is another thing entirely. There are rather few butterflies that can handle it. A couple do, however, and they are candidates for January's butterfly of the month.

In runner-up position is the plucky little mountain blue, a widespread resident of Tasmania's higher peaks and plateaus, including Kunanyi-Mount Wellington and Hartz Peak, the Central Plateau and Turapina-Ben Lomond; it's also found in high country in the south-eastern part of the Australian mainland. The mountain blue is a tiny butterfly, no larger than my thumbnail, and easily overlooked because, despite its name, its wings are largely suffused with dark, bronzy-brown scales. These are all the better to absorb heat, I assume, as with many other dark-coloured mountain insects. In certain lights, and at close quarters, you might discern a vaguely purple sheen in the forewings of a freshly emerged male, but nothing that I would count as blue. In any case, the butterflies seldom settle long enough for close inspection. They are more likely to be seen fluttering feebly over the alpine shrublands, searching for mates, reliable sources of nectar, or heath plants on which to lay their eggs. Even then, in the harsh sunlight typical of high altitudes, they are hard to follow against the backdrop of the dark greens and browns of the alpine shrubs; although if you get your eye in, you may find that they're actually quite common. To give the mountain blue due credit, it is an incredibly tough butterfly, braving the extremes of heat, cold, rain and shine that are characteristic of our alpine areas during summer. That's even before we consider the appalling weather conditions that its eggs, larvae and pupae must endure throughout the rest of the year. It doesn't bear thinking about. In fact, the more I do think about it, the more I feel that mountain blues should, by rights, receive my unconditional endorsement as butterfly of the month.

Yet my heart yearns for another. I am drawn rather more towards our second candidate, and duly anoint it with this accolade. It can't trump the mountain blue in the endurance stakes, but it is such a pretty butterfly. It associates with gentler terrain, at slightly lower altitudes, where I have a better chance of encountering it during my own wanderings.

My first encounter with the leprea brown—also known as the Tasmanian xenica, the delicate xenica and the elia brown (which must be a record number of names for a Tasmanian butterfly)—was one January, many years ago, while staying at Waldheim (close to Cradle Mountain). It was our first family visit to the area, and, with young kids in tow, we opted for a short potter around, to explore our local surroundings, rather than embarking on one of the fully-fledged bushwalks that would become the hallmark of our stays there in later years. The Weindorfers had chosen a lovely spot for establishing their 'forest home', back in 1912. They set it amongst a grove of myrtles and towering King Billy pines, occupying the mid-slopes of an otherwise open, buttongrass-filled valley through which the tannin-stained waters of Ronny Creek gently meander. The location offers a homely combination of shelter, dry ground, access to fresh water and, nearby, commanding views over open country. You'll see the same elements in countless landscape paintings from around the world, including Glover's Tasmanian artworks, which depict both colonial and Aboriginal scenes evoking a rustic, rural idyll. Across cultures and across time, these elements seem to be universally appreciated in a landscape, even if we can't always articulate why that is. I can't know what the Weindorfers would have thought, but they almost certainly didn't appreciate that these are also the conditions preferred by the leprea brown butterfly.

To be more ecologically precise, leprea browns occupy upland rainforest glades or edges where forest grades into natural grassland or shrubland, which is where their larval foodplant, hook sedge, thrives and where the butterflies can find both flowers for nectaring, and open, sunny conditions for courtship and basking. You could say that leprea browns have ridden on the back of the wombat, since these accomplished and single-minded herbivores do much to stem the rainforest's advances, preventing it from shading out the herbs and shrubs. There are wombat-maintained glades and scalloped rainforest edges within a hundred metres of Waldheim, and it was in one of these glades that we unintentionally intruded on a thriving colony of leprea browns, during one of our short walks. They were all males, patrolling the sunny edge of the glade and ascending high into the canopy of the myrtles, before zigzagging back down again, to ground level. Our first impressions were of a small, rather delicate, cream-coloured butterfly; but as I took a closer interest in them (while Chris and the kids marvelled at a pile of cubic wombat droppings), I began to make out more details. The wings were divided into cream and tan sections in equal measure, but it was impossible to discern the intricacies of any patterning while the butterflies continued their frenetic fluttering-by. By chance, one descended in front of me and settled-down to probe the tubular flowers of a late-flowering scoparia—that beguilingly beautiful but spiky shrub that is the bane of off-track bushwalkers. The cream of its wings perfectly matched the colour of the flowers, rendering it all but invisible, had I not seen it land. With the wings now closed, I was able to observe the intricate patterning of their undersides, which combine undulating bands of cream and dark tan with white chevrons and an arc of dark-centred, white eyespots. I struggle to find the right word to describe their overall appearance. 'Frilly', or 'lacy' don't quite sum it up, but they are the best I can do. Suffice to say that it was love at first sight. The butterfly and its intimate, homely setting enchanted me in a way that no other Tasmanian butterfly has done, either before or since.

I have had many similar encounters with leprea browns over the years, and they never fail to enchant just as much as my first encounter at Waldheim. Among other places, I have watched them fluttering along the rainforest margins at fagus altitude beside the Lake Dobson Road, high on Mount Field; among the subalpine scrublands fringing the Lake Osborne track in the Hartz Mountains; and, most recently, in similar terrain on Adamsons Peak. They also inhabit rainforest glades around the margins of Kunanyi-Mount Wellington, including at quite low elevations, although seldom in the numbers that can be experienced at these other localities. In my totally unbiased opinion, if you haven't spent quality time with leprea browns, then your appreciation of a Tasmanian summer is incomplete.

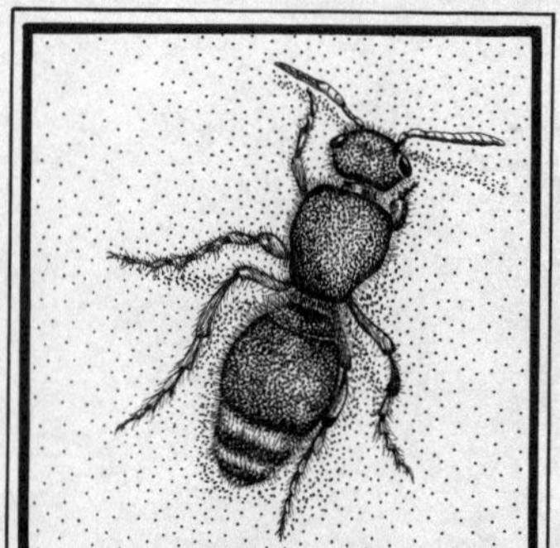

Cuckoos of the insect world

I'll have more to say next month about cuckoos of the avian kind, but here is another kind—insect cuckoos. The proper term for these is actually 'kleptoparasites', a term I threw into the narrative earlier, without explanation. So here's an explanation: whereas klepto*maniacs* can't help collecting things that belong to others, klepto*parasites* survive by stealing what others have collected. In the insect world, that's usually the food reserves a diligent mother insect has stashed away for her offspring. Think solitary bees and their stores of pollen, or solitary wasps and their stockpiles of paralysed caterpillars or spiders. The provisions put away by other species are a tempting target for kleptoparasites, which save themselves the bother of preparing their own stash of food for their young by sponging off some other insect's food reserves. Where there's an opportunity in nature, there's usually one or more species with the skillset to exploit it. Tasmanian food webs may not be as intricate as those in the tropics, but there are still plenty of niches for some highly specialised and intriguing kleptoparasites—and January is a good month for observing some of them.

Many native bees excavate holes for underground brood chambers in sandy ground. They gradually provision these with pollen, before laying their eggs on the food and then sealing the hole with fresh sand to deter interlopers. Although they aren't social like honeybees or bumblebees, they may nest in aggregations. This might be because there are only so many places where the soil is suitable for digging; or perhaps they gain some benefit from nesting together, such as safety in numbers. But an aggregation of native bee burrows is a tempting proposition for insect cuckoos, and if you hang around and watch the goings-on, you will have a good chance of spotting some. My favourites are little relatives of the carrion- and dung-feeding fleshflies. They have no common name but are known as miltogrammines. There are three species known to occur in Tasmania. One feature that these elongate, housefly-sized insects have in common is a face entirely covered in silvery-white, microscopic scales, which lie flat against the fly's integument. These enable them to hide in plain sight. A female miltogrammine will loiter around the fringes of a nesting aggregation, awaiting her opportunity. Once she spots a bee disappearing into its nesting hole with provisions, she will fly down and position herself right next to its entrance, facing the hole. The emerging bee will look intently around, checking that the coast is clear, before flying off to gather more provisions. If the fly is lucky, the bee will look straight at her without seeing her because her silvered face blends in perfectly with the sand all around the nest entrance.

Once the bee has departed, the miltogrammine reverses into the bee's hole, and deposits one or more live larvae, which she will have been keeping within her body, ready-hatched, and ready to get to work on the bee's larder before the bee's own offspring have even hatched. I wish that I could say that's the only thing the fly's larvae do, and that they share the pollen stores with the bee's larvae; but, once they have finished off the pollen, they move onto the next course—the bee larvae. Weeks or months later, it will not be a fluffy little bee that digs its way out of the burrow, but one or more miltogrammine flies.

The miltogrammines don't have it all their way, though, because there are other cuckoos waiting in the wings. Among them are the velvet-ants. These are antlike wasps, amongst whom the females have dispensed with wings. Their name comes from the velvety appearance of many species, although it's hard to discern this when most local species are only a few millimetres long. Males have wings and, although common, are hard to spot as they zoom back and forth along sandy paths, just a few centimetres from the ground. Up close, a female velvet-ant is a thing of beauty. She has a deeply pitted, often metallic integument, decked with bands or spots of silvery, orange, or maroon hairs (usually a combination). Velvet-ants are good diggers, and it seems that their method of invasion is to simply open up the bees' burrows so that they can enter and lay their eggs in the larder. Like the miltogrammines, the velvet-ants' larvae ultimately become predators of the bee's larvae.

And then there are the bee-flies, so named because some of their kind look like bees, even though they are flies. All are kleptoparasites, with different species selecting different hosts. The adults include some strikingly marked species, with black-and-white pigmented wings. Others have furry, stripey bodies; or the tip of their abdomen cloaked in silvery, reflective hairs. Most species have long tongues, which they use to delicately sip nectar from flowers while hovering, a bit like a hummingbird. Locally, there are several very fluffy, bee-like, bee-fly species, and these can often be seen hovering around ground-nesting bee aggregations. One in particular, the golden bee-fly, is well worth following. Her forté is precision hovering, plus bowling. (Yes, bowling.) Once a golden bee-fly's senses tell her that she is hovering in front of a nest hole (her big, bulging eyes probably enable her to discern the dark opening), she flicks her abdomen forward in a series of rapid bursts of movement. Each flick liberates one or more tiny eggs, coated in sand-grains that she has picked up with the tip of her abdomen during an earlier sortie. Her goal is to aim true enough for at least some to roll down the hole and lodge close to the bee's larder. It's a bit like the bouncing bomb made famous in the *Dambusters* movie. The effect can be just as destructive. When the eggs hatch, the very active larvae make a beeline (or flyline) for the bees' pollen stores. There they will shed their skins, becoming sedentary grubs that, like all the other kleptoparasites we have seen, finish off the bees' pollen, and their larvae.

Some bees are cuckoos of other bees. There is at least one Tasmanian species of leafcutter bee that has spurned leaf-cutting or resin collection in favour of cuckolding the nests of other leafcutters or resin bees (the latter are lumped in with the leafcutters, taxonomically speaking). I rarely see the host bees, and have only once seen their cuckoos; so they must be quite a rarity. Their presence probably indicates a particularly high-quality patch of habitat.

Wasps also get in on cuckolding behaviour. Here I'm talking about native, solitary wasps, not the black-and-yellow European wasps. One family of native wasps is known simply as the cuckoo wasps, since all members of the family are kleptoparasites. Some cuckold native bees, others native wasps, particularly those nesting in holes in logs or tree trunks. Our cuckoo wasps are diminutive but gorgeous in appearance, sporting intensely malachite-blue or jade-green bodies, from the tips of their antennae to the ends of their toes. They are also very quick-witted and have excellent vision, so you'll be lucky if you ever get to observe one up close before it clocks you and makes a hasty departure. Then there is the cuckoo spider-hunting wasp. It belongs in the same family as the rest of the spider-hunters, which provision their nest cells with spiders they have paralysed with a sting. Wrestling with a live and potentially lethal spider is a risky business. It would be so much easier to leave that to others, then to sneak in and lay your own eggs on the paralysed spider. Of course, that strategy is not risk-free, either, since the provisioner's sting is also effective against cuckoo spider-hunters. As far as I know, only a single species of cuckoo spider-hunter calls Tasmania home, but it's not uncommon in sandy areas where other spider-hunters are likely to be nesting.

If you are inspired to go looking for your own insect cuckoos, I suggest heading for sandy country, because that's where you can expect to find the greatest diversity and abundance of ground-nesting bees and wasps. A sandy path through coastal heathland would be perfect. Pinmatik-Rocky Cape, Narawntapu, Musselroe, and the Remarkable Caves to Crescent Bay walk on the Tasman Peninsula come to mind; but there are plenty of other promising areas. That said, I have had most encounters along the fire trails of the Peter Murrell Reserve, near Blackmans Bay; although perhaps that's because it's local to me and I visit often. I can quite easily spend hours there on my hands and knees, covering no more than a few metres of sandy path, just quietly observing the goings-on among the provisioners and their cuckoos. It's a very absorbing pastime, so long as you're prepared for the strange looks from passing walkers and their dogs.

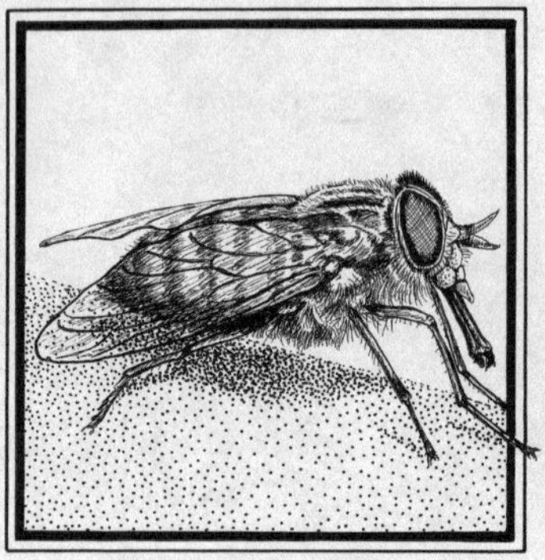

Warra days

A few years prior to my arrival in Tasmania, a 15,000-hectare parcel of the southern forests alongside the Huon River, upstream of Tahune, along with adjacent mountainsides, had been delineated as the Warra Long Term Ecological Research (LTER) Site—part of a global network of LTER sites in which research efforts are focused and pooled. You may have heard of Warra in another context, since data from its weather station regularly appear on the daily weather report that follows the ABC TV news (often it is the wettest, and sometimes the hottest or coldest, place in Tasmania). Forestry Tasmania, my employer at the time, drove much of the research at Warra, often in partnership with students and academics from the University of Tasmania or other institutions. One of the core projects was the Silvicultural Systems Trial, which compared the ecological, silvicultural and economic consequences of various alternatives to clearfelling in wet eucalypt forests. Each replicate (a single coupe) of each alternative treatment was monitored before harvest, in the first year after harvest, at three years and then at various subsequent time intervals. My role in this project was to oversee the zoological monitoring and collate, analyse and publish the findings, with co-researchers. One of these colleagues, Sue, extended the research beyond these initial analyses, and continues to do so. (I have Sue to thank for sending me that cutting from the *Mercury* alerting me to the conservation biologist job vacancy at Forestry, when I was still writing up my doctoral thesis in Queensland). We primarily monitored birds and beetles. Birds were surveyed monthly, using timed counts along pre-defined transects through the various coupes and unharvested control areas. Arrays of pitfall traps were set for beetles at multiple plots within each coupe, which operated continuously but were collected monthly. Back in the lab, we would sort, pin, label, identify and database the thousands upon thousands of beetle specimens arising from this huge trapping effort.

Over time, our findings helped demonstrate the ecological benefits of retaining stands of unharvested trees either within the harvested area or close by. Coupled with the findings of silvicultural and economic research, it led to the adoption of 'aggregated retention silviculture' as the preferred practical alternative to clearfelling. The rather ungainly term is a coinage from the forests of the US Pacific North West, where the system was first implemented. The practice requires demarcating, in advance, areas of forest that will be left unharvested within a delineated coupe area. Insofar as is feasible, the areas are selected to optimise the retention of features of particular conservation significance, such as stands of old eucalypts or rainforest. The idea is that they will be retained for at least one full silvicultural cycle (of harvest, regeneration and regrowth: notionally 80 years), at which point the next harvest may see the same or different areas retained, depending on the assessed conservation priorities and their spatial distribution at that point.

There is no set threshold for the percentage of the coupe that must be retained under this silvicultural protocol; but, crucially, the arrangement of retained forest must ensure that at least half of the area to be harvested will remain within a distance of one tree-height (usually about

40 metres) from retained, unharvested forest. This broadly corresponds to the distance over which eucalypts can naturally disperse their seed, yet it allows for sufficiently extensive areas of harvested forest for a safe and effective post-harvest regeneration burn to be undertaken. The approach leads to the retention of 'aggregates', or patches of standing forest. Sometimes these form 'islands' surrounded by harvested forest, but usually they comprise 'peninsulas' or deeply scalloped coupe edges. Generally, they occupy 15-30 percent of the original coupe area. In practice, these coupes are usually aerially resown with locally sourced eucalypt seed after the burn, as happens with clearfelling, to ensure sufficiently dense regeneration throughout, approaching what would eventuate in the aftermath of a bushfire.

Our research demonstrated that the aggregates would function as 'lifeboats' for plants and animals that can't survive a regeneration burn—mimicking the 'fire skips' that occur naturally, when wildfires rage through forested country. The aggregates also act as local sources of seed, spores and animals, which can disperse into the adjacent regenerating forest as it matures. These two functions should make sure that the regenerating forest will bear a stronger ecological resemblance to natural forest of a similar age than would be the case after clearfelling. Compared to clearfelling, the method requires a different mindset and skills. It usually costs more, too, and the window of opportunity for conducting the autumn regeneration burn will be narrower—because burns can only be carried out when there is a sufficient humidity differential between the harvested and retained forest, to prevent a fire from encroaching on the aggregates, or beyond. It's an art as well as a science.

Even a simple experimental design like the Silvicultural Systems Trial represents a huge undertaking; but it was by no means the only largescale, long-term study at Warra. One that my predecessor had established was called the Warra log decay project. It aimed to document the beetle fauna in fallen logs as they gradually decay on the forest floor, and how this might differ in old-growth versus regrowth-age logs. The study involved felling six old-growth stringybark eucalypts and six nearby regrowth trees of the same species, and then periodically enclosing three-metre sections of the resultant logs in aluminium-framed netting cages fitted with collecting heads at their upper and lower corners, to sample the beetles and other insects emerging from the logs. Each cage was in place for one year, then removed for three years to allow for further colonisation (while adjacent sections were in turn enclosed), then reinstated, and so on. The project continued for ten years, which covered the first of five decay stages for the logs in question. The aim is to repeat the process in subsequent decay stages; assuming that researchers of the late twenty-first century and beyond can maintain the momentum—these logs are expected to take close to two hundred years to fully decay. I got to analyse and publish on the first decade of sampling, which revealed differences between the old-growth and regrowth logs. It seems that the old-growth logs hit the ground with a ready-made beetle fauna, while the regrowth logs relied more on colonisation from the outset. Over time, a succession of beetle species colonised and emerged from the logs, amounting to some 350 species. At that rate, I wouldn't be surprised to see that total rise to a thousand or more beetle species over the full course of the logs' decomposition.

For some of our later projects, Warra wasn't big enough; so we developed the concept of the Southern Forests Experimental Forest Landscape, anchored on Warra but extending eastwards through state forest, almost as far as Geeveston—a distance of some 30 kilometres. In that landscape we embedded a series of wildfire chronosequence plots, in part to extend the range of forest age classes in which we could monitor birds and beetles beyond the age of the silvicultural treatments at Warra, and so that we could compare silviculturally regenerated forest of a given age with similarly aged forest regrowing after wildfire (the main documented bushfires in this landscape dating from 1898, 1939 and 1967). It turns out that wildfire regrowth and silvicultural regeneration of the same age are reassuringly similar; which perhaps shouldn't surprise us, given that occasional, sometimes stand-replacing, wildfire is a defining feature of the ecology of wet eucalypt forest. That said, we observed differences, reflecting the valuable ecological role played by structural legacies from the pre-wildfire stand, such as dead standing trees, logs and treeferns, that typically are not replicated through clearfell silviculture.

Our most ambitious project entailed surveying or sampling birds, beetles, and plants at well over a hundred temporary plots that we established across this landscape. These plots were chosen both for the age of the forest within the plot, as well as the degree of ecological, forestry or agricultural disturbance in the landscape 'matrix' surrounding the plot. Finding, surveying and sampling plots to meet these criteria was no small feat; but doing so enabled us to tease out the extent to which the landscape surrounding a bit of forest—its *context*—influenced what was able to live within it. In the resultant data-sets, we focused on species known, from our previous research, to be particularly associated with either mature or young forest growth stages. Reassuringly, the study found that both old-growth forest and forest regenerating after harvest are remarkably resilient to changes in landscape context, in terms of the diversity and composition of the flora and fauna found in relation to the growth stages on which we focused. These species must have either continued to survive and reproduce in our plots as the landscape changed around them, or were able to recolonise from, or through, the matrix despite these changes. However, we found that resilience breaks down at the highest levels of landscape disturbance, such as where large areas have historically been given over to plantation or agricultural land. Thankfully, those levels are unlikely to prevail under current forest management and reservation practices, because of the greater present-day emphasis on maintaining a network of formal and informal reserves throughout the landscape, as well as the shift towards alternatives to clearfelling. As a result of our study, landscape-context planning (aimed at keeping well within these ecological thresholds of landscape change) is now embedded in how forestry operates, on public land at least.

Once Warra became known among the ecology and forestry communities, other researchers began to be drawn to the site; so the forests quickly became crisscrossed with transects and informally tagged paths to and from various temporary and permanent plots. It became clear that Forestry Tasmania would need to keep track of all these projects, so I set up a database and website. In a bid to be more strategic about what research we would like to see conducted at Warra, I developed and published a conceptual ecological model, which aimed to show (in a rather complicated flowchart often referred to as the 'Warra horrendogram') how various aspects of Warra's ecology—such as log decomposition, fire frequency, soil respiration, climate, vegetation structure, carbon sequestration—might relate to one another and contribute to forest function. The model enabled Forestry Tasmania to prioritise research projects to progressively fill in the most significant pieces of the Warra ecology puzzle. We offered small grants to students and attracted postgraduates and postdoctoral researchers, who also brought their own expertise and funding. Eventually, over a hundred projects were added to the database and website. One day I would like to synthesise all the findings from this intensive period of ecological research into wet eucalypt forests. Suffice to say that it was an extraordinary decade of activity, and a privilege to have been in the thick of it.

I used to look forward to fieldwork at Warra with a mixture of excitement and trepidation—especially in January. Why the trepidation? Because Warra can be stinking hot in high summer, and often rather humid. The combination can be exhausting if you're trying to do hard physical work *and* wearing the obligatory high-visibility vest and hard hat. January is also peak marchfly season, and marchflies seem to be irresistibly drawn to high-vis. Even though they rarely settled on any exposed skin long enough to bite, the incessant buzzing and dive-bombing, often by hundreds of Jackson's marchflies (a Tasmanian endemic species), is part of the Warra experience that drives even hardened entomologists to distraction.

We held regular open days at Warra, bringing busloads of interested punters from Hobart to hear about our research and see the forests for themselves. My party trick at these events involved inviting visitors to give me a hand at unrolling a printout listing the species of plants and animals that we had so far recorded at Warra. The list had parallel columns for reptiles, mammals, birds, vascular and nonvascular plants, fungi, and beetles. It wouldn't take long for the reptile and mammal lists to be revealed in full, because there are so few species of these. Birds took a bit longer, and the plants longer still. But even at this stage, fully four-fifths of the roll still remained to be unfurled, because there were so many species of fungi and beetles on the list—over a thousand of each. Whoever was holding the end of the roll would have to take several steps backwards to completely unfurl the list. What became known as my 'Warra toilet roll trick' was always a terrific way to demonstrate how much more there is to a forest than the trees and other elements of biodiversity that usually come to mind; and how fungi and insects really run the whole show. The Forest Education Foundation still employs a version of the Warra toilet roll to this day.

Warra lives on as part of a national and international network of ecological Supersites, although, sadly, the scope of research has been scaled back since much of the original LTER site has been absorbed into an expanded World Heritage Area. Today, the focus is on understanding how the forest switches from being a carbon sink to a carbon source and back again through the seasons, and how heatwaves exacerbate this effect. Warra hosts one of very few carbon flux towers in Australia, or indeed in the Southern Hemisphere. Instrumentation on the tower takes continuous measurements of temperature, humidity and the concentrations of carbon dioxide and oxygen at various heights, from ground-level to high above the canopy. From these streams of data, it is possible to work out when the forest is, effectively, inhaling, and when it is exhaling. It turns out that the eucalypts at Warra most efficiently photosynthesise and sequester carbon during cool weather, and don't cope at all well with heat, when they switch to being net emitters of carbon. Heat stress can eventually kill these trees, even in the absence of fires; so these forests face an uncertain future as the climate continues to heat up.

Sadly, a few years ago the flux tower's sensor infrastructure was damaged in the Riveaux Road bushfire, which impacted on much of the Warra research area. Though it was repaired afterwards, it suffered a second blow more recently, when a windstorm felled a neighbouring tree. This broke the tower's tethering cables and caused the whole structure to come crashing down. It was a stark reminder of the dynamic nature of our forests, and of the many perils and challenges in conducting long-term ecological research. It's always worth the effort, despite all of the possible setbacks; but I must say I'm glad that I don't have to cope with such trials and tribulations in my museum job.

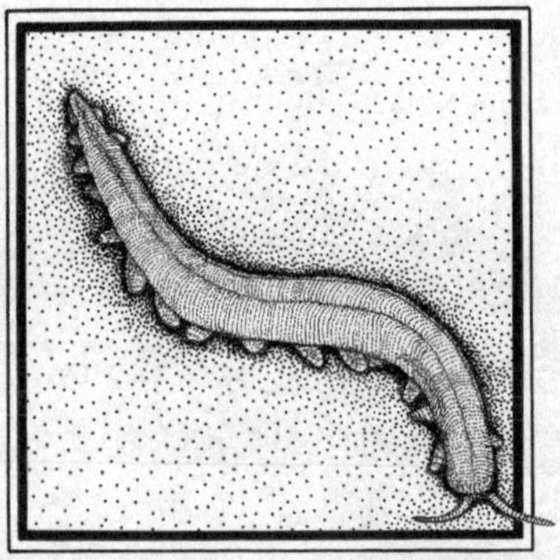

Velvet worms

In the distant Cambrian past, almost half a billion years ago, a diversity of strange, almost-segmented, worm-like creatures called lobopodians stalked the seafloor. They had stumpy, hydrostatically operated legs tipped with claws, and grazed on sponges or filtered particles of food from water. We know this because their fossils appear in the famous Burgess shale, from the Rocky Mountains of Canada, as well as in slightly older rocks in China. Many of these animals' contemporaries have long since vanished, with more recently evolved forms taking their place. But animals remarkably like lobopodians do exist today; it's just that there's no point in looking for them on the seafloor. Velvet worms, as today's versions are called, are entirely terrestrial. How and why their ancestors made a complete transition to dry land, yet changed so little in appearance, is a mystery; although they are not alone in having done so. Scorpions did the same, tens of millions of years later.

Tasmania is one of the best places in the world to see velvet worms. I deliberately wrote 'to see', rather than 'to find', though. It's often simply not productive to go looking specifically for velvet worms; but if you're out and about, fossicking for other creatures in velvet worm habitat, you might chance upon one. I was over the moon when I discovered my first one. It was before I moved to Tasmania, in the rainforests of northeast Queensland. At the time, I thought of velvet worms as entirely tropical beasts. Admittedly, this was because I was used to seeing them in Attenborough-style documentaries on the Amazon rainforests, where they were usually referred to as peripatus ('the walking-around one'). I was, therefore, delighted to learn that Tasmania hosts several known species of them (and probably many more unknown ones); and that, working in forestry, I was highly likely to cross paths with them, since most species are forest-dwelling.

And cross paths I did. It almost always happened serendipitously, although I have to admit that first impressions were rarely enthralling.

Generally what one sees is a dark, amorphous blob, about the size of a thumbnail, cowering under a log or in amongst a mass of rotting leaf litter. It could be a slug, or a stick; or it could be a velvet worm!

Velvet worms shun the light and, if disturbed, prefer to curl up and wait for dangers to pass. If you give one a gentle nudge, it may slowly uncurl as a prelude to wandering around (peripatetically) in search of cover. This will be your best chance to get up close and take a good look, and it's well worth the effort. Doing so will quickly dispel the idea that they are some sort of worm. Velvet worms actually have well-coordinated, stumpy legs; fourteen pairs of them, in fact. Their skin is superficially velvety in texture, but, under magnification, its surface can be seen to comprise a close-packed carpet of raised papillae, each of which bears microscopic ridges and grooves—almost like intricately cut gemstones. The background colour—that is, the colour of most papillae—is nearly always a deep plumbeous-blue (a lead-grey-blue); but other papillae on the same animal may be orange-brown, pale blue or off-white. These other colours are usually arranged in blotches or in oblique crosshatching, sometimes giving the impression that the animal might be wearing a tight-fitting tartan blanket. Nearly every forested part of Tasmania hosts one or more species of such small, dark, egg-laying velvet worms. I have photographed a fair number over the years, and on processing the images each one turns out to sport a slightly different colour combination. Blow the image up to fill the computer screen and each will be revealed as a creature of beguiling beauty.

Beautiful they may be, but velvet worms are predatory, and rather unconventional predators at that. Some might even call their mode of attack unsavoury. Sitting below and slightly lateral to the animal's main sensory tentacles, on its head, are two organs best described as glue guns. When a velvet worm senses insect prey within striking range, it shoots out two streams of watery, proteinaceous glue to ensnare and entangle its victim. The protein molecules in the glue come in a mix of shapes and sizes, which means they work as a glue by getting tangled up with each other. This only happens, however, once the water in the glue evaporates, which it does when thinly spattered over an insect's waxy, water-repellent cuticle. With the prey incapacitated, the velvet worm will then begin its leisurely approach, following and slurping up its trail of glue, which leads, ultimately, to the main meal. Termites are their favourite food (hence their association with rotting wood); but it is likely that other small invertebrates are also considered fair game. A velvet worm will also shoot glue at perceived predators, as my fingers have experienced, on many occasions.

If velvet worm feeding is a little weird, sex is weirder still. In most species, a male will attach a sperm packet to the female's skin, from which the individual sperm cells drill their way into her body cavity and then swim through her body fluid in search of her ovaries. Some males forcibly spike the female to insert the sperm packet under her skin. Only a few go in for conventional penetrative sex.

Though most of Tasmania's velvet worms are egg-layers, globally, the majority are live-bearing. Which brings me to our own, rather special, live-bearing velvet worms. These belong to quite a different lineage from the egg-layers. For no apparent reason, they also bear an extra pair of legs.

Each live-bearing species has a very restricted range, with two being confined to the northeast of the state. My first opportunity for an encounter with one of those came early in my Forestry tenure, when I was invited to join a hunt for the blind velvet worm, under the auspices of the Forest Practices Board (now Authority). This threatened species is confined to a small area of forest close to St Marys, where it lives in the interior of rotting logs. It was only discovered in 1987, and its entire range is some 200 square kilometres—it sounds a lot but, if it were a square, each side would be only 14 kilometres long. Looking for the creature is destructive of its habitat, so our permit came with instructions to strictly limit the number of logs examined. Fortunately, we found one in almost the first log that we looked at (despite my earlier comment about not looking for them).

As its name suggests, the blind velvet worm has forsaken sight as surplus to the requirements of its inner-log home. It has also foregone pigmentation, so it is a deathly white colour. Strange and enigmatic though it is, I must confess that I was a little underwhelmed at its appearance, although my reaction was undoubtedly not helped by the fact that it stoically refused to uncurl itself for our viewing pleasure. We eventually reassembled its log around the strange creature as best we could and returned to the car.

The range of the blind velvet worm abuts that of a much more charismatic species, the giant velvet worm. During my time at Forestry, I had several occasions to get to know this beast, which is also considered threatened and also confined to a small part of northeast Tasmania (although its range is three times greater than that of its blind neighbour). The giant velvet worm is also about three times as long as its neighbour; which is to say, almost as long as your index finger. It can see; and it is fleshy-pink above and off-white below. And the reason I have included velvet worms under January is because the first time I had an encounter with the giant velvet worm was in the aftermath of a severe bushfire, which, one January, tore through a substantial part of its range in the hills behind Scamander. I took the event as an opportunity to see how well the giant velvet worm had coped with the fire, given that it, too, had an association with rotting logs, although it is known to wander afield more than the blind velvet worm.

I drove to the northeast with my Forestry colleague, Marie, who works as a conservation planner and shares my interest in log-dwelling invertebrates, having also researched them for her PhD. There doesn't seem to be any obvious reason, such as a physical barrier or change in habitat, to explain why the range of one velvet worm species abruptly gives way to the other; but that's what happens with these two. It is another aspect of their mysterious appeal to people like me and Marie.

We pulled over at a spot where giant velvet worms had been recorded before the fire, and set about our search. It wasn't looking very promising. The canopy had been burnt to oblivion and all that remained of the trees were bare, ashen skeletons. The breeze stirred up miniature tornadoes of ash, tainting the air we breathed with the smell of burnt wood. There were still plenty of logs around, but the fire had been so hot that many had been burnt right through; tell-tale lines of ash and charcoal on the bare woodland floor revealed where other logs had been completely incinerated. We turned over a few log fragments and found nothing but the charred remains of snail shells. Dispirited, we returned to the vehicle and drove through another few kilometres of blackened bush, until we reached a dip in the road where we could pull over again. This spot, sitting low in the landscape, looked a bit more promising. Some of the trees retained foliage, albeit scorched and, apparently, dead. There were more logs here, too. Some were still smouldering. I had learnt that wrecking bars make handy tools for lifting and rolling logs, and they proved their worth here, saving us from burning our fingers. Log after log yielded no sign of velvet worms, though. Half an hour passed, and with no velvet worms to be seen, we were on the verge of conceding defeat. I dispiritedly kicked a small, charred shard of wood—all that was left of a hollowed-out log that had succumbed to the flames. As it budged, there, curled up in a loose coil on the surface of the soil, was a giant velvet worm! It recoiled slightly from the sudden exposure to the light, but otherwise appeared unharmed. How it had survived the fire in that insubstantial refuge, I really don't know. Perhaps it had moved from some other refuge after the fire had passed and simply made do, having failed to find better shelter, or perhaps it had been caught out by the sunrise (they only go wandering at night).

Surviving the fire is one thing, of course; the velvet worm and others of its kind would also have to endure this devastated forest long enough to reproduce, if the species were to persist there. Finding this individual gave us hope that the species has some inbuilt resilience though, in the face of apparently disastrous fires. That optimism was bolstered when we found more velvet worms in similarly scrappy habitat later that day. By the day's end we had concluded that the giant velvet worm was down but not out—with a bit of luck, it would be able to recover and gradually recolonise lost ground. Our prognosis remains contingent on there not being another severe fire for a good few decades, though, for the trees would need to have time to recover; and these trees would also need to live long enough to generate more logs to replace those burnt. In the current era that may be a tall order, but I live in hope.

© K. DAVIS 2023

February

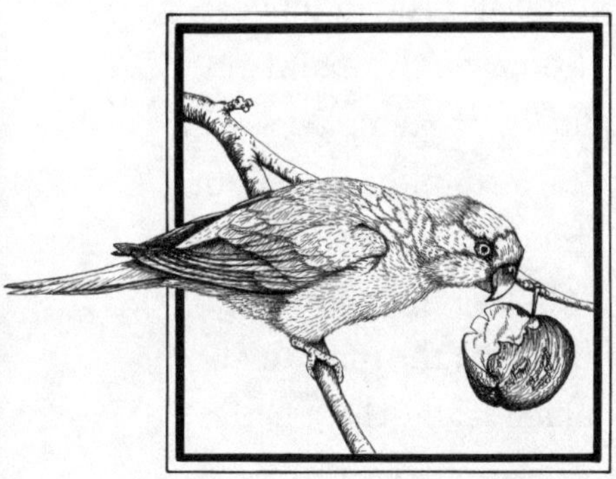

Enduring warmth

Those lazy, hazy days of summer continue, often with little interruption, throughout February. High-pressure systems are perhaps at their most assertive now, often parching the land and calming the seas. In the garden, tomatoes and veggies need daily watering; and apples and pears ripen, leading to constant squabbling between the wattlebirds and lorikeets, who battle for garden supremacy even though there would be plenty to go around if they would all just share nicely. In the lowlands, the earth looks spent: dry, dusty, even desolate. It's a different story up in the hills, though. They start the summer with more moisture than the lowlands, and then dry out more slowly. Many upland plants are at the peak of their flowering season in February, and mountain insects time their development to take advantage of this later summer peak; which is why I find that February is often the best month for getting into the high country.

Blitzing the bush

February, as it happens, is a poor month for fieldwork on much of the Australian mainland. It's just too hot, and either too dry or too wet, depending on location and year. As other entomologists or invertebrate enthusiasts often like to escape such conditions on the mainland and visit Tasmania at this time of the year, I quite often tee up trips with them. The benefits seem to be mutual: I get to learn from them and put my own local experience and knowledge into a broader, national context; while the visitors get an insider's insights on good places for collecting, camping, feeding and watering. They also usually make the most of an opportunity to check over our museum collections. Ideally, the museum will receive identified specimens collected by the visiting expert, too, bolstering the scientific value of the state's collections.

Another consequence of mainland summer heat is that the Bush Blitz program is likely to slot in a Tasmanian expedition during this period. Bush Blitz is a national, rolling program of expeditions that aims to document the plants and animals of poorly-known parts of Australia's national estate of protected areas and to discover previously unknown species. Each expedition comprises specialists from their respective state's museums and herbaria, as well as from other states, if they will help to fill some gaps in expertise. The whole thing is backed up with the logistical resources needed to successfully run an expedition, such as catering, four-wheel drive vehicles, helicopter hire when access to particularly remote areas is required, accommodation and support staff. A case could be made that, compared with vast swathes of inland Australia, even the most far-flung parts of Tasmania are already well explored and documented; but my museum colleagues and I are hardly going to make that case. We are, in fact, always delighted when a Tasmanian Blitz is mooted. These are our best opportunities to do some targeted collecting in out-of-the-way parts of the state, with like-minded people who understand our particular needs and whims. It's a highly successful and productive model, boosting the scientific value of our collections, expanding knowledge of the fauna and flora of particular areas and the state in general, and advancing taxonomic research on poorly known groups. Over the past decade, Bush Blitz expeditions have taken us to the Five Rivers reserve on the Central Plateau, the Tarkine, Flinders Island, Bruny Island, the Far South and, most recently, Stony Head, on the northeast coast.

In non-Blitz years, we mount museum expeditions of our own, grandly labelled our *Expeditions of Discovery*. The Friends of the Tasmanian Museum and Art Gallery and other benefactors have generously supported these expeditions, which are like home-run versions of a Bush Blitz. Under this banner, we have surveyed the Wind Song property near Little Swanport, the Musselroe Wind Farm in the far northeast, Spring Bay Mill at Triabunna, the Turapina-Ben Lomond plateau, and, most recently, the Woolnorth Wind Farms.

The concept behind all these expeditions is to 'blitz' a defined area with all the surveying, sampling and collecting effort that we can muster. For me, that involves covering as many of the major habitats and microhabitats as feasible during the week or so of the expedition, leaving no stone unturned—in a metaphorical as well as literal sense. Fieldwork is definitely not a walk in the park. I'm generally decked out to the eyeballs with an array of different collecting equipment, each of which brings the prospect of unique discoveries. I'll always have my sweep net with me, of course. Additionally, I'll often take a beating tray to examine insects dislodged from overhanging bushes or branches by my beating stick or net handle. Also, a wrecking bar for rolling over logs and rocks; a tomahawk-style hacking implement and hand-held rake for investigating the interior of rotting logs; a set of graduated sieves for sifting leaf litter; cans of knockdown insecticide and a white sheet for sampling the bark fauna. And (yes, more): a long-handled dip-net, kitchen sieve and white plastic dish—these are for sampling puddles, ditches, ponds and creeks. Then, of course, I'll have hundreds of specimen vials, assorted plastic bags and sample pots on me, as well as my notepad, camera, radio, first aid kit, repellent, sunscreen, food and water. And did I mention that I will also go out at night? If the weather's suitable, I'll go spotlighting for insects or checking on other expeditioners' light traps (my zoology curator colleague, Cathy, is a lepidopterist, or moth specialist). On the first day of the expedition, I'll establish some traps which I'll collect on the final day. We'll dig pitfall traps (plastic cups or sample pots) into areas likely to be suitable for trapping ground-dwelling insects and spiders; and we'll lay out arrays of pan traps (brightly coloured plastic picnic bowls filled with water and a dash of detergent). These go on patches of ground selected for potential to be visited by flying insects likely to be foolish enough to drop in for a closer look. Somewhere among all this sampling, I'll need to find lab time to process my catches, so that, when I return to the museum, I can be assured that the specimens will be in good order, and, most importantly, I will know what came from where, how and when.

If you think this is the extent of our sampling, I should also mention that we generally conduct a reconnaissance earlier in the season. The recce allows us to get a feel for the site, which helps in planning the main expedition. It's also an opportunity to sample at a different time of year, and to set up Malaise traps that can be left to passively sample flying insects until our return for the main expedition.

Each expedition has its own unique feel, but for a taste, I'll recount some of the highlights from Turapina-Ben Lomond—as I write, I am still working through the collected specimens and drafting a report for this one.

It's always a gamble planning an expedition in mountainous country, since good weather is paramount for sampling insects and other invertebrates. That's partly why February is a good bet: because of a reasonable prospect of stable weather. Even so, the Ben Lomond plateau sits at an altitude of around 1500 metres, so the conditions really need to be warm as well as settled. As it turned out this time, however, the day that we all drove up from Hobart, in several vehicles packed to the gunnels with field gear, was hot and sunny. For this journey, and indeed the entire expedition, I had the pleasure of excellent entomological company. Lynne is one of our honorary curators and an outstanding entomologist and arachnologist. She had jumped at the chance to take a few days off writing her doctoral thesis, to spend the week entomologising on Ben Lomond with us. Lynne and I go back a long way—we both worked in the same field of research at Forestry. I don't think that there was more than a brief pause in our conversation during the whole journey, nor for most of the time that we were out on the mountain for the ensuing week.

On our drive north, the mountain massif, or at least its southern extension, Stacks Bluff, was clearly visible from the Midland Highway beyond Campbell Town, rising defiantly through the haze. Leaving the busy highway, the most direct route took us along quiet country backroads; so quiet, in fact, that I didn't see a single vehicle from the highway turnoff until we reached our destination—one of the lodges in the ski village—a good hour later. Dusty agricultural vistas, featuring clouds of cabbage white butterflies, gradually gave way to dry woodland and pine plantations, until we turned off the main road to begin the long climb up Ben Lomond Road. The dirt road passes through a succession of forest types—brown stringybark, gum-topped stringybark and alpine cider gum—before the trees peter out completely at the foot of Jacob's Ladder. Elsewhere in Tasmania, the trees would continue upslope for several hundred metres more, but here on Ben Lomond, there are no snow gums, so the treeline is correspondingly lower.

Jacob's Ladder comprises a series of dramatic switchbacks that mount the almost-sheer northern face of the plateau's dolerite ramparts. It beggars belief to imagine how that road was ever constructed, and I doubt it would have gained planning approval today; but our expedition would have been out of the question without it. At the top of the Ladder, I found myself letting out a sigh of relief as the road veered away from the ramparts and levelled out, passing through boulder fields and flowery alpine shrubland. A few minutes later, the ski village came into view. It is nestled in the lee of Legges Tor, which, at 1,572 metres, is Tasmania's second-highest peak (after Mount Ossa, which is a mere 45 metres higher).

The ski lodge is not normally open at this time of year, so we had the place to ourselves. Indeed, we appeared to have the entire ski village to ourselves. Notwithstanding its vaguely unsettling similarity to the Overlook Hotel of Stanley Kubrick's 1980 horror movie *The Shining*, it proved a perfect venue for us: comfortable, spacious, and with splendid views out over the broad sweep of Meadow Vale and across to Hamilton Crags, the ridgeline that forms the north-eastern limit of the plateau. The air up here was chilly and shadows were creeping steadily up the Crags from the valley floor. With no time to lose, Lynne and I packed a bag with our pan traps, pitfall traps, bottles of water and trowels, and ventured out. We decided that there was no need to go far to set up our first arrays of traps, since one patch of shrubland around here looked much like the next. Having crossed the creek, we found a suitably flat and sheltered spot and set to work, digging small holes for twenty pitfall traps around the bases of various shrubs. The ground was too stony to sink our plastic cups, so we resorted to smaller sample pots instead, making sure that the lip of each was level with the surrounding ground surface so that insects and spiders bumbling along at ground level would simply drop into them. Job done, we then distributed green, blue, white and yellow plastic picnic bowls—five of each colour; nestling these in amongst the shrubs and rocks so that they stood less chance of being blown over, while remaining visible to insects flying overhead. The idea is that the bright colours would act as super-stimuli for the insects, which might perceive them as giant flowers or particularly attractive foliage. We topped up each pan with water, and a dash of detergent to reduce the surface tension if an insect fell in. Yes, we were hoping to drown our insects. It's a sad reality that, to identify insects, you usually need to kill them. Furthermore, you need to retain the proof of your identifications in the form of preserved specimens, ideally in a museum collection. This being science, all species concepts are hypotheses and all identifications are provisional, pending further knowledge—which may come decades or even centuries into the future.

We repeated the setting-up process in different habitat, on flat, boggy ground down in Meadow Vale, a few hundred metres from the ski village. With dusk now well advanced, we returned to the lodge for a quick meal with the rest of the museum expeditioners—zoologists and botanists, curators and collections managers, numbering up to ten of us. We always greatly appreciated the meals prepared by Charlotte, our much-loved expedition caterer and cook. Heading out again, this time with headtorches and puffer jackets, Lynne and I were met with cold mountain air and a sky full of twinkling stars. Spotlighting for insects and spiders can be very absorbing, but it can also be quite disorientating, especially when you're tired. At least here we were in no danger: we only needed to look up to see the lights of the lodge a mere hundred metres away.

It was really a bit too chilly for much insect activity, but some beetles seemed not to have read the weather forecast and were out anyway, albeit mostly just sitting around on rocks looking cold. In fact, rocks are probably a good place to sit in cold weather, since they would have been radiating some of the day's heat. Most of the beetles were darkling beetles, like those commonly found under logs in lowland forest; but here they are a different, montane species, whose wing cases sprout stiff hairs (rather too sparse, I thought, to serve as insulation). I bent down to collect one, and spotted another dark shape at the fringes of my torch beam. It was close to the base of a prickly scoparia bush. It wasn't moving, and I was about to conclude that it wasn't animate at all, but I gave it a poke anyway. It moved, but only under the influence of my poking. Looking closer I vaguely discerned a beetle-like outline. I scraped it up into the palm of my hand and my suspicions were confirmed: it was, indeed, a beetle, albeit a dead one, and close to two centimetres long. It was not any old beetle though. It appeared to be Walker's flightless longhorn beetle! This is a species otherwise known only from Mount Wellington and a few other high peaks. I had never seen one before, but it has a lowland congener that I was familiar with. I couldn't believe my luck, and nor could Lynne.

Tired now, we were starting to meander homeward when I spotted a different beetle on a pathside stone. It was much smaller than the darklings, and shimmered a metallic blue-green in my torchlight. I recognised it as one of the leaf beetles—of which there are many types of this size and colour—but when Lynne saw it, she announced that it was *the* leaf beetle; the one we had been most keen to find here! Of course, it has no common name, but let me call it the sieve-necked leaf beetle, since that is a fair translation of the specific name given to it by Arthur Lea (whom we will meet again in May), when he described it as new to science, in 1929—apparently on account of the many little dimples in the beetle's pronotum or thoracic shield. They are a bit like the holes in a sieve (I suppose). Lea described the species based on a single known specimen, which had been collected on Mount Wellington. Our find wasn't the first time it had been recorded from Ben Lomond, since one had been found in more or less the same spot as ours, back in 1999; but it's clearly a very rare montane beetle, currently known only from these two mountains at opposite ends of the state. Unfortunately, our find doesn't add much to what little we know about its habits, since we weren't able to associate it with any particular plant species; unless it eats stones rather than leaves, of course. We spent much of the rest of the week trying our hardest to find another specimen, but drew a complete blank.

Our first full day on the plateau was blessed with weather every bit as fine as the previous day. Lynne and I decided that we had better explore the highest points while the going was good, so we headed up towards the saddle behind the ski village, with a view to climbing Legges Tor and the adjacent Giblin Peak. For a bushwalk to such lofty peaks, it felt like a bit of a cheat to be starting from the ski village; but since we had all our collecting paraphernalia with us, we weren't complaining. Though we could have walked the distance in under an hour, it took us the best part of the day, because we kept finding interesting spots to fossick for insects. Our snail's pace would have driven most walkers mad.

Even before we reached the saddle, we had found some special insects: Hardy's hardy nannochoristids were active, montane relatives of the ones found on Hobart's western fringe, as I described back in November. We swept several from the shrubby margins of little peaty pools and trickles. To the best of my knowledge, these are the first records of montane nannochoristids from the northeast of Tasmania. This is significant because, like other naturalists, we had been under the impression that the montane fauna (and flora) of Ben Lomond and other mountains in the northeast was characterised as much by what was missing (compared to similar areas west of the dry Midlands) as by what was present. But, one by one, we found ourselves ticking off species long considered to be absent, and reducing the differential between west and east. Perhaps it just takes a concerted push, such as our expeditions, to find these things out.

Lynne and I rounded the back of Giblin Peak, poking around among the frost-heaved shards of dolerite that littered the feldmark (feldmark is a Norse-derived term for the characteristically lichen- and cushionplant-dominated, ground-hugging vegetation that gives the exposed, higher parts of the plateau its alpine feel). The detectable insect fauna seemed to comprise just a couple of hardy species of ground beetle and the occasional earwig (not the garden variety, but a larger species endemic to Tasmania's mountains). Leaving the feldmark, we began scrambling up the steeply inclined boulderfield towards the peak. Some sort of living missile glanced the side of my head and tumbled into a crevice. I couldn't see the animal concerned but it was followed soon afterwards by two more, which narrowly missed me. It was raining grasshoppers! One landed on the boulder beside me, and I knew then that these were our first sightings, on the mountain, of the talus grasshopper. They are chunky insects and strikingly coloured. The body is a deep blue-black, the legs bright red, and the antennae are clubbed, in contrasting black and white. They lack wings and spend their days among the boulder fields (talus is the geological term for these) of Tasmania's higher mountains, basking when the sun shines and secreting themselves away in crevices when it doesn't. They could equally be called the tumbling grasshopper, since that seems to be their default escape mechanism: launch into the air in some random direction and crash-land on a boulder, bounce off a couple of times, and eventually end up in an inaccessible gap between the boulders. They must eke out a living feeding on encrusting lichens, since there's precious little else to eat on a talus slope.

The summit offered spectacular, panoramic views in all directions, and I reflected on just how compact the island of Tasmania is, despite its diversity. To the northeast we could see all the way to the high peaks of Truwana-Cape Barren Island, situated off Tasmania's northeast tip; while in the opposite direction we figured we were looking at the outlines of Mount Field and, in the extreme distance, Kunanyi-Mount Wellington, just visible in the haze. But, mostly, we were looking out over the vastness of the Ben Lomond Plateau. It struck me that the plateau is Tasmania's largest expanse of alpine country, by quite a wide margin, and perhaps the only expanse big enough to generate its own weather. The summit of Mount Ossa may be higher, but the bulk of that mountain sits way below the level of Ben Lomond's plateau, and its weather is heavily influenced by what's going on at lower altitudes. Here on Ben Lomond, the plateau is almost a world unto itself, sometimes sitting above clouds that blanket the lowlands at its feet; while its shallow valleys are prone to cold-air drainage and year-round frosts. It's also the only part of Tasmania, I told myself through heavy breaths, where I feel justified in blaming the low oxygen levels of altitude for my apparent unfitness.

What the summit of the Tor lacked, though, was hilltopping insects.

Perhaps the day was just a tad too cool for that sort of behaviour. In fact, there were very few insects at all. I did find a millipede, under a stone—it turned out to be an undescribed species that is only known from these few tens of square metres of rocky grassland just below the top of Legges Tor, and nowhere else in the world! Perhaps it occurs elsewhere on the plateau, but if so, nobody has yet found or recognised it.

After a stop for lunch on the summit of the Tor, we headed back downslope, in the general direction of the saddle that we had left earlier in the day. A line of pylons for one of the ski lifts looked rather unloved at this time of the year. It was hard to imagine the same scene in winter, following one of the increasingly rare dumps of snow. The ski lifts would be working at full pelt, ferrying gaudily apparelled skiers to the summit. We noticed that some of the taller shrubs immediately below the ski lift cabling had been cut some years ago, presumably to reduce the chance of skiers getting scratched or, worse, snagged. Our thoughts turned to the dead wood from these shrubs, and the prospect of finding more Walker's flightless longhorn beetles, since their larvae are likely to feed on rotting wood, like their lowland congeners. We set to, breaking apart the dead stems in search of damp pockets of rot, which is where we figured the beetles, or their larvae, would be living. To cut a long story short (the activity kept us engrossed for a good hour), we didn't find any longhorns. We did, however, find stag beetles. They, too, live in dead wood (both as adults and larvae), and are flightless. The species in question is widespread in the east of Tasmania, from the coast all the way, it seems, to the mountaintops. It strikes me as quite amazing that these flightless beetles—longhorns, stags and darklings—can somehow successfully seek out these tiny patches of dead wood on a bleak mountainside. From what we could see, the density of such patches was extremely low; so, even if they could smell their target, they would still have to walk hundreds of metres to reach it; and all without feeding, since the longhorns and stags, at least, do all their eating as larvae.

That evening, low cloud rolled in from the northeast, spilling over the plateau's ramparts and enveloping the ski village in thick fog. Visibility dropped to mere metres, and the temperature plummeted. Every bush and tussock glistened with dew-spangled spiderwebs. Though I couldn't, in fairness, rate conditions as inclement, it was a reminder of the fickleness of mountain weather, and the hardiness of the creatures that live in such places.

I could continue to detail what we got up to every day, but I'll skip to day four, by which time our surveys had taken us further afield across the plateau, as well as back down Jacob's Ladder to the treeline around Carr Villa, which we worked during a cloudy spell when the plateau above was fogbound. Our species list had increased steadily, and we had continued to tick off 'western' montane species that had not previously been recorded east of the Midlands, though plenty remained elusive. One standout was the scratchy jewel beetle (my translation), a showy, red-and-black insect whose larvae are thought to feed in the wood of narrow-leaf orites, a common montane shrub. Though Lynne and I had been keeping an eye out for it (since it was already known from the plateau), the sole specimen of the expedition chose to land on the fluoro vest of one of our botanical colleagues, where it was snapped up by our quick-thinking vertebrate zoologist who had the nous to collect it on our behalf. Showy beetles aside, we wouldn't be able to formally identify much of what we had collected for another few months, once we had had the chance to work through our finds back at the museum.

Day four started out bright and crisp, after a frosty night. Lynne and I decided on an offtrack walk, focusing on terrain to the south and east of the ski village. By the time we had reached the saddle, the sun's warmth had begun to awaken insect life. Tiny mountain blue butterflies had emerged from their overnight roosts among the shrubs, to bask on bare rocks and cushionplants. Marchflies sunned themselves, too, noisily vibrating their bodies until they were warm enough for take-off. We left the saddle, heading south towards a line of crags that overlooked a broad, shallow valley—a side-arm of Meadow Vale. The alpine shrubland was looking particularly glorious: a wondrous patchwork of compact, wind-sculpted, and prostrate clumps of mountain rocket, orites, montane heaths, scoparia, teatree, pineapple grass and others. I was glad that someone had very thoughtfully spaced the shrubs widely enough apart so as not to impede our passage. It was a beautiful landscape; breathtakingly so. *'This is the life'*, I thought. I felt on top of the world.

There were many good finds that day, but the real entomological highlight came while we had stopped for a mid-morning rest and a bite to eat at the foot of the crags. It's remarkable how often stories unfold this way—the big discovery arrives while you're taking a break from looking: having a quiet cuppa, or going for a pee. In my case I saw a metallic, shiny beetle flying straight towards me, or perhaps towards my fluoro vest. With my left hand occupied by a sandwich, I stretched out my right arm to intercept the beetle, plucking it from the sky with my hand. Truly the gods of serendipity were on my side this week! I knew it was a jewel beetle, but I didn't know what sort it was. There are dozens of species to choose from, although few are likely to be found at this altitude. This clearly wasn't our scratchy jewel beetle, such as the one that had homed in on my colleagues yesterday. Once it was safely enclosed in a specimen vial, Lynne and I took a closer look. It was a beautiful insect, about 15 mm long, elongate, entirely metallic green-blue but with purple hues in some lights. On its longitudinally ridged, almost corrugated wing cases it bore two wavy orange transverse bands. The front one additionally branched into a narrow, longitudinal band of the same colour. We racked our brains, trying to place this beautiful beetle, and concluding that it must be the one whose name translates as the crested jewel beetle. It is a relatively common species in the western mountains, but yet another species that was not previously known from Ben Lomond. That said, we weren't convinced—for a start, those ridges seemed way too prominent.

I photographed the beetle and posted the picture on iNaturalist, which is a global platform where naturalists record their observations. Overnight, a German entomologist specialising in jewel beetles came up with an identification. It wasn't the crested jewel beetle at all. Rather, it was a much rarer species, McQuillan's jewel beetle, a species named in 1988, in honour of University of Tasmania entomologist Peter McQuillan (though the species epithet misspells his name). The year before its naming, Peter had collected several of these beetles from a mountain in the remote west of Tasmania. There had been only a single sighting since then, also on a western Tasmanian peak, until our find on Ben Lomond. Discoveries like this are what our expeditions are all about. I couldn't have been more delighted.

I'll wrap up this section with a brief rundown of our final morning on the mountain. It was time to collect all our trap samples and dismantle the traps. On our recce, earlier in the season, Lynne and I had set up two Malaise traps on the plateau, in Meadow Vale, about 800 metres from the ski village. We had been in thick fog at the time, with temperatures hovering not far above freezing. We had navigated by following the poles marking the route of a cross-country ski circuit—we could just about see the pole ahead before the one behind disappeared in the fog. We set the traps up at a distance from the route, and from the ski village, where we figured they wouldn't be visible, or at least not intrusively so. It was only on a subsequent visit to check on the traps, a few weeks later, that I realised how laughable our attempts to hide them had been. In clear weather they stood out like sore thumbs. I had bigger issues to deal with than their visibility, though, because both traps had been beaten and battered by wild mountain weather, their metal poles snapped, and their netting shredded. One was completely beyond repair. I gathered up all the bits and packed them into my rucksack. The other wasn't quite as badly damaged, and I was able to patch it up so that it could at least resume sampling until our expedition took place. This was the Malaise trap that Lynne and I had to dismantle on our last morning. I was relieved to find that it had been sampling somewhat effectively: the ethanol in the collecting jar contained a range of insects. I could even discern, in the insect soup, yet another jewel beetle species—the myrtle jewel beetle. Since there is no rainforest, and no myrtle, on the plateau itself, I figured that it must have been caught while dispersing from one side of the plateau to the other.

The pan traps were also thick with insects— mostly various flies, but I also spotted a nannochoristid in one. The pitfall traps were rather unexciting, although quite a few spiders—one of Lynne's specialities—had wandered in. Two further Malaise traps, which we had set at lower altitude—one at Carr Villa and one down near the campground—were full to the brim with insects, but in different proportions. The one at Carr Villa was dominated by flies, although quite a different suite of flies from those up on the plateau; the campground one was richer in beetles and moths. Given the sheer numbers of insects in the samples, I knew then that I would have to be very selective about which ones I chose to extract for pinning and further identification. After all, people have written entire doctoral theses purely on the contents of fewer trap samples than these.

When Lynne and I eventually examined our specimens back in the lab, and compared them to others in the museum collections, we were disappointed to realise that our flightless longhorn beetle wasn't the mountain endemic we had convinced ourselves that we'd found, but its widespread lowland congener. It's not the first time I've been caught up in, and caught out by, the euphoria of a moment, like this. On one level it's disappointing; but on another, I celebrate the joy and excitement that our 'discovery' brought us as the time, even if it didn't carry over into a scientific breakthrough.

Cider gums and silver xenicas

The slopes of Ben Lomond aren't the only places where alpine cider gums grow. The same species, sometimes known as the apple cider gum, is common along the northern ramparts of the Central Plateau, while further south, on the plateau proper, it intergrades with a closely related species, known simply as the cider gum (though a translation of its scientific name would have it called Gunn's gum). In turn, the typical form of *this* cider gum morphs southwards into a subspecies, better known as the Miena cider gum, since it is around Miena at the south end of Yingina-Great Lake that these trees grow. The common names are very telling. Gumtrees are gums because many of them ooze gum. Cider gums take exudates to the next level and produce cider, or at least an alcoholic ooze, the frothy fluxes of which vaguely approximate the eponymous fermented apple beverage. The First Peoples of Lutruwita deliberately wounded trees to elicit sap flows, which would naturally ferment *in situ* into a fruity, alcoholic drink known as wayalinah. Europeans learnt the same trick, and rather rashly rebranded it as cider. But even in the absence of humans, cider gums are prone to exude. Stresses in the trunk, which are often caused by the trees' tendency to twist as they grow, can be enough to induce fissures; as can wind or heavy snowfalls, if they take out limbs. More commonly, the culprits are likely to be the larvae of wood-boring longhorn beetles and goat moths. These tunnel extensively under the bark, until they eventually leave, as adults, via gaping emergence holes. As you can see, all of my stories come round to invertebrates in the end.

Of the three types, Miena cider gums are the toughest. They favour (if that is the correct word) frost hollows and the intermittently waterlogged fringes of upland mires, at around 1000 metres of altitude. Perhaps they would prefer to grow on better ground in the vicinity, but other trees, such as the gum-topped stringybark, won't let them; so being more resilient allows them to occupy areas other trees can't survive in. In any event, their domain is a challenging place for a gumtree to thrive in, and many of the older ones look as though they have lived lives of great endurance rather than bounty. Yet, much as a hard-pruned bonsai pine is a thing of beauty, so, too, is a veteran cider gum. Its short, gnarly trunk and undulating branches often seem unsure as to whether they should be reaching for the ground or the sky.

I have spent a fair bit of time among the cider gums of Miena, but no visit has been more pleasurable than one February day, about a decade ago, when Chris and I went for a ride on our mountain bikes along a backroad just to the south of the lakeside shack settlement. It was late in the afternoon of what had been a hot, sunny day. The shadows were lengthening but there was no breeze, so the air was pleasantly warm. The road, which had degenerated into a rutted dirt track, eventually emerged from the shade of low, gum-topped stringybark woodland into the glare of a treeless expanse of boggy frost hollow. This was lined with ancient Miena cider gums. The track hugged the eastern side of the hollow, which meant that the late afternoon sun now struck the trackside cider gums almost side-on.

Since leaving the woodland, we had disturbed quite a few butterflies, which had been basking on the track. Suddenly, we found ourselves engulfed in an orange-brown cloud of fluttering wings. Scores of butterflies simultaneously erupted into the air all around us! Demanding closer investigation, we parked our bikes and backtracked to where all the action was. The butterflies were settling down again, onto the track. At first, I thought that there must have been some salty deposit they were probing: perhaps deer urine, since we had seen a lot of feral fallow deer during our ride. But the butterflies—various species of browns—didn't seem to be feeding on anything at all; just loitering, flicking their wings open and closed, and generally looking rather vacant.

It was Chris who discovered why all the butterflies were there. While I was intent on studying the insects on the track, she had cast her gaze towards a cider gum beside the track. She quickly called me over to have a look. At about head height, dozens more butterflies had settled on a fissure in the main trunk, each intently probing the crack with its uncoiled proboscis. Warm, frothy sap oozed from the fissure and dribbled down the trunk, pooling in an irregularity in the bark before petering out a little further down. Honeybees, blowflies and European wasps had also crowded around the pool, jostling with each other, and the butterflies, for prime position. Judging by their poorly coordinated, swaggering movements, none looked particularly sober; but perhaps they were more so than the insects loitering on the path nearby—yes, insects get drunk too.

Chris and I could smell the cider now; in fact, I'm not sure how we could have missed it when we had first passed the tree. I wouldn't describe it as particularly cider-like. To me, it seemed closer to the smell of fresh, yeasty dough, with a hint of alcohol, and some subtle, toffee-apple overtones. With all those insects dipping their tongues into the foamy tipple, I can't say I was tempted to taste it myself. I decided to try to count the butterflies instead. *Ten, twenty, thirty, forty … no, did I count those already? Forty, fifty, sixty … Oh, I give up!* There *were* an awful lot of them, and counting was made all the harder by their constant jostling, and by some falling off, drunkenly. Furthermore, if one was spooked and took off, its flight would startle others, causing what looked like a Mexican wave of butterflies to launch into the air. Following this, each would settle down onto nearby parts of the trunk, prior to awkwardly wending its way back towards the fissure, on foot.

Several species of butterfly were present. A lone and very handsome yellow admiral remained largely aloof, unwilling to engage in the main jostling and, seemingly, preferring to seek out the odd drop of sap that had dribbled to the ground. At the opposite extreme, several rather faded common browns were part of the main melee, but they didn't look like hot-headed youths; rather, they were clearly in the autumn of their lives. Alongside them, though, were many of the slightly smaller, and more intensely hued, bright-eyed browns. Most butterflies, however, were diminutive silver xenicas, each roughly half the size of the larger browns.

Which brings me, at last, to my butterfly of the month: the silver xenica. This is not a rare butterfly, but it's always a treat to find oneself in silver xenica country. While, with some butterflies you'd be lucky to see more than one or two at a time, silver xenicas often occur in their hundreds. If it happens to be peak season (February), and you're in prime habitat (the fringes and glades of upland open eucalypt woodland), you can expect to see a dozen or more in a single vista. Their larvae feed on native tussock grasses, so prime habitat must include plenty of these; and the grasses must not have been too heavily grazed by wallabies, wombats or deer.

The 'silver' refers to markings on the underside of a silver xenica's hind wing, although their legs are also silvery-white. To be fair, the crescent- and wedge-shaped patterning that overlies cream and fawn background colours is bright white, not silver; but the effect is striking. Someone must have decided that 'silver' conveys this better than 'white'. In addition to the white overlay, the entire wing outline of freshly emerged butterflies is finely fringed with white, on both the upper- and underside. Unfortunately, though, this will be the first part of the wing to fray and lose detail, once the butterflies have been active for a few days. The underside of the hind wing bears two prominent eyespots. Each has a white 'pupil', thickly encircled with a thick black 'iris' which, itself, is more thinly rimmed with a cream 'sclera' (to continue the eye analogy); and then there is a fine brown outer circle enclosing the whole lot (mascara, perhaps?). One of the crescent-shaped white markings that I first mentioned forms an 'eyebrow' around one side of the eye. A single, fuzzier version of these eyespots, lacking the mascara and eyebrow, adorns the underside of the forewing. The upper sides of the wings are rather less distinctive, although they also bear a single white-in-black eyespot, which is overlaid on an orange-brown background with chocolate-brown checks and flecks.

Silver xenicas are found widely across south-eastern Australia and

Tasmania, and exist as several subspecies that differ somewhat in their colouring and habitat preferences. Only one subspecies is found on the Australian mainland, whereas Tasmania hosts the remaining three. The ones that Chris and I watched, as they imbibed intoxicating liquor at Miena, were the montane subspecies, which is characteristic of the Central Plateau. If I had had my net with me, and if I had been so inclined, I could probably have swept up a dozen with each swish and dispatched them for the museum collections; but I have long held an aversion to killing butterflies. I just can't do it. It's not based on science, unless it can be argued that we already know enough about butterflies and don't need to collect more; but this isn't strictly the case. It's more of an emotional attachment to the idea of butterflies, perhaps to Psyche herself (whom we will meet again next month). It may seem strange, considering all the drowning, dredging, log-breaking and other activities I have carried out in the name of Science; but killing butterflies is a line in the sand that I'm not prepared to cross.

We spent so long in the company of the Miena xenicas that we cycled no further that afternoon. Eventually the sun dipped behind the crowns of the other cider gums, on the western side of the clearing; and, suddenly, the air temperature dropped. For the butterflies, the six o'clock swill was over and, one by one, they left the pub. Many rose high into the cider gum's canopy, which was still warmed by the very last of the sun's rays. Others fluttered erratically across the track and settled in amongst the grass tussocks for the night. We, too, turned our bikes around and trundled back to Miena to re-join our hosts, Mark and Gabes, at their shack. And what did we do when we got there? We headed straight to the pub.

Where I grew up, our neighbours planted a eucalypt in their garden, and it soon overtopped the hedge between our garden and theirs. Within a few more years, it grew taller than the nearby oaks and chestnuts. This, I have since learnt, was a Gunn's cider gum, a species so hardy that their kind will grow quite happily in suburban gardens throughout England, at a latitude equivalent to that of Macquarie Island. As far as I know, the Miena subspecies hasn't been domesticated to the same extent. Perhaps a tree with such spartan tastes doesn't fare well in the decadence of suburbia; or perhaps the likelihood of it dripping sticky sap or fermented froth onto parked cars or garden furniture rules it out. I think I prefer it that way. It's a special tree, with a special relationship with its particular environment and the other unusual species that live there.

Sadly, Miena cider gums have an Achilles' heel. While they can tolerate extremes of cold and wet, they cannot abide heat and drought. And this is the basis for an unfolding tragedy on the Central Plateau. Drive from Bothwell to Yingina-Great Lake today and, between The Steppes and Miena, you will pass hundreds of dead or dying cider gums that have fallen victim to recent droughts and heatwaves. Wood-boring or defoliating insects sometimes deliver the *coup-de-grâce*. In death, their Tolkienesque, ent-like forms may engender a sense of unease in the observer; at least they do with this observer. And so they should. Modelling suggests that, within a few decades, no part of the existing range of Miena cider gums will fall within their climatic envelope of tolerance. Beyond domestication, they will have nowhere else to go.

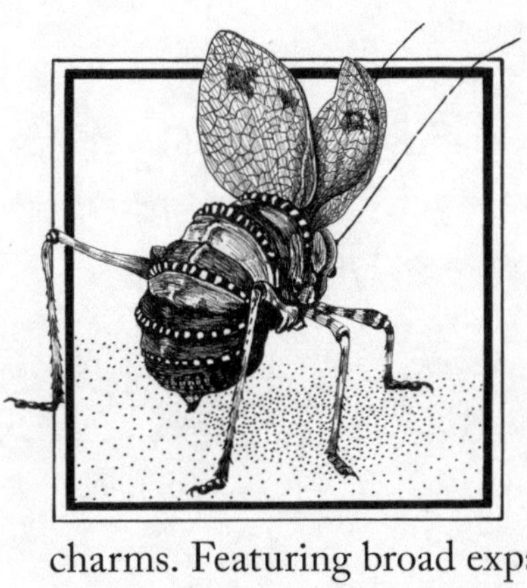

Three insects, one Plateau

Since we're up at Miena, let's stay on the Central Plateau a while longer. This region can't boast the rugged peaks of the mountains to its west, but it has its charms. Featuring broad expanses of undulating moorland, broken by wooded knolls and dotted with lakes and tarns, the Plateau is inviting and intimidating in equal measure. Arriving by road from Bothwell and The Steppes, to the south, the ascent to the plateau is gradual; but northern approaches must contend with towering dolerite ramparts, reminiscent of those on Turapina-Ben Lomond. The approach is similar for bushwalkers accessing the Walls of Jerusalem, from the west. Vast parts of the Plateau are essentially trackless, but drive or cycle along the potholed dirt road from Liawenee to Lake Augusta and you'll get a sense of the plateau's terrain, its wildness, and the brutal climatic conditions that plants, animals and people must endure if they are to make a living here. Liawenee (population, in the most recent census: two) often features on the television weather reports because of its record-holding coldness, especially in winter, when sub-zero conditions and icy winds can last for weeks on end. Even in summer, frosts are common. The Plateau is not for wimps.

In the right weather, though, the Plateau is *the* place to be. The northerlies that, periodically, bring extreme summer heat to southern Tasmania, must first pass over the Central Plateau. At this stage in their journey, they lack the ferocity and desiccating potential that will become their hallmark, once they gain momentum on their southward descent. On the Plateau itself, they offer the conditions that require its insect inhabitants to get active and 'make hay while the sun shines'.

Three special insects regularly draw me to the Plateau in February. With luck, I can sometimes find all three of them in a single day. Often, however, the best days occur when I don't try to bite off too much; so let me recount two separate days, from a few years back, when I sought and spent quality time with my three Plateau favourites.

We'll start with what is probably the Plateau's most famous insect inhabitant (though you would be forgiven for not having heard of it): the Miena jewel beetle. Jewel beetles occur worldwide, but they are especially prominent (in number of species, size, and showiness) in the tropics. Tropical Asian cultures have long used their shiny, brightly coloured, durable wing cases for making jewellery and clothing—it's a tradition known as beetle-wing art. Tasmania's many colourful jewel beetle species would be too small and rare to be used in such a way; but that makes close encounters with the animals in the wild all the more thrilling and, I would say, valuable.

The Miena jewel beetle was described in 1934, based on a specimen collected somewhere near Miena; however, after its initial discovery, decades passed during which no further specimens came to light. For a while the species was considered extinct; but then, early this century, a couple of chance finds rekindled hopes that a population might have been hiding in plain sight. (If memory serves me correctly, one of the finds was in a trout fisherman's tinny, on Yingina-Great Lake; and the other was in the back of a ute returning from the same area). This proved to be the case. Concerted searches by members of the Tasmanian Field Naturalists Club and others soon revealed that the beetles could be encountered on the flowers of kerosene bush (an aromatic shrub in the genus often known simply as 'everlastings')—but only within a wedge of country centred on Liawenee and along the shores of Yingina.

My own moment of discovery occurred on one such Club trip. The day was fine, and the kerosene bushes were in full flower. (We had checked to make sure they would be, when planning our trip). Their sprigs of tiny, off-white, papery flowers sit atop stiff, compact-leaved branches, wafting vaguely vanilla scents across the shrubland. On close inspection, we could see that the flowers were swarming with hungry insects, including native bees, tiny day-flying moths, metallic-green long-legged flies, pin-tailed beetles, and many others. Cameras at the ready, we dispersed among the tall shrubs, each of us with an image of our quarry in mind. A passer-by on the Lake Augusta Road would scarcely have known that we were there, although they might have spotted the occasional sunhat popping up above the shrubland.

A cry of '*got one*!' from one of our party soon signalled that we had timed our visit well, and the jewel beetles were doing their thing. '*Me too!*', came another shout. '*And me!*', answered another. It seemed that we had picked a good year for our target, for it later transpired that they alternate between years of plenty and years of scarcity. Eventually I found a Miena jewel beetle, too, noticing it in flight before it landed on a flowering sprig of kerosene bush. Like many jewel beetle species, it was bullet-shaped and bug-eyed, with a background colour of bright, metallic green. Longitudinally-ridged wing cases bore three paired blobs or partial bands of yellowy orange. The wing cases on this particular individual tapered to sharp points, signifying that it was a male. In females, the tips are blunter. It seemed beautifully poised for a photo, so I tried my luck but, no sooner had I pointed my camera at it, than it turned to face me, took an instant to figure out that I couldn't be trusted, raised its wing cases, and buzzed off, out of sight. I was disappointed; but the feeling soon turned to elation when I spotted two more, a male and a female this time, on a nearby kerosene bush. They seemed to be performing a sort of *pas de deux*, each facing the other and alternately advancing and retreating, jerkily, across the flowerheads. At some stage during these proceedings, they must have decided they were compatible because they were soon *in copula*, the male mounting the female, doggy style. I decided to leave them to it, but their mating got me wondering where the female would lay her eggs. So far, all we knew was that the beetles had an affinity for the flowers of kerosene bushes, but that didn't necessarily mean that the shrub was their larval foodplant.

Most jewel beetle species lay their eggs in the woody stems of living plants, and the larvae tunnel beneath the bark, feeding on the sapwood. We had figured that the full-grown larva of a Miena jewel beetle would be a couple of centimetres long, so it would be unlikely to live in a stem narrower than that. Therefore, I needed to find 'old-growth' kerosene bushes with thick, woody stems. It wasn't hard: there were many within just metres of me; so I got down on my hands and knees. I wasn't looking for the larvae as such, but for the emergence holes adults would make as they left their nursery. Lots of wood-boring beetles make such holes, but those of jewel beetles tend to be vaguely D-shaped, since that is the cross-sectional form of the adult beetle.

It didn't take long to find a likely-looking hole near the base of a senescent kerosene bush. Perhaps the larva's borings had contributed to the dying-back of this particular stem, since that commonly happens when other jewel beetle species are abundant. The hole in question was shaped like a rather tubby D, with just the right dimensions to have allowed the exit of a Miena jewel beetle. Circumstantial evidence, to be sure; but I checked some more kerosene bushes, and found further, similar, holes, as well as many other non-D-shaped emergence holes, which would have been made by other beetle species. I then checked some of the other shrub species and found no D-shaped holes, but some of other shapes and sizes. More circumstantial evidence. Others in our party had also been looking for holes and had developed the same suspicions.

Then came the moment when it all fell together. I spotted a female Miena jewel beetle as she landed atop a kerosene bush. Rather than feed at the flowers, she marched determinedly down the stem towards its woody, lichen-clad base. I crouched down to follow her progress, prising apart the wiry branches that otherwise hid her from my view. She deliberated for a while, turning this way and that, and then began probing the fissured bark with the tip of her abdomen, perhaps feeling for weak points. As I watched, camera in hand, she extended the tip to its full length, inserting it into some microscopic pore or indentation that she must have detected. I couldn't see an egg pop out, but she was clearly ovipositing; and I had the photo to prove it. Job done, she contracted her abdominal terminalia, rested a while, and then repeated the process further down the stem. After another rest, she clambered back up to the top of the shrub. The entire process must have taken some five minutes or so, although in those precious moments of insect intimacy, I was scarcely aware of the passage of time at all.

The association between beetle and kerosene bush is now common knowledge (in certain circles, at least). Others in the Club, with more patience than I, have gone on to document much of the beetle's life history in great detail. Its known range has been expanded and tightly delineated. The species' dependence on long-established stands of kerosene bush renders it vulnerable to fires; but fire is a double-edged sword: it kills the kerosene bushes but creates ideal conditions for new seedlings to establish. Presumably, the beetle has long lived with boom-and-bust conditions, such that local extinctions are in some sort of equilibrium with recolonisation events in other parts of its range. Indeed, Liawenee Moor may have become *the* place to find Miena jewel beetles because of its land-use history, which at some stage must have engendered massive regeneration of kerosene bushes—only lately had these reached the perfect stage of maturity for hosting the beetles' larvae.

Let's move on now to the other two special insects that I look out for on the Central Plateau, in February. One is going to be a hard sell, as it's a cockroach, but bear with me. My visit this time was with my museum colleague, Kirrily, and our destination was the Carter Lakes area. More precisely, we were heading for the inland dunes that form a peninsula, which more or less divides the eastern and western parts of Lake Augusta. Therein lie the smaller waterbodies known as the Carter Lakes. The dolerite bedrock in this part of the plateau is blanketed with drifts of windblown sand (known as loess), much of it dating to a previous periglacial period, when the region would have been even colder and windier than it is today; and drier too—almost a cold desert. Sand from the shores of the present-day lakes Ada, Augusta, and others, is still whipped up by the wind, especially when lake levels are low because of water releases for hydroelectricity. The inland dunes that all this sand has created support sparser vegetation than the surrounding shrubland, which means that they are a good place for some of the larger insects, since these need to bask to warm up. The dry, shifting sands also encourage short-lived plants such as fireweeds, which, as we shall see, are important for some insects.

At around this time, I was co-writing a chapter on Tasmania's mountain insects for a forthcoming book on Tasmania's Abels (mountain peaks over 1100 m), and we needed some photos of two of the species that I wanted to write about. One is the mountain katydid; the other is the Tasmanian sun-cockroach. The Carter Lakes area seemed to offer us the best chance of finding and photographing both. It was a day of rather ordinary weather: weak sunshine and average temperatures; but it was the best we could manage. At least it wasn't raining, or snowing. We arrived at Liawenee around mid-morning, stopped for a quick cuppa from our flasks and an unsuccessful search for Miena jewel beetles, and then drove the ute out on the Lake Augusta Road, before branching off towards Ada Lagoon. The increasing dustiness of the track, and the prevalence of roadside fireweeds, shouted 'windblown sand' well before we saw any signs of a lake; so, for safety, I slowed the vehicle down. I was also keen to avoid loose rocks and stones; therefore, I had my eyes peeled for those. A dark shape on the road struck me as not quite stone-like, so I slowed right down to walking pace, then pulled up to its left, so that I could look at it through my side window. My initial suspicions were well founded, because the shape sprouted long legs and began waddling towards the side of the road. It was a mountain katydid! I stopped the car, and we got out for a better look.

Female mountain katydids—or mountain bush-crickets as I prefer to call them—are strange creatures. Though I first likened this one to a stone, they more closely resemble a lump of wallaby dung, both in their rounded outline and in their mottled brown colouring and almost fibrous texture. The likeness was even more obvious now, because dung from a 'passing' wallaby littered the roadside right where our cricket was heading. The main contributors to the bush-cricket's disguise are its front wings, which are hardened and are elevated at just the right angle to create the desired form. I say 'wings', but these appendages are useless for flight. The hind wings are reduced to mere buds, too; so the animal cannot fly. Which is all very well, as long as her camouflage works, but what if her cover is blown? It turns out that she has a couple of options: the mundane one is to walk away, which is what she was attempting to do with us; but we were about to witness one more trick. Even though I knew what to expect, it still gave me a start. With Kirrily now in front of the animal, and me behind, she clearly decided that walking to safety was no longer an option, so she stood up on tiptoes, lifted her abdomen into the air, and raised her front wings, revealing a shockingly coloured abdomen. It was alternately striped in luminous blue and vermilion, *and it shuddered*, rhythmically! If one hadn't been startled by the static display, the moving one would surely do the trick.

Was this extraordinary performance all bluff, or is there some substance to it? The answer is, she's not bluffing. Not only can she kick potential predators with her long, spiny back legs, but her body fluids are loaded with toxic alkaloids sequestered from the fireweeds she feeds on. Those shimmering colours send an unequivocal signal to any would-be predator, that making a meal of her would be a big mistake. The male mountain bush-cricket also feeds on fireweeds and has the stripy abdomen, but he is long and slender and looks nothing like a wallaby dropping. But, then, he doesn't have the same need for camouflage, since he *can* fly. His deep brown forewings provide him with some camouflage though—he resembles a dead leaf—but if he's sprung, he has the option of flying away.

Mountain bush-crickets are neither rare nor entirely confined to mountains; but, in Tasmania, at least, they seem to be most common at higher elevations; usually in open, shrubby woodland. Their abundance peaks after fires, because of their liking for fireweed and open ground. They don't like areas of high rainfall, though, and appear to be absent from the west.

It was all rather convenient: we had our bush-cricket, and now we had photos of her doing exactly what I had wanted her to do. So, while we await our next move, let me make a little diversion to explain my dislike for the word 'katydid', and preference for 'bush-cricket'. Both are descriptive, but you'd be hard-pressed to know that with the first one. 'Katydid' approximates the sound made by one North American species; just one species, out of thousands in the world. I know it's unlikely to get me anywhere, but I don't think that's a 'sound' reason to call the rest katydids. And, if I'm to push it a bit further, 'katydid' also sounds rather too scientific, a bit like 'hominid' or 'nannochoristid'. It's the 'id' at the end, which, in taxonomic speak, denotes a member of a particular family, such as 'Hominidae' or 'Nannochoristidae'. There's no such family as Katydidae, though (katydids are in the family Tettigoniidae). I know I have my own biases, though; bush-cricket is, after all, the British term for tettigoniids (where a bush is merely a shrub, and not an expanse of native vegetation). And I have no such qualms about calling birds in the family Cuculidae cuckoos, even though 'cuckoo' comes from the call made by just one European species. Australian cuckoos don't make any such sounds. As it happens, many of our bird names are derived from ancient attempts to transliterate their calls. Pipits, tits, crows and ravens are all in that category. Meanwhile, 'cricket' (the insect version) comes from an Old French word meaning 'little crackler', referring to the sound that some crickets (but not our mountain bush-cricket) make. And while we're on my pet subject of etymological entomology (or is that entomological etymology?), vermilion (the colour of our mountain bush-cricket's abdominal stripes) is also from Old French, where it referred to a crimson dye; but the sense had evolved from an earlier Latin word for a 'little worm' or maggot. The 'maggot' in question was the cochineal insect (a type of scale insect, so not a maggot at all!), from which those crimson dyes were obtained. And cochineal? That came, via Spanish, from a Latin word for the redness of berries. And berries … I think I'll stop here.

Back to that dusty roadside on the Central Plateau. I was still photographing our beautiful bush-cricket when Kirrily, who was only a couple of metres away, poking around in the roadside shrubland, called out '*what's this?*' I coaxed the bush-cricket over to the side of the road, out of harm's way, thanked her for her tolerance and cooperation, and went to investigate. And, you guessed it, Kirrily had found our next special insect—the Tasmanian sun-cockroach. Rather incongruously, for such a heavy insect, it had climbed a grass stem and was hanging precariously from the seedhead, weighing it down so that it swayed in the breeze. My sudden movement caused it to drop to the ground and skitter off into the middle of a grass tussock. Although its rear end was still plainly visible, it seemed to have put its faith in the theory that, 'if I can't see you, then you can't see me'. I grabbed a sample pot and we nudged the sun-cockroach from the front so that it backed itself into the container.

If you're imagining a nocturnal household, hospital, or restaurant cockroach (which you may well do, if you're from the warmer parts of Australia or have holidayed in tropical Asia), think again. Sun-cockroaches are diurnal (day-active), and not at all inclined to hide away, nor to invade our buildings. This one looked more like a trilobite in bronze armour, with half a dozen rounded, interlocking segments decreasing in length from front to back. At about 35 millimetres long (excluding its long, filamentous antennae) it only just squeezed into the sample pot. I tentatively tipped it out, into the palm of my hand, and it sat there for a while, merely twitching those antennae. Eventually it started to shuffle off, and so I put my hand on the ground so that it could walk off, into the bush. In due course, it found a patch of bare sand and sparse grasses, and seemed quite content to sit there, basking in the weak sunshine, while I took some more photos.

We had found our two special insects for the day, and we had the photos I wanted; but we hadn't even seen the western arm of Lake Augusta. To remedy that, we drove on another few minutes, until a rough side-track told us that the lake was nearby. I parked the ute, and we walked the last couple of hundred metres through the low dunes, to the lake shore. The sun had now retreated behind a veil of high, hazy clouds, and we didn't see any further mountain bush-crickets or cockroaches on our walk. The lake itself was a bit of a disappointment, perhaps because the backdrop lacked grandeur, or perhaps because the lake has long been put to work as a hydroelectric reservoir. The water level was down, exposing a shallow, sandy beach, strewn with small stones. There was a strandline of sorts, perhaps from the most recent period of high winds, which would have whipped up waves and temporarily pushed the shoreline upslope. Dead leaves and filamentous algae made up the bulk of the strandline; but there were also hundreds of dead leaf beetles—victims, I think, of mistaken attraction to the sky-like, reflective surface of the lake. I knew that others had found dead jewel beetles here, too, on previous visits; but we had no such luck today. What I did find were a couple of drowned sun-cockroaches, but they were in poor condition, so we left them, rather than bringing them back for the museum collections.

On the long drive home, I couldn't help thinking about those drowned cockroaches. Why would such a fate befall an insect that seemed to thrive in the driest parts of the landscape and avoid going anywhere near water? I didn't know the answer at the time, but I have since learnt that the cockroaches had become zombies. Each had been parasitised by a horsehair worm that had taken up residence in the cockroach's body cavity. Once the worm had matured, it needed to get to water to disperse its eggs. To do that, it had to take control of the cockroach's brain. It's not that it physically controls 'levers' to tell the cockroach to turn left or right; it's more devious than that. In a related species, and probably in this one, the worm exudes chemicals that alter the cockroach's light-detection organs, effectively telling the cockroach to seek out horizontally polarised light. And where does that take the cockroach? Straight to the nearest waterbody. The cockroach wades in, drowns, and the worm is liberated into the water. Isn't nature marvellous?

At the time of our visit, the sun-cockroach had no agreed scientific name. It had been variously lumped with mainland species, but not consistently so. Entomologists long suspected that it might comprise its own unique species; and that Tasmania might even host more than one species, because sun-cockroaches are also found in sandy country near the east coast (from wind-blown sands atop the cliffs at Cape Pillar, in the south, to coastal sand dunes near St Helens, in the north). It took doctoral research by Shasta Henry, at the University of Tasmania, to resolve the matter. Perhaps surprisingly, Shasta found that all Tasmania's sun-cockroaches comprise a single species, and that all are distinct from mainland Australian species. In 2020 she anointed our species with the specific name *Yingina*, a tribute to their association with the inland dune country west of Yingina-Great Lake, and to Lutruwita's First Peoples.

Since the visits I have been telling you about, calamity has struck parts of the Plateau. The 2019 Great Pine Tier bushfire ripped through a tinder-dry landscape of moorland, peatland, shrubland and subalpine woodland, between Miena, in the southeast, and the Great Pine Tier, to the northwest. It incinerated the vegetation across much of Liawenee Moor, and the Lakes Augusta and Ada area in the process. On the Moor, the devastation it left behind made it all too apparent why the kerosene bush is so named. To the west, hundreds of slow-growing, fire-sensitive pencil pines on Great Pine Tier were incinerated. Three years earlier, a similarly intense fire took out pencil pines and moorland in the vicinity of Lake Mackenzie, near the north-western limits of the Plateau. The Plateau country is not meant to burn; at least not at such an intensity or scale. The scars will linger for years, but much will eventually heal over, as the moorland vegetation returns. There will be a new flush of growth, fireweeds and other ephemerals giving way to longer-lived shrubs. A new cohort of kerosene bushes will arise from the ashes. The same cannot be said for the pencil pines and other plant species in their communities, though, as they have not evolved to survive fire or to recolonise landscapes in its aftermath. The pines' sun-bleached and weather-beaten skeletons will stand as testimony to our poor stewardship of the land—and, indeed, of the planet—for decades or centuries to come.

The current weather

Planning a trip to the Central Plateau is a reminder that there is no shortage of weather in Tasmania, and it pays to remain responsive to its changing moods, whether to avoid a drenching, or to make the most of a cooling summer breeze. After a while, even without a forecast from the Bureau of Meteorology, one comes to see patterns in the weather, and to be able to predict what will come next, based on the clouds, winds, humidity, temperature and so on. The next step in my learning was coming to appreciate the connections between the weather on land and what is happening out at sea—not just in the atmosphere, but in the water itself; because, as it turns out, our weather and climate are very much shaped by what's going on in the seas. And everything in nature is connected: the climate plays a crucial role in determining what lives and grows where on this island. So, here's a brief overview of how the oceans influence climate, and the effects of this on Tasmania.

Whether from a cloud, a river, or a tap, water's flow under the influence of gravity is plain for all to see. But water doesn't stop flowing the moment it reaches the ocean. There, winds and tiny differences in temperature or salinity trigger local pockets of turbulence and eddies. These build up, cumulatively, into the great ocean currents, which move massive volumes of seawater in serpentine, undulating streams, from one side of the planet to the other, and back again. It sounds like a perpetual motion machine; but it isn't because, aside from the forces creating our tides (which I described back in January), two main drivers—the spin of the Earth (the Coriolis Force) and the sun's energy—conspire to achieve this mighty feat. The spin of the Earth helps to slosh water around ocean basins, causing it to bank up against continents or undersea obstructions in its path, from which gravity pushes it 'downhill' again. Meanwhile, the sun's influence plays out in the second law of thermodynamics: the tendency for physical systems to revert, constantly, to 'maximum entropy', or to structureless disorder. In essence, you could say that the physical world abhors order and structure, and physics doesn't like the idea of some parts of the planet hoarding more heat energy than others. So, when the sun pumps more heat into tropical waters than into waters at temperate or polar latitudes, the second law of thermodynamics kicks in and tries to redistribute that heat so that it is shared equally among all the oceans' water molecules. It never completely succeeds, but the world's major currents and countercurrents are a manifestation of its ceaseless attempts to do so. The amount of energy involved is staggering. If you've ever rued the additional work involved in carrying an unnecessary extra litre (just one kilogram) of water on a mountain bushwalk, imagine how much more work is involved in constantly moving cubic kilometres of ocean water around the planet, on multiple fronts. As I noted in January, the same physics are at work in the atmosphere; and the two systems—

atmospheric and oceanic—work in tandem. That said, they don't always work in a well-coordinated way because each is also a chaotic system with inherent unpredictability. The currents may also be out of phase with the seasons that gave them life, if seasonally variable wind patterns work against them, and because water warmed by intense summer sunshine in the tropics and subtropics may take months to wend its way south.

For the most part, the oceans are decidedly chilly at Tasmania's southern latitudes. Even in summer, the strong westerly winds whip up waves that continually mix the sun-warmed surface waters with cooler subsurface waters; so the sea surface seldom gets the chance to heat up as much as it would without these waves. In turn, the air above the oceans remains cool and relatively dry. But what happens close to Tasmania's west and east coasts differs from what occurs off southern Africa and South America—the latter both have warm surface currents on their eastern flanks and cold countercurrents on their western flanks. Australia bucks that trend, with warm surface currents that push south from the tropics along both its western and eastern flanks. In the west, the Leeuwin Current brings tropical fish south to Perth, and then laps around Cape Leeuwin and snakes along the southern edge of the continent. Rather than continuing eastwards, into the shallow waters of Bass Strait, it follows the continental shelf south-eastwards, towards Tasmania, where it has a name change and becomes the Zeehan Current. Sometimes it nudges right around Tasmania's south coast and starts heading north-eastwards, passing Storm Bay before petering out. Its peak flow, at our end of the Great Australian Bight, is in winter; which explains western Tasmania's equable climate of relatively cool summers and mild winters. It also accounts for the region's wet winters, because the westerly airflow gains warmth as it passes over the Zeehan Current, and warmer air can hold more moisture; that is, until it piles into Mount Read, the Tyndall Ranges, and other elevated ground along Tasmania's rugged western fringe. That's why median annual rainfall at Mount Read—Tasmania's wettest place—is about three and a half metres, peaking between May and October. It's also why fire-sensitive Huon pine trees there, and on the nearby Teepookana Plateau, can thrive out in the open, well away from their normal waterside haunts, because everywhere on higher ground is sodden enough to act as a fire refugium.

Meanwhile, in the east, Nemo's dad, Marlin, and his fishy, animated friends (from the Disney movie) were able to drift their way steadily southwards from the Coral Sea, along Australia's eastern seaboard, albeit a little out to sea. In fact, they would have been hard pushed to avoid being carried south once they had left the security of their coral reef. The East Australian Current is a humungous, southward-moving conveyor belt of heat and tropical marine life. Round about the latitude of Brisbane, part of it forks left and heads across the Pacific towards northern New Zealand. The rest pushes on, southwards, spawning massive, spiralling eddies that dissipate heat and life into the cooler Tasman Sea. By February or March, the current has reached its maximum rate of flow, delivering the larvae of brightly coloured coral reef fish to the coastal waters of southern New South Wales, and sheering off further eddies that spin southwards, past Tasmania and into the Southern Ocean beyond. At this time of year, undulations in the high-altitude belt of eastward-moving air (the jet stream) may spawn low-pressure weather systems. These 'cut-off lows' drift over the Tasman Sea and gain moisture and energy from the much warmer waters. Whenever the reinvigorated lows funnel easterly winds towards Tasmania's east coast, we know to expect drenching drizzle near the coast and torrential downpours on higher ground. So, we have the East Australian Current to thank for daily rainfall records of over 300 millimetres, set by places such as Gray and St Patricks Head, in the northeast. Occasional summer downpours and low cloud from the east sustain cloudforest and relict rainforest on east-facing higher ground all along the east coast, from the high cliffs of the Tasman Peninsula to the Blue Tier. But the rains seldom push far inland; which is why drier, more open woodland types dominate much of the east.

La Niña years are relatively cool over the Coral Sea, weakening the East Australian Current. In these years, the humid easterlies typical of summer in New South Wales extend further south, down to Tasmania's latitudes at times, increasing the chances of rain. In contrast, during El Niño years, the skies over the Coral Sea are sunnier and the ocean heats up more, driving an extra-vigorous East Australian Current, whose eddies extend further our way. Though the chances of rain in the east are generally lower, because more of our weather comes from the west, the warmer waters can also contribute to massive rainfall events when conditions are right (such as when those East Coast lows form). The conditions also depend, to some extent, on what the belt of westerly winds to our south is doing, since these may interact, to a greater or lesser degree, with weather systems from the warmer north, which are more under the El Niño/La Niña influence. When the belt of westerlies retreats southwards (the Southern Annular Mode's positive phase, which is more likely in La Niña years), it results in a higher chance of easterly weather in Tasmania, resulting in more rainfall in summer but less in winter. When the belt moves northwards, over Tasmania's latitude (a circumstance referred to as a negative phase of the Southern Annular Mode, which is more likely in El Niño years), easterlies are kept at bay and summer rainfall lessens, while winter rainfall increases. The system flips between these modes over a period of weeks, but it can sometimes get stuck in one or the other of them for months.

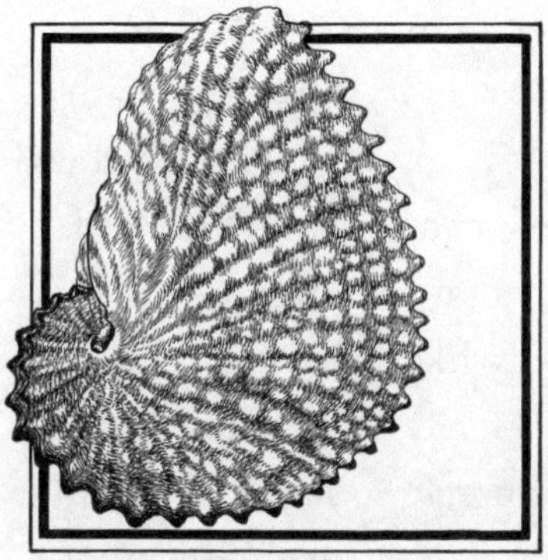

Molluscan mariners and other spineless drifters

Late-summer eddies from the East Australian Current bring with them the promise of strange marine life washing up on East Coast beaches. This prospect was what first lured me to Schouten Island. Schouten is a picturesque, largely wooded outcrop of granite and sandstone, off the southern tip of the Freycinet Peninsula, some twelve kilometres out from Little Swanport, and about halfway along Tasmania's East Coast. Although in earlier colonial times it hosted whalers, sealers, sheep farmers and coal miners, today Schouten is uninhabited; except in summer, when volunteer rangers take up residence. Our sons were quite young when we experienced a weeklong sojourn there. We had joined two other families and chartered a boat from Little Swanport, which would drop us off on one of the island's beaches, with all our camping paraphernalia; and pick us up again, a few days later. Our visit took place in an El Niño year and coincided with one of those periods of stable, sunny weather that are at their most prevalent and dependable during late summer; and we were spared rain-bearing easterlies of the kind that, from time to time, punctuate this stability and can ruin a good seaside holiday.

Hot and sweating from lugging our gear up the beach and setting up camp, we decided to grab our snorkels and masks and head to the beach for a cooling swim in the limpid sea. Calm it may have been, but it was a *cooling* swim—bone-achingly so (although the kids barely seemed to notice). Despite all my talk of tropical eddies, the warm water usually loiters many kilometres offshore, while the coastal waters remain stubbornly cold. As my shocked body gradually acclimatised, my breathing slowed, and I was able to survey my surroundings through the window of my mask. Below the waterline, the beach profile was quite steep, consisting of the white and pink quartz grit typical of a granite shore. Sunbeams danced across the seafloor to the tune of the ripples and wavelets above, glinting off flecks of granite-derived mica suspended in the water. Despite the beauty and tranquillity, there wasn't much marine life in evidence, until I paddled my way shoreward, to the point where my mask almost dug into the beach grit. It was then that I spotted a beautiful little animal that I recognised as our largest species of margin-shell. Its scientific name translates as 'fly-like'. I have never seen the resemblance to a fly, myself, and really wondered what whimsy had caused the French naturalist, Jean-Baptiste Lamarck, to name it thus, back in 1822 (he also named a smaller local species 'ant-like'). Regardless, my fly-like margin-shell was the first one I had ever seen alive (rather than cast ashore, as an empty shell). Through the mask, it looked egg-sized; but at a mere 15 millimetres long, the shell itself is diminutive, although beautiful. It is solid, polished, pearly-white, and enamelled with yellow and orange tinges on its main body whorl. Its spire is short, and its aperture is narrow, and bears three folds on the inner lip and a much-thickened outer lip—this is the 'margin' that gives the margin-shell family its name. The animal, too, is handsome, as I now discovered. It was a translucent fawn colour, mottled with darker brown and white, and peppered with orange spots. It sported a large 'foot' (the part of the animal responsible for the 'pod' in gastropod), a long, snorkel-like siphon, and a fleshy mantle that cloaked much of the shell. These molluscs are predatory, seeking out small clams and chemically drilling through their shells to reach the flesh within—for this they use acids secreted by the siphon.

For minutes I watched this one, on the sand a few centimetres in front of my mask, as it struggled to maintain its sense of direction in the surf, despite the fact that the waves that were buffeting it were only a few centimetres high. How, I wondered, would it ever have the composure to home in on a clam? But I was shivering by now, so I didn't wait around to find out. The surf zone is the species' exclusive domain, so I figured that they must be able to cope somehow, even with conditions a lot worse than today's.

After the shock of the cold dip, I resolved that, for the rest of our stay, I would stick to beachcombing; letting the molluscs come to me rather than immersing myself in their frigid world. Besides, I had developed a stubborn gum infection, which was making snorkelling uncomfortable. It would later transpire that the infection was the first outward symptom of the leukaemia that was taking hold in my bone marrow and blood; but that's a story for another day.

A couple of days passed before an opportunity for a multi-family beach walk came around, and I was impatient to get going. With an onshore wind in our hair, and the sun in our eyes, we lugged the kids' boogieboards over fluorescent-orange, lichen-splashed boulders, and scrambled around sandstone bluffs; at length arriving at a wide beach of soft, golden sand where a receding tide was casually depositing streaks of fawn-brown foam parallel to the waterline. A suitable picnic spot was found, and I took off my shoes and socks and left the kids contentedly scooting across the wet sand on their boards, while I went for a physical and mental wander along the beach. The sand surface had been moulded into sinuous ridges and furrows by the tide—while visually appealing, they made for uncomfortable walking. Closer to the sea, where the sand was wetter and the ridges flatter, the sand oozed soothingly between my toes. Its surface was dotted with squirty little piles of lugworm castings—a polite way of saying 'worm poo'—since lugworms, which live in u-shaped burrows, ingest the sand *holus bolus*, extracting diatoms and edible detritus from it before egesting it from their rear ends.

A triplet of pied oystercatchers pe-peeped in unison as they passed overhead, on their way to the far end of the beach, where they would soon be probing the sand for those lugworms with their comically long, vibrantly red beaks (I've never seen one catch an oyster, by the way). More than any other shorebird, oystercatchers seem to evoke a powerful sense of wild beachiness; perhaps because they announce their presence so vocally and hauntingly that they're hard to miss, compared to, say, a well-camouflaged and predominantly silent hooded plover.

Closer to the water still, legions of purple-hued soldier crabs took a break from nonchalantly clawing sediment into their mouths, pivoting their bodies, all the better to follow my movements with their stalked eyes. At my unrelenting approach, they scuttled slowly and deliberately seawards, in a choreographic ripple of motion. The crabs closest to me recognised this evasive action as being insufficient relative to the threat of being eaten, or crushed, by a giant, lumbering biped. With an intensified pivoting motion, they drilled themselves into the sand until only their eyes were visible, protruding from the centre of a spiral of wet sand. The moment I moved on, they would un-drill themselves and continue feeding. I know, because I looked over my shoulder and caught them in the act.

An amorphous lump of jelly on the sand caught my eye, and I crouched down to take a better look. Up close it wasn't so amorphous. About the length of my foot, cylindrical, almost transparent and with the texture of tripe, or perhaps one of those silicone shower mats, it wasn't a jellyfish but a colonial, ocean-going, filter-feeding sea squirt known as a pyrosome. The name translates as 'fire-body' and refers to the ghostly green-blue bioluminescence produced by the colony when alive. This one was a baby, compared with some ocean-going pyrosomes, which reach lengths of up to twenty metres and may have an internal cavity wide enough for a diver to enter. I can imagine the awe or fear that the sight of a luminous, monster pyrosome might have evoked in the mariners of old, had one bumped up against a sailing ship during the night (in the times before the animal's true nature was understood). Having spent one university summer vacation sorting plankton samples choked with globs of fragmented pyrosome, I don't need to imagine the swearing that went on aboard an oceanographic research vessel when their midwater trawl net emerged from the depths completely clogged with one of these monsters.

What was an ocean-going pyrosome doing washed up on a Schouten Island beach? To me, it could mean only one thing: the East Australian Current was running hard and fast, and one of its eddies had recently edged close enough to the coast for some of its living ingredients to be propelled shorewards. And if that were the case, well, there could be all sorts of other strange creatures on the beach! Had I thought about it more, I might have come to that realisation sooner, since the beach foam, too, was probably of oceanic origin. I could see, now, that it held the remains of tiny orange crustaceans within its matrix of bubbles. I pondered how the juices from these decomposing creatures would soon be seeping into the sand, helping to sustain the lugworms and their oystercatcher foes; and how their destinies, here on a Tasmanian beach, were entwined with what the weather was like in the Coral Sea many months previously—weather that had also affected the hunger of fruit-eating cassowaries in the Daintree rainforest, which lacked its usual rain; whilst on the Pacific coast of Ecuador, El Niño had brought drenching downpours …

With a renewed sense of purpose, I scanned the beach for other blobs that were not foam, wormcasts, or soldier crabs. I felt a bit like a detective at a crime scene, scouring the ground in front of me for the murder weapon. Before long, a sharp jab of pain on the sole of my foot told me that I had stumbled upon the fishing tentacle of another ocean traveller, a bluebottle; and that, clearly, I wouldn't make a good crime-scene detective. Fortunately, the animal had been dead for a while and most of the tentacle's special cells, those that house the poison-filled darts, were either already discharged or beyond being triggered by me treading on them. I sent the pain to the back of my mind, and gingerly scooped up the bluebottle in an empty scallop shell, so that I could have a closer look. In the bright sunlight, its deep indigo colouration fluoresced, as though made of pure essence of tropical ocean.

Like the pyrosome, a bluebottle is a colonial animal, but it's a siphonophore—a hydrozoan relative of the lowly jellyfish—rather than a sea squirt (sea squirts are much more closely related to us). Unlike the sole-operator medusae of their jellyfish cousins, siphonophore polyps have spurned individualism and autonomy in favour of working together for mutual advantage. They form a community of specialists, dividing up the labour of life, so that each member contributes some unique function to the union. This one had a gas-filled float that was about as long as my thumb and shaped a bit like a stomach. The taut sides of the blue-tinted float looked as though they had been crimped together at the top, forming a low, wrinkled crest, not unlike a Cornish pastie. The gas (which is largely air, but with a trace of carbon monoxide) had expanded in the heat, and it threatened to burst the bloated float at any minute. The whole float of a bluebottle is a single, specialised polyp, an individual member of the colony. Other polyps have taken on other roles: beneath the float, a mass of tightly coiled blue tentacles—each one a colony member—surrounded the pinkish stomachs (yes, plural) and gonads; while that single, long, blue fishing tentacle, which trailed a good metre across the sand, was another polyp.

Defying Australia's reputation for having the world's most venomous animals, our bluebottles are smaller and less potent than their Atlantic kin, the Portuguese men-o'-war, and only have the single fishing (and stinging) tentacle, as opposed to up to a dozen; even though, until recently, they were considered to be part of the same species. And yes, they do go fishing for fish—they can target prey up to the size of a mackerel, as they bob along on the warm ocean waves at the mercy of the wind and currents.

Now I had my eye in, I could discern dozens of bluebottles of assorted sizes scattered across the flats. Some were intact, like the one that I had scooped up; but the floats of others had already popped, and their tentacles had congealed into a dense, blue crust. The more I looked, the more I saw. Not all the blue blobs were bluebottles, though. I was excited to realise that two other ocean-going colonial jellies had also come ashore—not siphonophores but a related group of hydrozoans. Most numerous were blue buttons. Their name is quite descriptive. They are coin-sized glassy discs, fringed with azure-blue tentacles; here plastered to the sand, but, in life, floating on the sea surface. Among the blue buttons were by-the-wind sailors, which are similar in appearance but more uniformly blue, and they bear a stiff, vertical 'sail', bisecting the disc (a bit like the part you pull on to remove the seal under the lid of a screw-top plastic milk bottle). Like the bluebottles, both these jellies live their lives out on the ocean waves; but they feed on smaller planktonic animals, so their stinging cells don't need to pack the powerful punch of the bluebottle.

It's tempting to think of these jellies as passive drifters, completely at the behest of wind and current. That is largely the case, but nature has bestowed both bluebottles and by-the-wind sailors with a means of propulsion that can turn an ill wind into a fair one; at least for some colonies. It's all down to the orientation of the sail (or gas bladder) relative to the longitudinal axis of the rest of the colony. The orientation is always at a slight angle relative to the axis, but is fixed for a given colony, which means that they can't tack this way and that like a sailing boat might. But here's the thing: in some colonies that angle is to the right of the axis at the front and to the left at the rear—say, one o'clock and seven o'clock; while in others it's the other way around—eleven o'clock and five o'clock. So, while all colonies can only be blown downwind, some will end up skittering off to the right of the wind direction, and some to the left. Thus, in the worst-case scenario of a continuously blowing onshore wind, only some colonies will be shipwrecked while others sail off into the sunset and get to live another day. In reality, wind direction changes all the time, enabling these amazing little animals to be scattered to the four winds and maintain viable populations right across the world's warmer oceans.

There are seasnails that share the same habitat as these jellies; indeed, they feed on them. Violet-snails, imaginatively named for their violet or deep blue colouration, make themselves little multi-bubble rafts and cling to the underside, going with the flow of the same currents and winds that propel the jellies. Their life strategy relies on the likelihood that sooner or later their raft will bump into a jelly. Contact galvanises them into action, and they clamber aboard the jelly and devour it, presumably keeping hold of their bubble raft at the same time, to avoid sinking (the bluebottle's float is the last bit consumed, for the same reason). I knew that there was a fair chance of finding violet-snails on the beach, since the jellies were being washed ashore in such numbers. Sure enough, towards its far end I encountered several violet-snails. Most were still alive and clinging desperately to their bubbles. These were the common violet-snail, the most widespread species, found in both tropical and warmer temperate waters. The delicate, globular shells were deep violet on their undersides but almost white on the upper whorls of the low spire; the animals, too, were violet. How sad, I thought, that the life-rafts, that had safely carried them perhaps thousands of kilometres over the ocean, had now become their death-rafts, driving them ashore.

You will have noticed, I'm sure, that all these ocean surface-dwelling animals have one thing in common: their colour. In the mid-twentieth century, the great British marine biologist, Sir Alister Hardy, was moved to coin the term 'Blue Fleet' to describe this suite of blue animals. Blue is quite an uncommon colour in nature and it is probably difficult to make, so it must have some crucial function among these creatures. One idea is that the pigments serve as sunscreens, helping to protect the animals from the damage to cells and DNA that intense ultraviolet light can inflict. But blue and violet aren't the best colours for this, so, at best, this may be a secondary function. The more likely reason is camouflage. A blue animal bobbing on a blue sea is hard for a passing seabird to spot from above. The colouring of our violet-snail lends support to this second idea, for it would look violet from above but much paler from below—matching its background either way. There's an ocean-going, predatory seaslug called the blue dragon that also feeds on bluebottles, and similarly goes in for countershading. It is dark blue on top and pale blue beneath, although when I say 'on top' I'm actually referring to the animal's underbelly, because they float upside down. Blue dragons do, very occasionally, turn up on Tasmanian beaches, but I've never been lucky enough to find one.

With the tide on the turn and my stomach rumbling, I turned, too, and headed back towards our picnic spot. I had walked much further than I had realised and had to hurry back across the sand, dodging bluebottles. I thought I might be met with gripes and grumbles, but as I approached, I could see that Chris and the boys were grinning at me and holding out some beach treasures of their own, awaiting my reaction. They, too, had found violet-snails, but they had done better—they had also found paper nautilus! This really was turning into a beach bonanza, I thought. The paper nautilus, it transpired, had been found up on the strandline, among the pebbles and boulders, rather than out on the sandflats where I had been. The fragile creations were a bit battered from the experience, but a wonder to behold, nonetheless. Each was hand-sized, comprising one main body whorl with a flared entrance; almost paper-thin, but stiff and brittle; pure white, other than some chocolate-brown staining along the double keel of the main whorl; and the whole surface decked in spirally arranged rows of bumps—inverse dimples—that give this species its common name of knobbly paper nautilus. These were the first of their kind that any of us had found, and we were transfixed by their beauty and at the thought of there being an animal capable of manufacturing such works of art.

Paper nautilus aren't like your normal seashell, which is progressively grown by the soft mollusc within as its long-term home. A paper nautilus is a brood chamber, constructed especially for the purpose by a female argonaut, which is a sort of pelagic, ocean-going octopus. She occupies the entrance to the chamber, facing inwards while she broods her eggs within. The whole thing floats near the ocean surface because she occasionally uptilts the shell, capturing gulps of fresh air within. How she makes her shell is also rather unconventional: she extrudes it from the tips of two of her eight tentacles, which nature has modified for this purpose. Octopuses are well known for their ability to coordinate their tentacles, each of which is semiautonomous, with its own 'brain'. The female argonaut puts this ability to use by dexterously moulding a perfectly symmetrical structure between her two webbed tentacles. As for the male, he is a tiny imp in comparison; but he, too, has a special tentacle—effectively his penis—which he drives into the female during mating. The tentacle then detaches from the spent male, who promptly dies, while his penis lives on. Whoever first said that males think with their genitalia would have been particularly close to the mark when it comes to argonauts.

Paper nautilus have long been a zoological enigma. They have been known as nautilus (or *nautilos*) since ancient Greek times: the name means 'little sailor', because Mediterranean mariners of old thought that those two expandable, weblike, shell-spinning tentacles were the animal's sails. For centuries, indeed millennia, it was assumed that the octopus within the shell was an interloper, which had usurped the shell of a seasnail in the manner of a hermit crab. That myth was only dispelled in the early nineteenth century, by a French woman, Jeanne Villepreux-Power, who lived by the sea in Sicily, where argonauts abounded. Spurning the conventional role of a wealthy businessman's wife, Jeanne indulged her natural curiosity instead, and lived a life of fruitful scientific enquiry, unparalleled in her day. The male argonaut, meanwhile, had been overlooked entirely, and his penis-tentacle (hectocotylus) was first scientifically, and patronisingly (matronisingly?), described as a parasitic worm. As for the name 'argonaut'—that comes directly from the animal's scientific name, which Linnaeus borrowed from Greek mythology. The Argonauts were the sailors of the good ship the *Argo*, who accompanied Jason in seeking the Golden Fleece (amongst other swashbuckling exploits).

Incidentally, argonauts are only very distantly related to the pearly, or chambered, nautilus, although both are cephalopod molluscs. The resemblance between the paper nautilus and the 'real' shell of the chambered nautilus is superficial. The chambered nautiluses (there are several species) all live in the tropical Pacific. They use the gas in their multiple internal chambers to migrate up and down the water column rather than to float. This means that their shells float to the surface when the animals die. They can then become caught up in the currents and carried long distances. Until recently, I would have rated the chances of finding one in Tasmania as next to zero; yet I have since been shown a big-navel nautilus that washed up on the southeast coast; and I have seen photos of at least one other, which was picked up in the northeast, near St Helens.

The beach—and the boys—had one last surprise for me. Among their collection of common violet-snails was a smaller, bluish shell bearing a more pointed spire and with fine, oblique, parallel striations crossing each whorl. This was my first sighting of a second species, yet one more blow-in from the Blue Fleet, known as the little violet-snail. At the time, I was only aware of a single Tasmanian record of this tropical species—from the far northeast of the state. Schouten would constitute the most southerly find for the species. On one level, I was delighted: it's always nice to spot a rarity and to recognise its significance; but on another, I well knew that it was an ill omen, heralding bigger changes arising because of global warming.

Our southernmost little violet-snail record would not go unbeaten for long. Within a few years, a lengthening and less-stable East Australian Current had pushed eddies of tropical water further and further south, down Tasmania's eastern coastline, bringing little violet-snails, chambered nautilus, and not one, but two species of paper nautilus to the southeast coast. Coconuts, sea beans and the seeds of beach barringtonias, all tropical in origin, have also been washing up in the far northeast; yellow-bellied seasnakes, longhorn cowfish, blue dragons, and green turtles, too. Lumps of floating pumice originating from an erupting underwater volcano near Tonga washed up all along the east coast, having first been swept into the East Australian Current off the Queensland coast. Planktonic larvae of tropical goose-barnacles, pearl oysters and pumice-snails had settled out on the pumice and were carried to Tasmania, too. While snorkelling at Taroona, I learnt first-hand that jimbles, a type of box-jellyfish not previously known to occur south of Bicheno, on the east coast, had moved into southern Tasmanian waters. Meanwhile, our giant kelp forests are almost gone because they can't stand the heat and lack of nutrients in the warmer waters. As the atmosphere heats up, so too do the oceans; in fact, they are absorbing most of the excess heat and carbon dioxide that human activity is generating. They are now giving the second law of thermodynamics an increasingly vigorous workout, and weird, dead, tropical animals on Tasmanian beaches are one consequence. In nature, everything is connected.

Out of time
an interlude

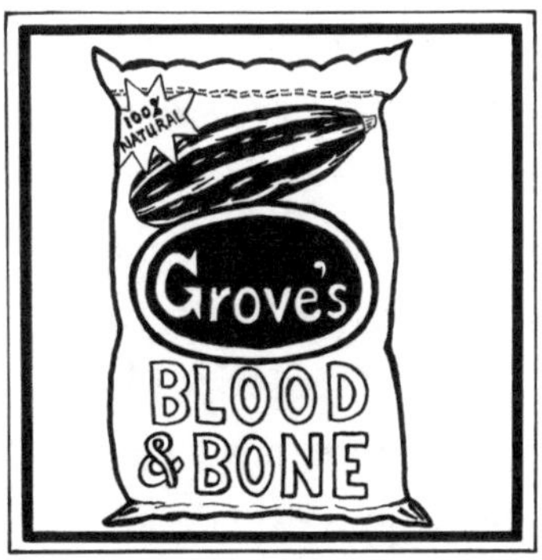

Blood and bone, *or* How to grow a better marrow

At the half-way point in this journey through the seasons, it's time for a pause. As I have already mentioned in passing, there have been episodes over the past two decades when poor health has necessitated pauses in my own life's journey; or, at least, has demanded that I live life in ways I would really have preferred not to. So, if the title of this section causes you to expect advice on how to grow giant zucchinis through the application of animal-based organic fertiliser, I'm sorry to disappoint you. Indeed, all prospects of gardening, of handling soil and compost, and even of eating fresh garden produce, were denied to me for months on end; and my encounters with nature were strictly rationed. I'd like to recount how things unfolded for me, and how the natural world was instrumental in pulling me through to the other side. And, if the accompanying illustration for this chapter intrigues or bemuses you, all will be revealed …

Migraine, malaria … leukaemia

Confucius wrote that we are born with two lives, and the second one starts when we realise that we only have one (or words to that effect). At some level, we all know how precious life is, and that we only get one go at it. But perhaps, as Confucius surmised, we need a reminder sometimes. My reminder came as more of a jolt than a nudge. I was on a post-conference road trip in Argentina, not long after I had returned home from the Schouten Island sojourn I have written about, in February. For a few weeks before the Argentinian trip, I had been suffering from intermittent, but increasingly intense, headaches. I had been treating these as migraines, which I had begun experiencing a few years earlier. I also had that annoying gum infection that had made snorkelling uncomfortable during our Schouten trip. The warning signs were building, but my attention was on other matters: work, mostly. I was finding my research at Forestry all-absorbing, and was pushing myself hard, which left rather little time for self-reflection. In retrospect, I was in a vicious circle, as is common in academia, where, with work taking precedence, mental health and relationships became neglected. The easiest way to paper over the cracks is to ignore the latter and work more. Argentina would prove a circuit breaker.

I had been in Argentina to take part in the *Southern Connections* conference in San Carlos de Bariloche, with my colleagues from Tasmanian forestry, Mark and Fred. We had presented some of our findings from long-term ecological research at Warra, and elsewhere, relating to forest management. Delegates had come from across the world's southern continents, having a common interest in our shared Gondwanan heritage.

Bariloche is a wonderful place for a conference. Arriving by plane from Buenos Aires, you descend over vast, rolling grasslands, with the high peaks of the Andes rising abruptly ahead. The region is sometimes described as the 'Little Switzerland' of Argentina. Certainly, the lakes and mountain scenery give the town a picturesque, almost Swiss, backdrop. Adding to this, many of its inhabitants have German heritage, which is evident in building styles and the delicious local beers. Many of the delegates were soon taking full advantage of the latter, including me. Other than the headaches, I found myself enjoying the conference immensely. A mid-conference fieldtrip deepened the warning signals that I should already have been taking notice of, though. We were on a guided walk in the foothills of the Andes, which took us through enchanting forests of *lenga* (a southern beech cousin of our own myrtle and fagus), where the entire forest floor was a waist-deep carpet of deep-yellow Peruvian lilies in full bloom. Onwards and upwards we went, through dry submontane shrubland, to a high, dusty col with magnificent views west over the snow-capped peaks of the cordillera. I was breathless; inordinately so, I felt, for such a moderate gain in altitude. I was surprised to find myself bringing up the rear of the party; but I put my unease aside when, on reaching the col, an Andean condor glided high overhead! If there was any ringing in my ears at that point, it was of Andean pan pipes and the tune *el cóndor pasa*. The bird's enormous wingspan and effortless traverse left us awestruck.

After the conference, the Tasmanian forestry contingent had been invited to co-run a workshop in conservation and forest management for Patagonian forestry professionals. The venue was not Bariloche but Ushuaia, some 2,300 kilometres to the south, on the island of Tierra del Fuego. The workshop would be held at the regional research station of Argentina's equivalent of Australia's CSIRO. We had a week or so to get there, and the research organisation's lead forest ecologist, the affable and irrepressible Guillermo (who had also attended the conference), would be our guide and companion for the journey. The road trip would take us southwards, down the eastern fringe of the Andes, via the famously wild and scenic National Route 40. At over 5,000 kilometres long, RA40 is one of the world's longest highways.

It took some effort to find a more-or-less roadworthy hire car capable of keeping up with Guillermo's well-travelled four-wheel drive; but, the morning after one last evening of Bariloche's fabulous beer and *asado*, we bid the town a fond *adiós*. What followed was a week of cultural camaraderie. The mild-mannered Tasmanian forest botanist, formerly known as Fred, transformed into Frederico, and took every opportunity to brush up on his Spanish, sometimes to the consternation of locals. To his credit, he persevered, and barely uttered a word of English for the duration of the journey. He and his Argentinian partner (now wife), Mercedes, drove the hire car, while Mark and I opted to travel with Guillermo. Mark, whose booming voice and no-nonsense approach to conversation is legendary (and much loved by the Australian forestry community), decided that the best way to be understood by the locals was to stick to English, bar the odd *por favor* or *gracias*, and turn the volume up. Those who know Mark will appreciate that he has a heart of gold, but I can imagine that some of the locals might not have known quite what to make of this *gringo*'s arresting manners. Nevertheless, Mark was the only one of us *tres australianos* who could give Guillermo a run for his money; and the two of them passed many an hour in excited, sometimes heated, but always amicable exchanges on matters of silviculture, forest ecology, culture, trout-fishing, and politics. As for me, I mostly sat meekly in the back of the car, enjoying the scenery, spotting birds, occasionally contributing something to the banter; but mostly feeling rather out of sorts.

The drive itself was both thrilling and a feat of endurance. Each leg of the journey brought different vistas, new walks, and variety in nature. Lush mountain beech forests and glacial lakes, one day, gave way to wind-blasted shrublands the next; then dusty semi-deserts, salt lakes, and rocky gorges appeared on following days. The ancient rock art in one gorge was outstanding, and poignant, given the dispossession (and worse) experienced by the indigenous peoples of the region. Groups of guanacos (relatives of llama) browsed nonchalantly on roadside shrubs; turkey vultures squabbled over roadkill; distant flocks of pink Chilean flamingos shimmered in the haze of the salt lakes; weather-beaten *gauchos* on horseback watched diffidently over their *patrón*'s vast herds of sheep or beef cattle. But, despite all of this, each day, I felt worse and worse. My headaches had become unrelenting, and now I was feeling an increasing sense of lethargy, which dampened my enthusiasm for any walk that took us more than a few hundred metres from the car. I was also developing a strange metallic sensation in my mouth, and any savoury food seemed unbearably salty. At length, we came to the intriguingly named roadside settlement of Bajo Caracoles (population thirty-one)—the name translates as 'Lower Snails', or perhaps 'Below the Snails', and refers to fossil seashells exposed in the limestone nearby. We booked into a hotel (which doubled as a general store and purveyor of rifle ammunition) and tried our luck with an evening meal. It appeared some two and a half hours later, by which time I was good for nothing. I took one mouthful of the boiled ox tongue, declared it inedible, and retreated to bed. Two days' travel further down the highway, we were to arrive at what should have been one of the highlights of our trip—a walk beside Lago Argentino, below the Perito Moreno glacier, with views to icefields high in the Andes where the glacier begins its descent. The truncated end of the glacier is truly awe-inspiring, constantly calving off great blocks of ice, which plunge spectacularly into the milky waters of the lake. I managed the walk, but I really felt rather morose, despite the magnificence of our surroundings. Already, the night before, in a motel in El Calafate (named after the local name for the barberry bushes characteristic of the region), I had broken out in fits of uncontrollable shivering and sweats, soaking my bedlinen. Although they subsided in due course, by the time morning came, I was feeling completely wrung out.

I didn't know what was going on, but, because my decline had been gradual, even now I didn't think that there was anything wrong enough with me to need urgent attention. Back in El Calafate, I had emailed Chris, describing my symptoms, and she passed my email on to a family friend who was also a general practitioner. The doctor suspected that I had malaria, and emailed to tell me that I should seek immediate treatment; however, I dismissed this as highly unlikely, given that I hadn't been anywhere near a malaria zone. In any case, there was nowhere within hundreds of kilometres of us, where I might find such treatment. So we continued south, crossing into Chile. In Punta Arenas I stocked up on paracetamol. Later, waiting for a ferry to take us across the Straits of Magellan onto the island of Tierra del Fuego, I enjoyed a few moments of mindfulness, picking up sombre-coloured limpets and top-shells from the shoreline and watching cormorants perched on the jetty pilings, drying their wings. With the strong westerly winds whistling along the channel, I felt I could almost have been in Tasmania, in the path of the Roaring Forties; but I realised that these were actually the Furious Fifties.

At last, we came to Tierra del Fuego. More herds of cattle and sheep, more *gauchos* left behind in clouds of dust as we sped onwards along the dirt road, across boundless, windswept grasslands. At our next stop, to catch up with some Chilean foresters over lunch, I didn't even leave the car until I was given a helping hand, which had been offered so that I could at least eat something to keep me going. After that we entered denser forests of *ñire* (a hardier southern beech), as we followed a deep, once-glaciated valley between snow-capped mountains. A final climb; then we dipped down towards the shores of the Beagle Channel for the last leg of our drive into Ushuaia, a roadside sign declaring that we had arrived at *El Fin del Mundo* (the end of the world). For me, it was feeling that way.

Guillermo introduced us to his lovely wife, Vanessa (also a forest ecologist), and their two young children, Conrado and Lucrecia. For our first night in town, we would be their guests. Vanessa had cooked us a beautiful meal, and she and Guillermo were the perfect hosts: Guillermo, proud and excitable; and Vanessa, a rather more calming influence. But all I wanted was to go to bed, so after dinner I made my miserable excuses. That night, the sweats returned, and the next morning I had to sheepishly confess to Vanessa that her nice clean bedlinen would be needing another wash. And this is where I owe my life to Guillermo, Vanessa, and Mark. After I had gone to bed the previous evening, they had stayed up, discussing my health; and had concluded that I really needed to get a blood test to find out what was wrong. Not long before, Lucrecia had been diagnosed with a blood disorder, so her parents were well attuned to such symptoms, and knew the right people to see in town, if I were to get tests done.

After breakfast, Vanessa took me to a health clinic where they took my bloods, and then we waited around for the result. I really had no idea what the words or figures on the printout meant; but Vanessa took it to the doctor who had treated Lucrecia. I couldn't understand a great deal of what was being said, but I could see that it was serious. The doctor was recommending that I have a bone marrow biopsy.

Ushuaia may be remote, but it is strategically important for Argentina; so it is well provisioned with government services, including healthcare. It turned out that the hospital could perform the biopsy the very next day. I have since had quite a number of these biopsies, and I grew to look forward to them, almost: more accurately, it is the woozy feeling afterwards, when the sedative is gently wearing off, that I came to enjoy. Nothing prepared me for this first one, though. I was instructed to curl up on the operating bed, in a foetal position, with my arms clasped around my knees and my backside sticking out—all the better to outline the pelvic bone, the surgeon said, so that she could clearly see where the marrow was to be aspirated (sucked out, that is; with a big, fat syringe). I had a sense that the surgeon was new to this procedure. Out of the corner of my eye, I could see her double-checking her notes before she manoeuvred into position, first holding a marker pen—X marks the spot—and then the syringe. If I was given any local anaesthetic at all, she had certainly been very frugal with it, because I couldn't help but cry out in pain. It was excruciating; and there were no nice sedatives for me either. As I was being patched up, I noted with alarm that the wall beside the bed was now spattered with my blood, which had erupted under pressure from my marrow. I knew, then, that things were bad, so I was not surprised by the solemnity with which the specialist doctor, later, read out the biopsy report. His advice to me was simple and to the point: get on the next flight back home, or I would, most likely, be leaving in a body bag.

For me, everything around this time is a bit of a blur. Erasure is the mind's way of coping with a stressful situation, I now realise. I remember Mark giving me a big hug. I remember that we found time for a trip with Guillermo, to a boulder-lined stretch of the Beagle Channel that served as a recreational area, just beyond the outskirts of town; I suppose this took place while we were awaiting the biopsy report. I recall that there were Chilean skuas and kelp gulls overhead, and black-necked swans, kelp geese, and various ducks in the weedy shallows; but no flightless Magellanic steamer ducks, a local 'special' that I had long wanted to see.

After the doctor's solemn edict, I submitted myself to the care of those around me. It was they who gracefully shouldered the responsibility for getting me safely home. Mark bravely volunteered to chaperone me, leaving our planned contribution to the conservation workshop in the capable hands of Frederico and Mercedes. My heart goes out to them all, as well as to Chris, back in Tasmania, who was also frantically trying to arrange flights and medical support through the Australian consulate in Buenos Aires. In the end, Mark and Guillermo drove to the airport to procure tickets for the next flight to Buenos Aires, for the two of us. It was due to leave in a couple of hours, leaving us with scarcely enough time to say our goodbyes, before we were back at the airport and boarding our plane. As it taxied down the runway and flew low over the harbour, I peered out of the window, hoping to spot a steamer duck; but they must have had other plans. While we were en route to Buenos Aires, the Hobart travel agents, through whom Mark had booked our original flights, managed to pull whatever strings needed pulling and procured premium economy tickets for our onward flight to Sydney. I remain forever grateful for this.

However, we weren't out of the woods yet. First we had to overnight in Buenos Aires, but we hadn't booked anywhere. Our taxi driver came to the rescue, eventually finding us a somewhat run-down, fifties-era guesthouse in the old part of the city. It would not have looked out of place in the movie, *Evita*. The next morning, we decided to seek additional medical advice at a nearby private hospital. We had hoped to procure a letter from a doctor that we could present to the airline in order to get me special treatment. However, the doctor took a cursory look at my medical notes, and dispassionately proffered the opinion that we shouldn't risk flying at all, since any slight knock to the head (for example, because of air turbulence) could be the end of me. For this he charged a hefty fee; and we chose to ignore his advice anyway. I felt a rush of relief as we stepped aboard the plane to Sydney.

Our flight home took us in a broad southerly arc, right back over Ushuaia; but we were, sadly, too high for duck-spotting. Then we were out over the immense white expanse of the West Antarctic ice shelf. It's funny the things you remember. I recall listening to a Supertramp biography on the airline entertainment system, allowing the music to take me back to a happier time and place. Rather than being high above the Antarctic, I was at a chalet in a meadow in the French Alps, thirty-one years earlier. I was reading an article in *Paris Match* about the band's new album, *Breakfast in America*. Meanwhile, back in the real world, Mark was simply counting the minutes until he felt it was safe to let the cabin crew know that I was unwell and in need of oxygen, without risking their turning the flight around.

One further night was spent at an airport hotel in Sydney; then, at last, Mark and I were on the home straight. Mark deserves a medal for withstanding the anguish he had to go through to get me back to Hobart. He hadn't slept a wink since we had left Ushuaia, remaining ever-vigilant in case I should suddenly keel over. I imagine that he must have been comatose within moments of arriving at his own home. As for me, I didn't get home for a few more weeks, because Chris whisked me straight from the arrivals hall to the Royal Hobart Hospital, where plans were already being made for my admission. Within an hour, I had been subjected to another bone marrow biopsy (this time, mercifully, with the benefit of anaesthesia and sedatives). Just a few hours later, my sickness finally had a name: acute myeloid leukaemia.

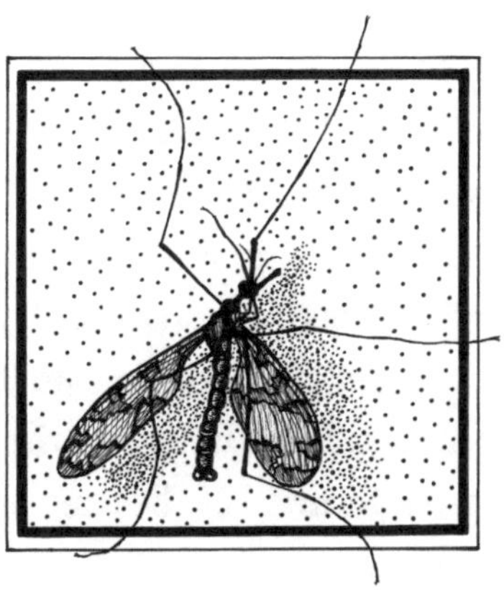

Chemotherapy, nature therapy

February, where we were when I broke off my narrative to subject you to this interlude, is supposed to be when we delight in the stable, blue-sky weather of the tail end of summer. So I'm sorry that your read has now become so dark. Still, since I've started describing my leukaemia journey, I feel I should keep going, even though not everything happened at this precise time of year. Please bear with me because nature really is at the heart of all this.

Winston Churchill once quipped that if you're going through hell, it's best to just keep going. Never were wiser words spoken, I felt, as I braced myself for three rounds of brutal chemotherapy. These took place over three successive months: it was not just plain hell—since it was blood cancer, I felt I had a right to call it *bloody* hell. I was an inpatient for each round, and remained largely bedbound during this time, drifting in and out of sleep and seldom fully conscious. I would recover enough to be allowed home for a week or so, and there I would just lie in the hammock, in the garden, and listen to the birds. Then the cycle would repeat. At least that's the simplified version of events. Chris and I, and our sons, are so lucky to live in such a kind and caring community. A meals-on-wheels roster was set up among local friends, and this kept Chris and the boys well fed and watered while Chris looked after me. Matters were made rather more interesting when Chris tripped in the garden—ironically, over a marrow. A zucchini, some might say, but this one had grown to truly gigantic proportions, while she had had her mind on other matters. She broke her ankle rather badly, requiring an operation to insert a metal plate. So, for a while, we were both in hospital, and then both at home, in wheelchairs. It was truly a mercy that my parents, in England, chose this moment to visit us, in our hour of great need.

Then, after the third round of chemo, brutal as all the others, came the news that we had all been holding out for: the leukaemia had gone! My chances of surviving the next five years, which had initially hovered on about forty per cent, were now substantially higher. For the next few months, I would be reliant on blood transfusions. I lost track of how many I was given; suffice to say that I am forever indebted to all those selfless donors out there. My six-month convalescence was slow and steady. I had been given special leave by my very accommodating employer, and scarcely gave my Forestry work a second thought, despite it having dominated my life just months before. In fact, I had little appetite for dwelling on the past, nor for contemplating the future. Life became a sort of suspended animation, a bubble of here and now.

As autumn turned to winter, I was less inclined to lounge around in a hammock, and took to gentle outdoor exercise instead. At first my atrophied muscles would scarcely carry me to the beach and back—all of four hundred metres—but, in time, I regained strength, and would practise walking up and down the steep steps leading to the clifftop path at the far end of the beach. Reaching the top, I would take a lengthy rest on a bench conveniently located there. I came to know the beach intimately, and those steps, and the bench and short section of coastal path above them. I found myself tuning in to every little change that marked the progression of the seasons, and equating these with steps towards my own recovery. There was always something to watch, smell or hear. The day-by-day, gradual decay of fallen leaves; the sudden appearance and distinctive, musty odour of toadstools; rainwater trickling down the steps; the subtle variations in the calls that songbirds used to express their own changing relationships with the seasons; the daily rhythm of the tides; the ever-changing seascapes and skyscapes, which kept time with the weather's moods through a never-ending cycle of ridges, troughs and fronts. These were commonplace events that most of us completely overlook in our busy lives; but for me, at this time, they *were* my life.

From my perspective, it was as though the beach, the sea, the cliffs, the woodland, all of nature was gradually embracing me and making me part of its daily existence, as it was part of mine. On the beach, the silver gulls watched my shuffling, daily progression through beady, red-rimmed, yellow eyes. They soon learnt that I was no threat to them, so they would sidestep out of my way rather than taking flight. A female pink robin took up residence for the autumn and winter, in the strip of undergrowth between the clifftop path and the cliff edge. She was seldom visible, but often audible, making a gentle *tuc, tuc*; whispered, rather than sung. I felt that she, too, had come to accept my daily presence as part of her own estimate of place. Her quiet calls seemed like ones of recognition: neither a *tut* of disapproval nor a *tic* of approval. One morning, I caught a glimpse of a beautiful firetail—there was no mistaking its crimson rump—as the bird exited a clump of rushes at the end of the beach and made a dash for the gully. On another walk, I had an uncanny feeling that I was being watched; a feeling triggered, perhaps, by an unusual lull in a soundscape that was normally filled with birdsong. I turned and glimpsed a pure-white goshawk eying me warily from a branch halfway up a blue gum.

I always carried my camera on these slow walks, and compiled a photographic record of the many kinds of insects and spiders that I spotted on the path while I rested on the bench, or on the handrails during my ascents and descents of the steps. Wooden handrails, it turns out, are wonderful places to look for insects. Perhaps the insects perceive them as branches, on which they would usually expect to be well camouflaged; but here they actually stood out like sore thumbs.

I wrote an article about my walks for *The Tasmanian Naturalist*. And, during this time, I also authored a book on seashells, but I'll leave that story for July. What I can honestly say, now, and say with great conviction, is that nature contributed enormously to my full and speedy recovery. *Leukaemia? Sure, I've had that, but I'm over it now.* How can I be so sure of nature's significance, with a sample size of one? Because, a few years later, the sample increased to two, when my leukaemia returned. And, this time, I didn't have such ready access to nature.

You may recall how, one December, I had been looking forward to hosting a group of visiting Russian entomologists. The day before they arrived, my oncologist phoned to say that she needed to talk to me and Chris about the results of my latest blood test. Could we come into the hospital right away? It didn't sound good. I had been on Rosie's books for five and a half years by that stage, and our regular reviews had generally been positive, as I passed various recovery milestones. Indeed, a few months previously, I had reached the five-year mark, after which the risk of dying from the original leukaemia fades into the background of one's overall mortality risk.

We had got used to Rosie's upbeat and cheerful demeanour, as the bearer of good news. Today we saw a rather more sombre side of her. Even before she spoke, everything was writ large on her face. A routine blood test had just confirmed that the leukaemia was back; and, this time, my only chance of survival would be to travel to Melbourne for a bone marrow stem cell transplant; that was, if chemotherapy could first bring me back into remission. On the plus side, we already knew, from screenings last time around, that my three siblings were perfect matches for a transplant. This meant that there would be no need to begin what is often a fruitless search of global databases for a compatible donor. There was no time to lose, though. I would already be vulnerable to infections, and my chances of pulling through would be better the sooner I started treatment. I would be readmitted the very next day.

I had been feeling perfectly fine up until this news—no breathlessness, no gum infections, nothing to indicate the rapidly proliferating cancer cells within. I was cycling to work most days, and, indeed, had cycled to our appointment with Rosie. I had been back to working fulltime for a good four years, firstly at Forestry, and then at the museum; and I had, long ago, put thoughts of leukaemia well and truly behind me. But it seemed that my world was about to come crashing down. Again. I had one last commitment to keep first, though: I had promised to drive the Russians to Mystery Creek Cave, in Tasmania's Far South, to start their Tasmanian fieldwork. Rosie was reluctant to agree to this plan, arguing that rainforest and cave air was just the sort of place where I was likely to inhale fungal spores; which, with my compromised immune system, could easily kill me. However, she could also see how much it meant to me. After all, it could, quite possibly, be my last outing into nature, ever.

So it was that I picked up Dmitry, Kiril and Elena from their hostel, later that very morning, and we settled in for the drive. They were excited about their forthcoming trip, and itching to get out into the field. I tried my best to give them a running commentary on the ecology of the country through which we were passing, as I usually do when hosting overseas scientists; but today my heart just wasn't in it. I don't think that my fragile predicament even registered with them, however—lost in translation perhaps; but they must have felt I was a bit distant at times. Treating them to coffee and pastries at the Dover bakery, it amused me to see that they weren't particularly impressed with such niceties. They had wolfed down their food and drink and beaten a retreat towards the door, almost before I had even settled into my seat. Oh well, I thought; my head hadn't really been in the right space for relaxing with a coffee, either.

When we arrived at the car park it was raining gently; and we were all a bit damp by the time they had unpacked all of the camping and collecting gear they would need until the next rendezvous, with my museum colleagues, several days later. We marched along the track at a cracking pace, through dripping-wet scrub and rainforest. Following the course of the creek, we soon arrived at the gaping hole through which it descends into the mouth of the cave. Rosie would have a fit seeing me now, I thought, as I followed the others into the cave. Our eyes gradually adjusted to the dimness within. Elena had come here in search of a primitive cranefly, called a tanyderid. Their larvae live in cold streams and aren't exclusive to caves; yet any larvae that do get carried into the cave would pupate in there, and then the emerging adults would find themselves unexpectedly confined to the darkness of the cave; this would make them easy to spot by torchlight, and to collect.

While Elena sauntered off to sweep the cave's recesses with her net, the rest of us enjoyed the spectacle of a galaxy of glow-worms. They were strung out across the roof of the cave, like our very own planetarium. Glow-worms are the predatory larvae of a species of fungus gnat. Each hangs a thread of silk from the ceiling, which is beaded with sticky secretions to ensnare small flying insects that have been attracted by the dim glow of the larva's rear end. Their natural habitat is the interior of hollow logs and shady overhangs; to them, a cave is just a magnificently spacious log. The same can be said of the Tasmanian cave spider, since it, too, more usually inhabits hollows in old trees and logs. I could see several at the mouth of the cave, each hanging, motionless, in its own sheet of webbing. These large, long-legged spiders live life in the slow lane. They may eke out a living for a decade or more, by ensnaring the odd cave cricket or cranefly every now and then. Males have special kinks on their front legs, which they use to prise open the jaws of the female during mating, and to keep those jaws agape. If all goes well, this prevents her from making a meal of him, gaining him time for more than one partner in life, which is more than can be said for the males of most spider species. The cave spider's closest living relatives are all in those Andean beech forests that I had visited on my Argentinian trip; so their journey since the break-up of Gondwana mirrors that of those vicariant moss bugs that I talked about back in December, and which were also a target of the visiting Russians.

A whoop from Elena echoed deep within the cave, making it clear that her cranefly quest had met with success. I had never heard of tanyderids and didn't have particularly high expectations, but as Elena held out her trophy in a clear plastic pot, I had to recalibrate. This was quite the most beautiful cranefly I had ever seen. It was dainty, rather than gangly, with its slender wings intricately marbled in tones of grey and brown. There is such natural beauty in the strangest of places, I thought, for those who know where and how to look. I decided on the spot that, should I eventually come out the other side of this horrible disease, I would mount my own tanyderid quest. Our mission accomplished, and the creek's waters rising, we figured that it was time to beat a retreat, and we scrambled out of the cave into the sudden, sullen, sodden greenness of the rainforest. Trudging along the muddy track, I couldn't help but wonder whether this was indeed my last-ever trip. Back at the car park, I left the visitors setting up camp for the night, and excitedly discussing their finds, while I had a rather less spirited drive back to Hobart, nursing thoughts about what lay ahead for me in the coming days and weeks. The next morning, I checked myself into hospital, with Chris by my side, helping me to settle in and adding some homely touches to my room.

The doctors explained to me and Chris that the chemotherapy regime would be a little different from the last time because my relapse demonstrated that the cancer cells had evolved resistance to that approach. Regardless of any differences, I found the experience similar. Essentially, several days hooked up to tubes that variously delivered poisons, painkillers, anti-emetics, fluids, and antibiotics. Days of semi-consciousness followed by a couple of weeks of reprieve and home leave, before the entire process repeated. I was glad that I had been living an active life before the treatment began, because it gave me a buffer of fitness that helped get me through the chemo. In turn, that stood me in good stead, emotionally, and I remained in quite good spirits, all things considered.

From time to time, I was administered opioids to help with the odd bout of pain related to my illness or treatment. Before I learnt which ones worked best for me, I experienced a number of hallucinatory trips. Most were rather dark; one was terrifying; but one still makes me smile. In my altered state of mind, I experienced the scanning room in the hospital basement, into which I had been wheeled on my bed, as though it were a large barn, complete with rusty old farm vehicles and a number of contentedly clucking chickens. Recalling the experience recently, I decided that I would see if I could turn it into an illustration to accompany this chapter, using an online artificial intelligence text-to-image generator ('Dream', by Wombo). The program spat out scores of very trippy images, most of which bore little resemblance to my own recollections of the experience, although they were comically engaging, nonetheless. One particular image did strike a chord though; with an eerie, uncanny concordance of almost-realism and patent absurdity. I hope my choice brings a smile to your face, as it does to mine. The recumbent, bald-headed waif in the hospital gown (that's me!) was cleverly painted into the scene for me by talented local digital artist, Andi Magnusson.

Friends and colleagues popped in from time to time, bearing news of the outside world. Their stories helped me hold myself together, even though I was gutted to be missing out on a cracker of a Tasmanian summer; at least I could experience the summer vicariously, through them. I also remained sanguine (and there's an etymologically fitting term, given my disease) about having to forfeit my place in a Bush Blitz. Then my day of reckoning arrived. With the final biopsy results in, Rosie again called me and Chris in for a consultation. And once again, her facial expressions told us that she was going to be the bearer of unwelcome news. The chemo had worked, she started, but not well enough. I wasn't in remission; so they wouldn't be risking a transplant.

I don't fully recall anything else Rosie said, for a while, because my mind was racing. When I came to, she was explaining that my main options revolved around what sort of palliative care I might want. I think Chris and I both burst into tears at this point. Rosie led us to a private room, where we could work through our emotions. I don't want, or need, to go through the details, but perhaps you can imagine what it was like to be told to get ready to die. Eventually, the box of tissues now half-empty, we regained enough composure to return to Rosie's consultation room. It was only then that she said there was one further option, which she described as maintenance chemotherapy. This might keep me going for months, maybe even years; but it was highly unlikely that it would bring me into remission. For those on death row, the adage about dying being very easy, and the waiting being very hard, is probably quite apt; but, given the choice between 'die now' or 'die a little later', who wouldn't plump for later?

So it was that I was discharged from hospital, yet returned every few weeks for subcutaneous 'maintenance' chemo injections (delivered straight into my belly), as well as frequent top-ups of blood products. Regular blood tests ensued, confirming that the regime was indeed maintaining me. Back at home, I felt nauseous much of the time, yet well enough to resume soul-nourishing walks to the beach and clifftop path. Summer gave way to autumn, and I tuned into these seasonal changes in nature, as I had during my previous journey of recovery, six years earlier. I also embarked on some museum work at the dining table, sorting and identifying samples of deep-water micromolluscs from Bass Strait. For much of the time, I was at home on my own; just me, my molluscs, and my thoughts. Although I knew I was lucky to be alive, I found it emotionally draining, and immensely difficult, to exist in this state of limbo. In reality, I was just waiting to be told that my time was up. Even writing about it now is hard. But, clearly, things didn't pan out that way. One day we got a call from Rosie, saying that the maintenance chemo had, unexpectedly, overachieved. I wasn't in full remission, but they were now prepared to take a chance on me surviving a transplant.

From that moment, things moved quickly. Once again, I was granted special leave from work, for the duration of my treatment; and Chris, likewise. Chris put the word out among friends that our teenage sons would soon be left at home on their own for a few months, while she and I decamped to Melbourne. She was reassured by plenty of kind people offering to look out for them, and to pop in with meals and shopping. In England, my elder brother, Jon, underwent further tests to confirm that his bone marrow stem cells were as good a match as had been determined, six years earlier. He, hurriedly, managed to arrange leave from work and a rapid departure for Australia. With the invaluable support of the Leukaemia Foundation, we fixed up accommodation, in a unit within walking distance of the Royal Melbourne Hospital. Chris would live there for the month while I was an inpatient; then we would both live there, for a further three months, during my follow-up treatment, and, we hoped, recovery.

Jon spent a week in Tasmania with us, before we transferred to Melbourne. It was a very special time—his first visit to Australia in thirty years, and his first ever visit to Tasmania. We took him to some of my favourite places around southern Tasmania: the giant eucalypts, in the Styx valley, which are impressive at any time of the year. They were all the more so now: their buttresses dripping with emerald mosses and their topmost branches reaching up to touch the swirling autumnal clouds. At Marion Bay, recent wild easterly weather had gifted us with drifts of fresh seashells. There were king scallops, wedding cake clams, pen shells, freckled cowries, wavy volutes and more, strewn widely across the beach. Jon shares the Grove family love of fossicking, and left with his pockets full of treasures.

Then we left for Melbourne. After some preliminaries, I was admitted to the stem cell transplant isolation ward, which was to be my home for the next month. Jon was given daily injections to prompt his stem cells to proliferate and, later, more injections to liberate them into his bloodstream. By the day scheduled for the transplant, I had been infused with chemicals to wipe out my own marrow and stem cells; including, hopefully, all the cancerous ones. On midwinter's day, 2016, a centrifuge-like machine painlessly (or almost so, I am told) extracted the stem cells from Jon's blood; an hour or so later, a milky suspension of those same cells was infused into what passed for my own blood. Not long afterwards, Jon returned to England, tired and aching, but otherwise fit and healthy. Meanwhile, Chris and I had a nervous ten-day wait to see whether the stem cells would settle comfortably into their new home. To cut a long story short, I was one of the lucky ones in that ward, because the success rate with these transplants is still rather low. Others on the same journey, at the same time as me, did not make it. To cut the story shorter still, in due course I became well enough to be discharged.

If you thought it was all over now, I need you to think again. Remember Churchill's quip about keeping going if you're going through hell? This was just the start of the hardest three months of my life; the mental health repercussions are still with me today. It's not that I had any major setbacks on my road to recovery; indeed, I was told that I had had a dream run. It's just a very difficult journey, and it took a toll on my mental health. The Australian health service excels at treating the physical aspects of acute medical conditions such as mine, and I owe my life to the dedication of all those health professionals. It does rather less well at looking after the emotional needs that arise from the trauma.

Where to begin? The immunosuppressive drugs left me constantly nauseous. Just the act of swallowing the many supersized, foul-smelling pills without gagging (the smell reminded me of snake poo) was challenging enough. Food anxiety came to rule my days, and the winter daylight hours turned into an often-fruitless quest to find nutritious food—indeed any food—that I could swallow, and keep down. Then, while all of this was going on, and we were still in Melbourne, Chris' father passed away, in England. Then her sister's husband died, tragically, from a brain tumour. Chris was torn between caring for me, and being with her father during his last days, and her sister in her time of need. In the event, my mother made another mercy dash from England, relieving Chris for a couple of weeks, while she returned to England for her father's funeral. Medals all round, I reckon.

For me, the hardest part of the whole post-transplant recovery period was not being able to escape into nature—restorative, nurturing, nature. During those long, long weeks in Melbourne, the best part of my day was falling asleep, and the worst part was waking up again, the next morning.

For weeks I suffered from severe fatigue, and scarcely had the strength to walk around our city block, let alone venture further afield. In any case, I wasn't allowed to do so initially, because of the risk of picking up an infection from airborne dust or spores. In due course, we hired a wheelchair so that Chris could push me along the busy streets to Royal Park, where we would do slow laps of the main jogging-and-dog-walking circuit. After a while I could walk the laps without assistance, and eventually I could walk to the park as well as around it, albeit with frequent rests.

Royal Park is expansive, and we could at least experience greenery and birdcalls; but skyscrapers always loomed large on the horizon. We were never out of earshot of road traffic and sirens; traffic helicopters hovered, more or less continuously, overhead; and planes rumbled, drones buzzed, and dogs yapped. Though these walks were the highlight of the day, they did not nourish me, and I found myself descending into a spiral of despondency, doom, and despair. In later weeks, as I was eased off the immunosuppressants, our horizons expanded. We got to visit the beach at St Kilda, and to meet up with friends living in the city's outer suburbs. The company, and destinations, were great; but the journeys, often along tens of kilometres of soulless, congested, urban motorway, were anything but. If Melbourne was considered one of the most liveable cities in the world—a gold standard for urban living—I deeply pitied the billions of people around the world for whom a city park (at best) was the closest they could get to nature. Perhaps you don't miss what you don't know, or perhaps I simply lacked the necessary resilience for city living; either way, I was enormously relieved when I was, at long last, deemed fit enough to fly back to Hobart and continue my convalescence at home. I'm sure that Chris did, too. She had been my rock for the past months, there for my every need; but she missed our sons and friends greatly, and was badly in need of some carer's respite and breathing space of her own. Our homecoming came not a moment too soon.

But something had changed in me. I re-trod my old paths, revisited old haunts, and saw, heard and smelled all that had been familiar and dear to me; yet every re-encounter, and every new *feeling*, now triggered a deep rush of anxiety, of panic almost, tinged with revulsion. Let's face it, there's nothing particularly scary, edgy or revolting about rounding a corner and seeing a familiar gumtree, or hearing the familiar call of a wattlebird, or even observing a billowing cloud; but in me these triggered hypervigilance, as though I were about to be attacked, or some other terrible fate would befall me. I knew that I should have been feeling nothing but gratitude for the gift of living, which had coalesced around me thanks to the blood, sweat and tears of my family and the hospital staff, and through my own months of suffering; but, every day, I found myself thinking that if this was to be my life post-transplant, I wasn't sure if I could bear it. I might be a walking miracle, but I felt like a damaged soul. I experienced everything through the lens of acute anxiety. It commandeered every other emotion and made all of them its own. I was broken, and I couldn't imagine ever being fixed again.

Others saw the signs of my distress, too, and helped save me from myself. Our neighbour, Pamela, enrolled me, for free, in her mindfulness classes; and Rosie lined me up with a psychologist and prescribed meds. I didn't get to see an official diagnosis, but my own take on the matter was that I was suffering from a form of post-traumatic stress disorder. Name it to tame it, they say. I hoped 'they' were right about that. I concentrated on healthy eating, rebuilding my physical fitness, and gentle nature therapy. It took all of these interventions, over many months, for the dark clouds to dissipate. After all, if you're down, the only way is up. I found that I could contain the worst of the anxiety within discrete 'anxiety demons', which I pictured hovering above my left shoulder, prodding and poking me, given the slightest opportunity. I just had to learn not to give them those opportunities. And to some extent I did learn. I have become adept at viewing my own emotions with a degree of detachment and equanimity. I would like to be able to report that the demons have been completely banished; but alas they are still there, impatiently waiting and ready to exploit any emotional upset, and occasionally testing the waters when I am least expecting them. I suspect that they will be with me for as long as I live. Consequently, I feel like damaged goods, and these days it takes me a bit more effort to find pleasure in living. It's as though life has lost its innocence, and I have discovered that beneath its lustrous sheen lies darkness, as well as light. But nature always steers me towards the light.

You may think that I am about to conclude this sorry tale, but it has one more twist. My new immune system seemed to have settled well into my old body, and I had been weaned off my immunosuppressants for several months, when I suffered an unexpected flare-up of what's known as graft-versus-host disease. This happens when the new immune system (the graft) decides that there are bits of its host body that it doesn't like after all. It attacked on three fronts, destroying my tear glands (which had already been damaged by the chemo), taking a painful shot at my leg muscles (a myopathy); and then changing tactics and eating into my peripheral nerves (a neuropathy). The latter started at my toes and fingertips, and progressively worked up my limbs and then my lower trunk, numbing everything in its path. At that point we stopped it in its tracks with a course of steroids, but some of the damage was permanent. To this day, in addition to requiring copious applications of eyedrops, nerve damage to the soles of my feet makes standing or walking on hard, flat surfaces painful. Tragically (he says, with tongue firmly in cheek), that rules me out for extended shopping sprees and supermarket runs. Mercifully, however, the pain is scarcely perceptible on rough ground, so I'm fine for bushwalking, birdwatching and photographing insects. It's almost as though my body is telling me that I need to get out into nature more. I won't argue.

And there *are* silver linings to the darkest of clouds. Firstly, I can report that contracting leukaemia is a great cure for migraines. Secondly, the whole experience taught me a thing or two about the importance of friends and relationships, and that working long hours to the exclusion of those relationships does nobody any favours. Thirdly, I've embraced the Roman poet and satirist, Horatius's, attitude to life (from more than two thousand years ago): *carpe diem, quam minimum credula postero.* This translates, broadly, as 'seize the day, since you can't place much faith in the future'; hence, I have adopted the personal motto: '*carpe the diem before the diem carpes you.*'

Finally, true to my word, on my recovery I mounted my own tanyderid cranefly quest, and I can report that I eventually met with glorious success on the banks of the icy-cold Arve River, high on the lofty plateau of the Hartz Mountains.

© K.DAVIS 2023

March

Summer in autumn

March is officially the first month of autumn. But just as spring, in Tasmania, can be a long and drawn-out affair, so, too, is summer; usually. Indeed, March often boasts the most stable weather of the year, and though the nights rapidly lengthen as the equinox approaches, days can still feel summer-like; at least towards their middle, when the sun is still high in the northern sky. In the garden, it's the month for generally tidying up, grubbing out, and salvaging the last of the summer veggies: withering tomato plants, whitefly-infested beans and mildewed zucchini. Then I prepare beds for the winter crops. Beyond the garden, much of nature's frenetic summer activity is now spent, and this always saddens me. Seasonal affective disorder clearly isn't just about shorter days. Flowers become few and far between; and, with the soil at its driest, much of the land takes on muted tones. Most adult insects will have expired; but there are exceptions, and perhaps because there are fewer of them, those which remain can be observed quite readily.

There is one reason to welcome the first intense cold fronts of autumn: it is the expectation (or hope, at least) of spotting swifts, the most aerially adapted of birds. Strangely, no species of swift breeds in Australia; though there are diminutive relatives—swiftlets—which occur in the north of the continent. But swifts in their thousands, or perhaps millions, can be found high above the wide brown land during the austral summer, where they forage for tiny insects carried on updraughts from below. They are white-throated needletails and Pacific (fork-tailed) swifts, passing their winter holidays far from their breeding grounds in northeast Asia. Swifts are renowned for being so adapted to aerial life that they sleep on the wing, and they may not land from one breeding season to the next; although it is now suspected that some do find a quiet spot, such as a hollow in an outback gumtree, to take an occasional, surreptitious break. It seems that some itinerant swifts undertake a grand tour of the Australian continent: they concentrate over Western Australia in early summer, before making their way south-eastwards as autumn approaches, shortly before heading north, back to Asia.

When cold fronts bring downdraughts, the swifts work these airstreams for insects, careening and tumbling out of the sky and descending to levels where they can be spotted from the ground. They don't have any particular terrestrial habitat associations, so they can appear anywhere. I've spotted them from our Taroona garden, on the Central Plateau, and while driving the Tasman Highway through Buckland. But you must be quick, for they may appear and then vanish again in a matter of seconds.

Then, one day in March, I notice an absence. I can't put my finger on it at first. When did I last see swallows swooping over the oval, or along the foreshore; or hear the pleasant chatter of tree martins as I walked my favourite clifftop path? This is when I realise that they have already left the neighbourhood, and suddenly I miss them, dearly. I can reason that the cold, dark days of winter are still weeks away, but their departure is one more sign of summer's passing.

King tides

Back in December, I recounted how my fellow seashell enthusiast, Rob, had taken to increasingly ambitious approaches in his quest to find more and more Tasmanian shell species. In parallel with his dredging drive, Rob also hatched plans to visit more out-of-the-way places. Those plans came to fruition one March, in the form of a week-long trip to King Island. The idea had ticked boxes for me, too, since we had little material from King Island at the museum. It also worked out that my younger son, Ben, whose keen eyes, shell knowledge and pre-teenage quirkiness made him great company, was also able to join us. So, together, we hired a car and a holiday house near a west coast beach, and flew to Currie on a scheduled flight out of Wynyard.

King Island sits in western Bass Strait, midway between the north-western tip of the Tasmanian mainland and Cape Otway, on the western coast of Victoria. Our flight took us over the low hills and muddy shores of Robbins and Three Hummock islands, before bumping its way over 80 kilometres of windswept water to King Island. The island bears the full brunt of the Roaring Forties, and must be one of the windiest places in the country—it certainly made for a hairy descent into Currie. The winds also bring ample rain, and King Island has long been renowned for verdant pastures that produce top-quality beef and dairy products. Peering through the plane's windows, it was clear that much of the island's extent (some 64 kilometres, from north to south, and 24 kilometres east to west) had long been given over to farming. Many land animals have been driven to the brink of extinction, or beyond. The island's emus, wombats, and quolls, for example, are long gone. But, despite such impacts, our flightpath revealed a spectacular coastline; this, surely, offered enticing prospects for shell discoveries. We couldn't wait to get started.

And we didn't. On the first evening, we were already strolling down to the beach below our holiday house, which gave us a gratifying taste of what to expect. The shoreline was piled high with wrack: chiefly the leathery hulks of bull kelp, torn, evidently, from a jagged line of rocks just offshore by near-permanently rough seas. Strong swells and a high tidal range made for a very active, high-energy shoreline, which was almost as good at pulverising or removing shells as it was at beaching them. We accumulated a reasonable list of species, but it was based almost entirely on shell fragments. So, although we welcomed the enlivening, salt-soaked wind and roaring surf after our long and tiring day's travel, we knew that we were going to have to work hard over the week ahead, to find better spots for shell collecting.

With those thoughts in mind, we made our first daytrip an outing to the rather more sheltered east coast, in the vicinity of Grassy Harbour. The tidal range is just as great there, but we were relieved to find that the wave action was much more amenable to the deposition of shells. And it was here that we got our first inkling of just how different King Island's seashells are, compared with those of the rest of Tasmania. I had thought that I knew Tasmanian cowrie species well—we have five smaller species of 'brown' cowrie, plus the larger umbilicated cowrie—but here I found that I couldn't rely on my usual rules of thumb for telling one brown cowrie species from the next. The arrangement and size of the brown blotches and spots towards the pale base of the dorsum (the shell's 'back'), the variable presence of continuous or interrupted transverse bands of dark pigment sweeping across the dorsum, and the background colours of the dorsum, all broke the usual rules for distinguishing brown from freckled, or from peppered, or Compton's. I found that the true brown cowries were a sumptuous deep purple-brown; but so, too, were some of the Compton's cowries. Elsewhere in Tasmania, Compton's tend towards orange-brown. Meanwhile, some of the small, slender, peach-coloured shells that elsewhere I would assign to peppered cowries, based on their interrupted transverse bands (more like chevrons), here sported continuous bands instead, rather like Compton's cowries do elsewhere. It was all thoroughly confusing, but also intriguing. I then began to see the same, not-quite-this-species, not-quite-that-one pattern among abalones, nutmeg shells, and kelp snails. It looked as though the waters around King Island were a melting pot of introgression; that is, where speciation goes into reverse and multiple closely related species interbreed. Though testing this theory would require genetic studies, I figured that the patterns we were observing were probably a reflection of fluctuating sea levels over recent glacial and interglacial cycles. These had drained and drowned Bass Strait, alternately, causing periodic isolation of marine life to the west and east of the dried-out strait, and then enabling intermixing in the vicinity of King Island, at the next drowning. In the end we were satisfied, however, that we had sighted all five species of brown cowrie (the plump cowrie being the fifth); and eagle-eyed Ben eventually added the umbilicated cowrie to our list.

I find it interesting that cowries are almost everyone's favourite seashell. Any seaside curio shop will have rows of tropical cowries for sale, each plucked live from a coral reef somewhere to satisfy our lust for their beauty. In turn, museums have boxes and boxes of them, many donated by people who fell for the urge to buy them on some seaside holiday and then, later, wondered what to do with them all. They pop up in antique shops, too; and, of course, a bathroom is incomplete without one or two sitting on the ledge of the little opaque glass window, next to the bath bombs and other gifted toiletries. I don't have an answer for their universal appeal, but I probably share everyone else's attraction to them. At some deep, emotional level, I occasionally find their sheer beauty almost overwhelming. There is certainly something uniquely sensual about a cowrie shell. I remember the strange emotions that I felt as a child, sitting for hours among my shells, on my bedroom floor, simply holding and stroking each cowrie shell in turn, or brushing it against my cheek. It was love! It's hard to put into words, but, strangely, I recall the emotion as a pleasant aroma, a little like that of a recently emptied box of After Eight mints. I'm not particularly synaesthetic, but strong emotions can have that effect for me.

Some would say there's something erotic about cowries. Take the derivation of the word 'porcelain': it comes from the Italian term for a cowrie; no doubt reflecting the tendency for both cowries and china to have smooth, lustrous surfaces. More literally, the word means 'little female pig' (so a 'porcelain piggybank' is a tautology). In prudish Victorian times, the derivation was said to be due to the similarity of the cowrie shell to the shapely curves of a pig's back. An earlier, more down-to-earth explanation has it that the cowrie's slit-like aperture earned it the name from a similarity to a young sow's genitalia. I have never looked closely enough at a young sow to offer an opinion; however, such things were not in my mind when I fell for cowries as a child. That said, it does make some sense that, in our predominantly patriarchal societies, cowries have commonly been regarded as symbols for female fertility. You may take your pick of which explanation you think is correct, but, as for the word *cowrie*, itself—it's Indian in origin. In traditional Indian societies, cowries were used as money—there is one species still known as the money cowrie, which might sometimes have been used for the purchase of brides. That would be cowrie dowry, I suppose.

But back to King Island. Our next day's outing took us to sandy beaches in the southeast of the island. There, we found the first of the shell species we had never seen elsewhere in Tasmania. Prize finds were wavy and broad-hinged Venus clams. Both are common along the coasts of South Australia and western Victoria, but they scarcely ever turn up in Tasmania, other than around King Island. Wavy Venus clams were the more common of the two; in fact, they were the most common bivalve on one particular shore, which contributed to our growing awareness that we were somewhere quite different from the rest of the state.

The background colour of the wavy Venus shells is dull white; but, overlaid on this are streaks of orange-brown, forming aesthetically pleasing, tent-like shapes, which radiate out from the shell's hinge area. The much rarer broad-hinged Venus clams are a more solid, rounded shell; off-white, with wavy, pinkish overlays.

Also on the beach where we found the Venus shells were oblong keyhole limpets. These are purplish-brown, saddle-shaped shells, with a keyhole-like opening in the middle. They are another example of a species that seldom occurs as far south as the Tasmanian mainland.

Day by day, Rob, Ben and I covered more of the island's coastline, and our shell list grew. All up, we visited twenty-one beaches, from relatively sheltered, sandy bays and coves, to exposed rock platforms. We traipsed barefoot over miles and miles of windswept sand along the northeast coast—silent and deserted now, but once teeming with boisterous, malodorous, and very vocal elephant seals, in the days before sealers slaughtered them all early in the nineteenth century. Our rewards on these haunted strands were some concatenated keyhole limpets, ramshorn squids and fragments of pen shells. Swaddled in rainwear, and buffeted by gales, we rock-hopped our way along the wild west coast, dodging king waves, drifts of kelp and wind-blown spume. We got down on hands and knees and peered intently at patches of salt-encrusted shell grit, piled up along the strandline fringing the Currie golf course. Ben had the advantage here, with his acute vision, physical flexibility, and complete lack of concern for comfort. In one cove, close to the southernmost tip of the island, he and I revelled in the unique experience of finding the cowries so thick on the ground that we had to scoop them out of the way to search for rarer species. A rocky cove near Cape Wickham, at the northern end of the island, turned up Spicer's spindle shell, another rarity in Tasmanian waters; while Fitzmaurice Bay produced our only specimen of King Island whelk.

Further evidence of glacial fluctuations in sea level and water temperature came from collecting at Naracoopa, where, amongst present-day extant species, we found subfossils of species of moon snail and rock whelk, which would have lived in the region when sea temperatures were a couple of degrees warmer than today. By the end of our week, we had clocked up about three hundred seashell species. Later, sorting through shell grit under the microscope, we bumped our total haul up to over four hundred species. Nearly a fifth of these hadn't previously been documented for the island, and some were new for Tasmania. A couple of the micromolluscs were, apparently, new to science. Despite our early misgivings, King Island had done us proud.

The flight out of Currie was as bumpy as our arrival, but by now we were somewhat inured to island weather and thought little of it. King Island may only be a short hop across the water, but it had felt like a world apart. I have every intention of returning, one day.

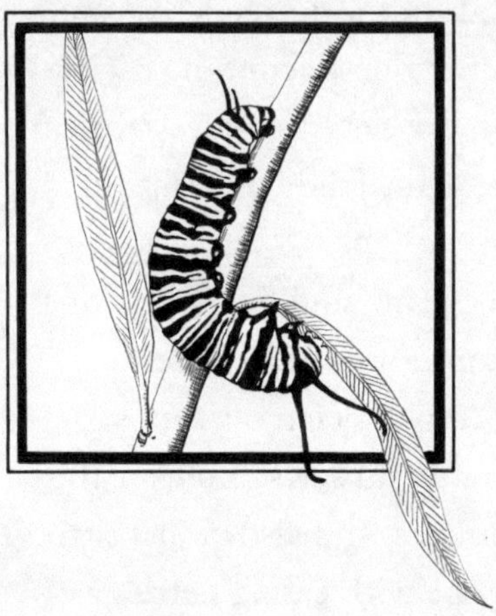

The wanderer

Butterflies are symbols of transience, fragility, and, these days, of vulnerability to our rapacious lifestyles, which place ever-increasing demands on the planet. Yet there are two species of butterfly that stand out as beneficiaries of humanity's colonisation of the globe.

One is in the title of this vignette, and we'll get onto that one shortly. The other is the much-despised cabbage white butterfly, known in the Old Country as the small white, to distinguish it from that country's other, substantially larger cabbage white butterfly—the excitingly named large white. Genetic evidence shows that our cabbage whites came from New Zealand stock, which came from North American stock; the North Americans acquired theirs directly from Europe. In Tasmania, cabbage whites can be seen on the wing from early spring to late autumn. March often sees a mass immigration of them, along the north coast of Tasmania. They are refugees from the increasingly parched agricultural areas of the Australian mainland, coming in search of a better life for their offspring.

I have yet to figure out why, once the small white made its debut in Australia in 1937, people took to calling it a cabbage *moth*. They're even called moths on posters of pest insects, and on the backs of cans of garden insecticide. Notwithstanding that all butterflies *are* moths (in an evolutionary sense), cabbage whites are clearly butterflies: they bear clubbed antennae, they are day-active, and they, well, flutter by purposefully, rather than making a panicked dash for the nearest dark nook. That last point, incidentally, is why I have been known to suggest that the main difference between butterflies and moths is that butterflies are clever and moths are stupid. It hasn't won me any favour among lepidopterist colleagues, and it may be a bit judgemental (especially since day-active species of moth buck the stupidity trend); but there's just something about butterflies. The ancient Greek word *psyche* meant both 'butterfly', and 'soul', since it was obvious to the ancients that our souls reside in those ethereal creatures. Psyche, whose feminine beauty outshone all (bar Aphrodite, who was a far more earthy goddess), became their guardian of the soul. Psychology, then, is the study of beautiful souls, or, perhaps, of butterflies.

I would be quite happy to have a butterfly as my soul, but I wouldn't choose a cabbage white. Perhaps a yellow admiral would be nice—independent and wide-ranging, not one for crowds, keen sense of sight and smell, able to find sustenance and succour in all sorts of out-of-the-way and near-at-hand places, very fond of flowers, and also of just sitting around in the sun. I would have needed to cope with eating nettles as a larva, but I think I could live with that. As it happens, I attribute my first opportunity for a visit to the tropics to having shown initiative on a selection weekend for an expedition to Costa Rica, held in the wilds of the English West Country, during which I supplemented our meagre food rations with soup made from nettles that I had picked from a handy hedgerow. That tropical expedition fired up my taste for adventures in nature, and ultimately led to my move to Tasmania.

But my thoughts are wandering; which, in a roundabout way, brings me to the wanderer butterflies of Flinders Island. I feel slightly uneasy about waving the flag for a non-native species, and I can't quite bring myself to crown it butterfly of the month. For that accolade you'll have to wait; but the wanderer has many redeeming features, which make it a close runner-up. It has wandered widely from its native home in North America, where it is called the monarch. There, it is famed for its yearly return migrations, which take it from as far north as Canada, southwards to mass overwintering sites (now, sadly, much diminished) in California and Mexico. A closely-related species occurs in the warmer parts of South America, where, bizarrely, individuals also migrate south for the winter. Perhaps what they are doing is searching for places where temperatures are low enough to induce energy-saving torpor.

On the eastern North American monarchs' southward migration, which takes them down the eastern seaboard, they sometimes become caught up in the tail end of hurricanes. These can whisk them away across the North Atlantic. That may be how they came to colonise the Canary Islands and Madeira, and, more recently, southern Spain. They may have always had this tendency, but it's only since their larval foodplants, milkweeds, became established as weeds that they have been able to settle in those areas. A female monarch can sniff out a milkweed plant from a distance of many kilometres.

Monarchs even turn up in Britain and Ireland, most autumns; but the climate there is too cold for the butterflies and their foodplants, so populations have not become established there. Meanwhile, in the North Pacific, monarchs somehow managed to colonise Hawaii, during the nineteenth century, and (rebranded as wanderers) they have since island-hopped across the Pacific to Fiji and New Zealand; perhaps with some human assistance along the way. They first turned up in Australia in 1871, around Brisbane. This was in the aftermath of the 1870 cyclone season, which could have blown them over from Fiji. From there it took them merely seven years to arrive in Tasmania, judging by the fact that one of their caterpillars features in an 1878 watercolour by Mary Allport. By the 1930s there were enough wanderers in Australia for people to start noticing their overwintering aggregations—scaled-down versions of their Mexican mass gatherings. Today, such gatherings have been recorded in more than thirty locations in Australia, including a small one on Flinders Island.

Throughout Australia, wanderers are totally dependent on introduced milkweeds and their relatives (commonly known here as swan plants, or balloon plants). Swan plants are sometimes grown by Tasmanian gardeners, and it seems that these are what attracts the wanderers.

My first visit to Flinders Island was many years ago, on an autumn holiday accompanied (and kindly underwritten) by my parents. They had come from England especially for this adventure, and we stayed, together, at Killiecrankie. Our flight to the island, from Launceston, was spectacular. Its path took us along the northern fringe of the northeast highlands, then high above the massive fossil dune ridges and sand-blows of Waterhouse, and over a succession of islets and islands emerging from vast meadows of seagrass, before skimming past the granite hulk of Mount Strzelecki and descending into Whitemark. From there, we hired a van and drove the 30 kilometres north to Killiecrankie, along deserted but potholed gravel roads.

I was in a seashell frame of mind at that time in my life; and from the plane, and the van, the beaches and coves looked every bit as enticing as I had dreamed. Consequently, many of our daytrips out of Killiecrankie were to beaches, and we gradually ticked off those which were both accessible, and made good destinations for picnics or walks. On our drive to one, in the northwest of the island, we would pass a stretch of roadside verge where the locals had encouraged swan plants to set seed and proliferate, specifically for the caterpillars of wanderer butterflies. I didn't rate my chances of spotting any butterflies; but I slowed down as I drove that stretch, and peered hopefully verge-wards, whenever it was safe to do so. I didn't actually know what a swan plant looked like, but when I started seeing strange plants with milky, blue-green, strap-like foliage (not unlike a diminutive oleander), I figured they were what

I had been looking for. So I slowed the van to a crawl and peered more intently into the vegetation. Then I spotted what I thought might be a chrysalis, suspended among the foliage. Vague protestations and mutters of disbelief came from the others as I brought the van to a halt and put the handbrake on. I could no longer see the putative chrysalis, though, so I got out and walked back a few metres, along the verge.

And, yes, there it was—a perfectly formed chrysalis of a wanderer butterfly! I had never seen one before, but I did have an image in my head, for an American book on animals, which I had owned as a child, had contained a painting of one. It had been spearmint-green and almost cylindrical, apart from its tapered extremities; an exquisite construction—just like the real one in front of me. This was about as fat as my index finger and almost half as long. Fine ridges outlined where the future butterfly's wings and abdomen would take shape, while several reflective gold spots and a further dotted horizontal line of gold and black adorned the surface of the chrysalis, serving no obvious purpose other than to add to the beauty; although perhaps they lined up with the sutures where the chrysalis would later split open for the butterfly to squeeze out. The whole thing hung from a short hook embedded in a pad of silk, which was attached to an overhanging stem.

I called the others over to have a look, which they dutifully did, unloading from the van while I hunted around for any further signs of wanderers. All I found, though, were some caterpillar-damaged leaves. Regardless, I was happy to have found signs of wanderers so late in the season.

The rest of our holiday revolved around more beach-wandering and bushwalking. We climbed the alluring and imposing granite edifice of Mount Strzelecki, snorkelled in the crystal-clear waters at Trousers Point, and paid our respects at the graves of the First Peoples of Lutruwita, who had been callously exiled to Wybalenna—ostensibly for their own good—as the early colony began expanding apace. For them, Wybalenna was a place of complete desolation and suffering. For us, Flinders Island offered exhilaration and joy, but also a haunting sadness, deeply felt, at Wybalenna.

Wanderers featured more prominently in my second Flinders Island visit, which was in March, a few years after our family holiday there. This time I was with my museum colleagues, participating in a Bush Blitz (like the one I wrote about in February). We made a special side trip to the patch of roadside verge where I had seen the wanderer chrysalis, in the hope of seeing the butterflies; and we weren't disappointed. Before even stopping the ute, I spotted several, lazily drifting back and forth along the verge. Wanderers are giants among butterflies, or at least among those of temperate latitudes. These were fifty per cent larger than female common browns, which are our next largest species (they are sometimes mistaken for wanderers). Like common browns, the wings of wanderers are suffused with orangey-brown scales; but in the wanderer, these are overlaid with a reticulated pattern of thin, sharply defined black lines, which radiate out from the wing base. The entire margin of the wings, apart from the leading edge, is thickly painted black and studded with white spots. The underside is not unlike the upper side, but with markedly higher contrast—the black lines are thicker, and the white studs are larger. The butterfly's body is black with white spots. Overall, its colours shriek 'danger!' to would-be predators; and with good reason—alkaloids in the swan plant are sequestered by the caterpillar and carried over into the body of the butterfly, rendering it extremely toxic.

Toxic or not, the wanderers were a delight to watch as they sailed over the swan plants, occasionally intersecting with others of their kin, but seemingly uninterested in liaisons. In fact, most seemed to be females intent on finding plants on which to lay their eggs. The trouble, for them, was that there were few plants that didn't already host eggs or caterpillars; and there would be little point in adding to the feeding pressure, if it resulted in all caterpillars starving. The eggs resembled tiny, spearmint-coloured barrels affixed to the leaf or stem of the swan plant. The caterpillars were far more spectacular, decked out in alternating concentric bands of white, black and yellow, with a pair of black, whiplike 'horns' at each end. Those colours, too, signalled toxicity, and the caterpillars made no attempt to hide themselves.

There were also pupae on the plants; so, in the span of a few minutes and metres, we could see all stages of the wanderer life cycle. Wandering among the wanderers, I wondered whether all of them would be able to continue their development to adulthood, this late in the season; or whether they would succumb to the cold, or perhaps go into torpor.

Perhaps, as in North America, only the adults can hibernate. If so, then this is the place, on Flinders, where they choose to do so.

Wanderers do turn up on the Tasmanian mainland, from time to time. When they do so, they are excellent trackers of their foodplants. During northerly weather, they have been seen flying in from the sea, along the north coast. In some years, they even make it as far south as Hobart and the Channel. As far as I know, though, the little patch of swan plant on that Flinders Island roadside verge is the only part of Tasmania in which they can lay claim to maintaining a permanent presence.

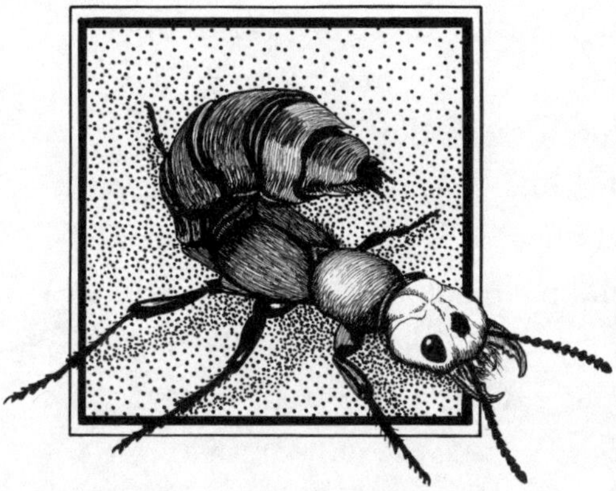

How to recycle a dead wallaby

One of my abiding memories of our Flinders Island holiday is the smell of dead marsupials. The island's rough roads were littered with the corpses of wombats and wallabies, and in our first day of driving our rented van, we managed to 'collect' a quantity of rotting wombat flesh. It had been scraped off a long-dead, road-killed animal by the van's underbelly, when I foolishly drove over it rather than slowing enough to go around it. No matter what we did to find the source of the odour, which soon began to infiltrate the air circulating within the van, it evaded us. Consequently, we had to put up with the stench for the duration of our holiday. And there's something about the whiff of carrion that gets right up the back of your nose and stays there, inducing a sense of unease and revulsion that is hard to ignore. Our revulsion over the stench of death—which comes from a compound called putrescine—is hardwired into us. Furthermore, it is, apparently, shared by our closest living relatives, chimpanzees. Such a response makes sense: it's not a good idea to hang around in places where others have met with death, because their misadventure could have been caused by some contagion, or a predator; and you might, at any moment, find yourself condemned to the same fate.

But if you want to witness the amazing process of carrion decomposition, and the army of fascinating insects that orchestrate it, you must override your gut reactions. You don't need to travel to Flinders Island, though—for better or worse, Tasmania is the roadkill capital of Australia. For worse, because of all those animal lives needlessly lost. For better, because it reflects the fact that the Tasmanian bush still contains many animals, compared with much of the Australian mainland (where dingoes and foxes—both absent from Tasmania—take a massive toll). Perhaps it's habitude, but when I'm driving, I don't tend to give roadkill much thought—it's all just part of the journey. That said, having learned my lesson well on Flinders Island, I avoid driving over roadkill, and I'll do what I can to avoid creating more; otherwise, roadkill just impacts on me as rather sad lumps of flesh, viscera, fur, and bone that you can't help but notice as you whizz past. It's a different matter from the vantage point of a bicycle, though. You don't merely glimpse roadkill in passing; you often smell it first, and sometimes you can even hear it—or at least the buzz of attendant blowflies. The overall encounter is a lot more, well, visceral.

A much younger version of myself once made the rather rash decision to cycle home to Taroona from Miena, on the Central Plateau—a distance of some 150 kilometres. It was after we had spent a long weekend with friends there, exploring some of the old Hydro tracks and generally pootling around on our mountain bikes. For reasons that escape me now, come Monday morning, I felt the need for a bit more exercise; so I was in fine spirits, waving Chris and the boys goodbye as they sped past me in the family car. Mountain bikes are fantastic for off-road cycling; but on sealed roads their greater weight and fatter tyres (relative to road bikes) make them hard work. While the ride involved an overall descent of 1000 metres, I was unprepared for the number of kilometres that were either flat or resolutely uphill, and I made slow progress, reflecting this. Still, you see, hear, and smell so much more when you're cycling slowly.

Encounters with nature came on thick and fast. Somewhere around St Patricks Plains, a raucous flock of some twenty sulphur-crested cockatoos (cock-a-twenties?) flew overhead, following the alignment of the road ahead. Several seemed to be making sideways glances at one of the flock members. It took me a while to twig to the interloper that they were all well aware of—a grey goshawk! Grey goshawks range from Cape York to Tasmania, and come in two colour morphs: grey, and pure white. All of those in Tasmania are white, and they clearly know that their plumage, and general size and shape, offers them the possibility of hiding in plain sight among cockatoos. Though they are no threat to the cockies, I suspect the latter get rather fed up with such deceit at their expense. The stealthy goshawk will choose its moment to peel off in pursuit of some unsuspecting pigeon, or another bird lulled into a false sense of security by the familiar sight and sound of cockatoos overhead. I had heard about this behaviour, but this was the first time that I had witnessed it.

A little further on, my own stealthy approach spooked a Latham's snipe from a roadside ditch. It zig-zagged low over the paddock, and dropped down out of sight, into a patch of rushes. A snipe is a remarkable bird, up close. Its streaky brown plumage provides excellent camouflage among the rushes, and its extra-long, straight beak is ideal for probing soft ground in search of worms and other invertebrates. The problem is that you don't often see snipes up close, as they are so quick to take flight, once they sense that their cover is blown. The erratic course of a snipe in flight seems almost like a preadaptation to avoiding getting shot; but presumably the behaviour stood them in good stead against goshawks, long before shotguns were invented. One night soon, this one would be leaving for Japan, which is where Latham's snipes breed. With luck, it would return to this same paddock in the northern autumn, our springtime.

The road continued its descent through beautiful open woodlands of gum-topped stringybark and mountain gum—the latter is a particularly picturesque tree, distinguished by the blotchy, grey bark of the main trunk. It was a few kilometres further still, where a concrete culvert rather ingloriously channels the intermittently flowing Blackburn Creek beneath the highway, that I got my first whiff of roadkill. Either the animal had not been killed outright and had staggered off the road to die, or someone had gone to the trouble of dragging it to the verge. Either way, I might not have spotted it but for the sudden buzz of blowflies disturbed by my approach. It was a small wallaby, old enough to have been out on its own but young enough not to have learnt of the dangers of roads and traffic. It was a lesson that it would now never get to learn. Perhaps it had fallen for the delusion, common among macropods, that the grass is always greener on the other side of the road.

Curious now, and in need of a rest from cycling anyway, I parked my bike on the verge, a few metres beyond the wallaby and out of range of the blowflies. Fighting a strong urge to turn, I crept towards the source of all the insect activity. The scene before me was as horrific as it was fascinating. The wallaby must have been dead for about ten days, judging by its state of decomposition and the fact that tufts of fur and hide had become spread far and wide across the turf. Already the carcase had lost most of its soft tissue, and stark white bones jutted out at odd angles, piercing the dark, mostly fur-free skin. For a while I couldn't look away; but since I was holding my breath, I eventually had to take a couple of steps backwards to exhale, and to reload with relatively fresh air—a ritual that I was to repeat many more times, since there was so much here to keep my curiosity alive.

Returning to the carcase, I watched as several bloated, fully-grown blowfly maggots wriggled their way out of it and started out across the turf, in search of pupation sites. They were clearly out of their element and, in their clumsy, undulating movements, like diminutive elephant seals hauling themselves up some Subantarctic beach. Female blowflies have such a remarkable sense of smell that they often arrive at a carcase within minutes of death, and some species lay not eggs but pre-hatched larvae, just to be sure to stay ahead of the competition. But fly larvae can only slurp, not chew; so their job here was, by now, largely done. From here on, it would be over to the beetles, as they can chew their food and tackle tougher tissues. Dozens of shiny, black, trilobite-like insects seethed over, in, and under the corpse. They were the larvae of carrion beetles. They often feed gregariously like this, which helps them monopolise the resource. However, they didn't have the carcase entirely to themselves. Hide beetles, dark on top but with white flanks and bellies, peeped out from parts where the leathery skin still carried tufts of hair, for that is what their larvae feed on. They, too, have tactics for monopolising their chosen resource—the adults give off a pheromone that attracts other egg-laying adults of the same species to the carcase.

The jerky movements of the hide beetles reminded me of little clockwork toys, but I was startled out of my ruminations by the sudden appearance of a bright red head through one of the holes in the skin. A lithe, red-and-black beetle emerged, menacingly waving its tail in the air. It was a rove beetle, this one known as the Devil's coach horse. The name comes from England, where a similar, but dark-coloured species, which also attends carcases (and cadavers), has long been associated with the work of the Devil. Apparently, if you crush one, God will forgive you for seven sins. I had no intention of crushing the one in front of me to find out, even though, having now emerged fully from the carcase, it seemed to be egging me on by raising its head and ferociously baring its jaws at me. The display is more usually aimed at would-be predators, warning them to back off or face the wrath of those jaws, as well as a dab of noxious chemicals from the animal's rear end. The jaws are not only for defence, though: they are also perfect for piercing and slicing through blowfly larvae, on which the beetles feed. It seems that it is the prospect of a feast of maggots that attracts these opportunistic rove beetles to carrion.

I was keen to find out what other insects might be engaged in recycling this wallaby, so I chose a stick from the adjacent woodland and returned to gently lever the carcase over (gagging ever so slightly). Bright, metallic-blue hister beetles dropped onto the ground as I did so. These beetles are like the Devil's coach horse: here for the fly larvae rather than the wallaby; but, as they are only a few millimetres long (and almost as wide), they're not in competition. They favour much smaller fly larvae.

An excavated pile of soil on ground that had been beneath the carcase piqued my interest. Using a slenderer stick as a digging tool, I worked through the loose soil, making my own pile next to the original one. What I at first thought was a small stone in the soil began to wander off, and I realised that under the caked soil and bits of fur was a different type of hide-feeding beetle; one related to dung beetles. The knobbly beast in question is built like a tank and camouflages its upper surface with errant bits of dead animal that it accumulates while ploughing through a carcase. The fact that it was burrowing beneath the carcase suggested that she was female, and preparing a chamber where she could safely lay her eggs in the knowledge that her larvae would have access to all the food they needed, right above their heads. I did my best to clear the debris from her chamber and brushed her in with the stick, before deploying my larger stick to return the carcase to its original position.

Quite a few other insects were hanging around the fringes of the carcase. Some were among the recyclers; others were predators of the recyclers; others may have just been curious bystanders. But, unless you follow the fate of a carcase from beginning to end, you only ever see a snapshot of the process, and a small part of the diversity of the players involved. The usual sequence is that the blowflies, fleshflies and scuttleflies arrive first, and their larvae make quick work of the putrefying liquid components. Some of them fall victim to predatory rove beetles and hister beetles, or to the larvae of other flies such as black carrionflies. Next come the carrion beetles, whose larvae eat the tougher flesh. Hide beetles colonise the pelt, with different species specialising on the skin or the hair; bright blue-and-orange ham beetles (so called because they are also partial to hams traditionally curing in the open) then move in to tackle the tendons and ligaments. Within a few weeks, all that remains is a pile of bones amid a patch of newly fertile, verdant vegetation. I had passed a good dozen such piles since leaving Miena, each a visible reminder of a life lost and a life successfully recycled. But even bones harbour marrow, and that, too, is a resource to be exploited. To see the flies responsible, you must sit quietly beside your chosen carcase until they no longer see you as a threat. That's what I now did, with respect to this dead wallaby. And that's how I came to see my first-ever bone-skipper flies.

The scientific name for these flies translates as fat-lovers, which is as good a name as any, given their tastes. The flies may arrive at a carcase at any time, but they need to wait until a good thighbone or similar is exposed before they can get to work. The males will stake out territories on or around such bones, and, once mated, the females will lay their eggs in the little pores that once held blood vessels, and now lead from the outside world to the marrow within. Once fully fed, the larva emerges via the same pore and then curls into a C-shape as a prelude to rapidly uncurling, thereby flipping itself away from the bone. It may need to repeat the procedure a few times, until it finds itself in a location more congenial for pupation. It's this skipping action that gives the flies of this family their name. If it all sounds rather disgusting, then you probably wouldn't enjoy one of the famous artisanal cheeses of Sardinia, in which bone-skipper fly larvae, rebranded as cheese-skippers, are an integral part of the gustatory experience. For a bone-skipper, fatty, proteinaceous cheese is like bone marrow; and one species has readily transferred its allegiance from one food source to the other. But don't worry, if you decide you don't want to eat the larvae, you can leave the cheese out for a while, and they'll eventually skip off. Alternatively, you can keep the cheese in an airtight container and they'll eventually suffocate—you'll know when by the silence, for they will have stopped rattling against the sides of the container. Easy.

The first bone-skipper flies arrived a few minutes after I settled into position, on a patch of fur-free grass about a metre from the carcase. By now I had become somewhat desensitised to the odours; but, all the same, it was not easy deciding whether to breathe through my nose or risk swallowing a blowfly by breathing through my mouth. (I chose the former.) The flies arrived as if from nowhere, but soon settled in on their chosen bones. For flies with such uncouth habits, they were rather lovely—much smaller than a blowfly and not much larger than a vinegar fly; bright shiny orange on the head, eyes and thorax, and dark blue-black on the abdomen and legs.

I didn't see any female bone-skippers, or at least none that I could discern. For the most part, the bones made the stage for a succession of males jostling for territory. Just as one settled in, another would pounce on it, and they would both disappear; only one reappearing to claim the vacated spot. On these flies, the top of the thorax bore a rearward extension like a triangular shield over the first part of the abdomen, and the shield was tipped with two long, stout bristles. I wondered whether these, and the shield itself, were really a means of protecting the abdomen from attack by other males, or perhaps by parasitic wasps. It seems as good a proposition as any.

A glance at my watch told me that I had been in the company of nature's undertakers for a good hour. With most of my ride home still ahead of me, it was time to move on. I'm sure I could still smell that carcase, or the memory of it, hours later, when I arrived home completely exhausted (I had battled an onshore sea breeze for the last forty kilometres). On closing my eyes in bed that night, I nodded off to images of bone-skippers and writhing masses of carrion beetle larvae. Carcase-watching, like cycling from Miena to Taroona on a mountain bike, is not for the fainthearted.

The frigid butterfly and the heartless wasp

There is a butterfly that makes the month of March its own, and that is the ptunarra brown. 'Going to see the ptunarra browns' has become a regular pilgrimage for me. The species in question waits until every other butterfly's time in the sun has passed before appearing; and that earns it my respect and admiration, and a place as butterfly of the month for March. The ptunarra brown is all the more impressive when you appreciate that it only lives at high altitude, where March nights often dip well below freezing. There are good reasons for its late emergence. One is that it gives the caterpillars the entire spring and summer to feed up, which is helpful for a coldblooded insect, because activity levels and growth rates are strongly tied to ambient temperatures. The other reason is that March often has the most stable weather, which is especially important for an upland butterfly.

Ptunarra browns derive their name from an Aboriginal word meaning 'frigid'. I take it that this alludes to their habitat and flight season, rather than to their temperament. They are only small, with a wingspan about the same as the width of two of my thumbnails (I know, because I once enticed one to sit and bask on my thumbnail). Their wing colours are also rather muted, and the impression they give when flying by, so low as to almost skim the grass heads, is of a grey or orange-brown blur. That's because, on the upper side, the male is grey-brown with cream-white flecks and blotches. The female is a faded orange-brown with smaller patches of grey-brown and cream. Each wing bears a black eyespot with a white centre and an orange outer ring. The underside is more strikingly and delicately marbled in silvery-white and brown, again with eyespots.

The usual destination for my ptunarra pilgrimage is Liawenee Moor. It's almost too easy—all you need to do is to pick your day well, park near the Lake Augusta Road turnoff, and look around. If the sun is out, the butterflies will be, too. If it's not, then you're out of luck. But at least there are other special insects to spot there—such as the inland jackjumper ant (a placid relative of Tasmania's most dangerous animal), the biting snipefly (not a common trait in snipeflies), and a grasshopper whose name translates (rather incongruously) as the Tasmanian Kosciusko-dweller. (The first species in its genus was described from specimens found in the alpine grasslands of Australia's highest peak.)

Liawenee is all well and good, but if you want to experience ptunarra browns in all their splendour, there is no better place than the Vale of Belvoir. I travelled there recently, one March, on my way (admittedly, a roundabout way) to the Tarkine. Relieved to be leaving the main road and tourist traffic behind, I had branched off into the Vale, down the potholed track to Lake Lea. There, I parked the van for the night, in the little trackside bay set aside for campers. The Vale is not far from Cradle Mountain, but it lacks the latter's rugged splendour, which attracts tourists, adventure seekers and bushwalkers from far and wide. The Vale's wide, sweet, grassy plains lie upon limestone and basalt, and are pleasingly hemmed in by part-wooded rocky ridges of sandstone and quartzite. It seems a place more suited to quiet contemplation or losing oneself among the butterflies. Come in February, and the Vale is awash with silver xenicas—close cousins of the ptunarra browns—as well as bright-eyed browns, shouldered browns, and mountain blues. Come in March, and you and the ptunarra browns will have the place to yourselves.

When I arrived, it was late afternoon and, though the day had been fine, menacing grey clouds had begun to block the sun's remaining warmth. They cast deep shadows across the moorland, muting the vibrant russet and ochre tones of the buttongrass and gone-to-seed rushes to a pale and insipid beige. I had arrived in high spirits, but they now began to sag as I considered the likelihood of becoming enveloped in buzzies, scoparia prickles and leeches, rather than butterflies. Nevertheless, I brewed a reviving cuppa and put my walking boots and gaiters on, while a stubbornly slow-moving cloud blanketed the sun. I decided that I would skirt the rainforest and scoparia scrub fringing the grassland first, to see if I could stir up any sheltering butterflies that way. Prickly scoparia is not a shrub that usually endears itself to bushwalkers; but, here, the individual bushes are neatly pruned by wind and wallabies, into scalloped glades and intimate dells that would do a landscape gardener proud. I advanced several hundred metres through this enchanted garden, but met with no butterflies. My doubts began to billow like the clouds as I wondered whether my equinoctial visit had been poorly timed. Perhaps the ptunarra browns' short flight season was already over.

Disheartened, I found myself turning around and making a beeline back towards the van, across the open grassland this time. I had only gone a few paces when the sun peeked out from behind the cloud. Almost instantaneously, butterflies started appearing, as if from nowhere. Not just ones and twos, but dozens of them! Within minutes, the entire grassland had erupted into a shimmering panorama of frenetic butterfly flybys. It was breathtaking. I had my camera at the ready, but none of the butterflies—all ptunarra browns—seemed to show even a remotest interest in settling anywhere for more than an instant. The males were quartering purposefully back and forth, and the females were mostly drifting in the same general direction as the breeze. With few flowers out, there was little to tempt them down to feed, other than a scattering of bright orange paper-daisies, which looked to have gone to seed. I figured that a camera wouldn't have been able to do the scene justice, so I returned it to its case and allowed myself to become fully immersed in the spectacle. Given the size of the Vale, I judged that there might easily be tens of thousands of ptunarra browns, all engaging in the same behaviour at this very moment. It was so wondrous that my spirits soared alongside them all.

It's no accident that the Vale of Belvoir is such a hotspot for butterflies. Much of it comprises one huge frost hollow, which is subject to such intense cold-air drainage that trees struggle to gain a hold. Such frost hollows are, presumably, the original natural habitat (in interglacial times, at least) of the ptunarra brown, the caterpillars of which feed on tussock grasses. Even so, the grasslands owe their current extent, structure, and composition to human management. For thousands of years before European colonisation, Lutruwita's First Peoples would have periodically and meticulously fired the grassland in a rotating or rolling patchwork, to encourage the sweet tussock grasses favoured by wallabies and wombats and to keep rushes, buttongrass and rainforest at bay. Europeans largely replaced fire with livestock. Cattle were driven up to the high plains over summer, where they enjoyed the same grasses once maintained for the benefit of the native grazers and their human hunters. For a while the Vale even hosted a dairy farm. In places where burning and cattle grazing ceased, the grasslands began to give way to tough buttongrass, scrub, and rainforest, and the ptunarra brown colonies sputtered out. In the Vale, patch-burning and cattle have become the allies of conservation—cattle are again summered there, intermittently; in low densities, and carefully controlled, to help keep the grassland sward open.

It was only when the sun went behind another cloud that I got a chance to see the butterflies at rest—frigid, you might say. I was lucky that I managed to follow a couple as they dropped down into the tussock grass, because, within a minute or so, it was as though there were no more butterflies. They had all secreted themselves away. The rule seemed to be: immediately alight on an elevated stem, close one's wings, then steadily inch one's way down, into denser vegetation. If the sun reappears, reverse the process *pronto*. Such stop-start, on-off behaviour presumably serves ptunarra browns well, given the prevalence of fickle weather in their chosen environment; but it makes photographing them a challenge.

The clouds were dissipating now, as they often do, in the uplands, towards the end of the day. Before long, the sun emerged again, and this time there were no clouds in line to spoil the show. Accordingly, the butterflies resumed their activity; however, over the space of the following half-hour, their behaviour changed. They became less frenetic and more inclined to perch and to bask; and, as the sun began sinking in the western sky, they took to basking on the western, sunlit sides of scoparia bushes. Sometimes my approach, and my long shadow, would spook them, and a cloud of butterflies would take flight. I soon learnt to approach from down-shadow. Thereby, I discovered all the observation and photo opportunities I could have possibly hoped for. Those greys, creams and browns, which had appeared a little underwhelming on the flighty butterflies earlier in the afternoon, now took on deeper, more saturated hues and contrasted beautifully with the equally saturated greens of the scoparia.

It wasn't long before an evening calm enveloped the valley. One by one, its residents prepared for the chilly night ahead: the basking butterflies closed their wings for the last time and crept deep into the bushes to roost, and I made my way back to the van to prepare my own roost. There is no evening chorus of birdsong at this time of year, but dusk progressed to the sounds of yellow-throated honeyeaters chupping and chooping mellifluously to one another among the snow gums, while yellow wattlebirds lowered the tone with their ludicrous coughs and splutters—like cats ejecting hairballs. The mountain soundscape would have been incomplete without the honking of black currawongs, and they duly obliged; and, somewhere deep in the adjacent rainforest, a pink robin jingled. The last avian word was had by a distant Bassian thrush, which performed a mournful solo piece, only just audible above the gentle rustle of foliage newly enlivened by descending, cold, mountain air. Out on the moorland, the bleating of Tasmanian froglets faded into the night.

Sadly, the scenes of marvel and wonder that played out before me at the Vale of Belvoir that afternoon are becoming a bit of a rarity; and the ptunarra brown is considered a threatened species. Habitat change, brought about either by a lack of active management, or by the wrong kind of management, is one of the drivers of the butterflies' decline. But they also face another insidious threat: predation by European wasps. These much-despised insects have only found their way into Tasmania in the past sixty years or so. Indeed, the *English* wasp, which is more tolerant of the cold and damp than its continental cousin, only turned up in the mid-nineties. Between them, they have rapidly become the bane of picnickers, forestry workers, gardeners, and bushwalkers. Although they're called social wasps, for humans they are anything but—they pester, despoil, and sting. What is less appreciated is their ecological impact. A mature wasp colony holds thousands of individual wasps, each of which has been reared on a protein-rich diet, comprising, primarily, the flesh of native insects and spiders. From spring to autumn, a colony grows exponentially and its demand for insect flesh ratchets up, peaking in autumn, when there may be thousands of larval mouths to feed. The adult wasp workers scour every nook and cranny within hundreds of metres of their underground carton nest, in search of prey. Roosting or basking ptunarra browns are an easy target, because they have evolved no behaviour for evading such an aerial onslaught. A wasp will approach from above, land on the unsuspecting butterfly, pin it to the vegetation, and snip off its wings; before transporting the hapless and now wingless victim back to the wasps' nest. They can even capture a butterfly in flight; and, earlier in the season, they will seek out caterpillars. There is no morality in the insect world—the wasps merely do what wasps must do to keep their species going. It is, nevertheless, distressing to witness their depredations on such innocent and innocuous-seeming butterflies; indeed, it's hard not to think of the wasps as heartless, callous, malicious, vindictive; maybe even evil.

I was pleasantly surprised, however, not to see a single European wasp at the Vale. Perhaps a baiting program is keeping them in check. At Surrey Hills, a vast rectangle of private land a little further to the northwest, which is largely given over to plantation forestry, the wasp problem is more severe. The remaining grasslands there are mostly protected, but the plantations are prime habitat for wasps. Where these abut grassland, the wasps take an immense toll on the butterflies. This problem is particularly severe after a plantation is harvested, as the stumps and roots are piled into windrows, in preparation for planting the next rotation. The windrows provide perfect nesting habitat for wasps. Fortunately, the land managers are now aware of the problem, so their practices have been changing, to give the vulnerable butterflies a fighting chance. But, without any easy fix, it seems that the butterflies' survival in the Hills will forever depend on intensive care. Death by wasp is a sad fate to befall such a tough yet fragile little butterfly; but in many ways it's symptomatic of an escalating mismatch between the demands that our modern lifestyles place on the land, sea and air, and the way nature would run things in our absence. By and large, we would not be missed.

© K.DAVIS 2023

April

Fine days, cold nights

Autumn is in full swing in April. Though tinged with sadness because of summer's passing, there is still much to enjoy. In most years, the stable weather continues well into the month and there's much pleasure to be had in simply sitting quietly in the garden, on a seaside rock, or on a woodland log, and soaking up the last of the sun's warmth without fear of sunburn and untroubled by marchflies or mozzies. Seek out a north-facing spot in the middle of the day and you may find yourself basking in warm sunshine, along with the last of the season's insects, which are, quite literally, in the autumn of their lives. But the days are shorter, and clear skies make for cold nights. Cold-air drainage soon brings a taste of winter to Hobart's western suburbs, which flank the mountain.

The sound of the season, for me, is that of little parties of silvereyes scouring the bushes for fruity titbits to fuel their imminent northward migration—most will overwinter on the Australian mainland, where they will mingle with silvereye populations that reside there year-round. Those from Tasmania will, however, retain a separate Tasmanian identity and appearance throughout the winter, and will feel compelled to return to breed here, come spring. I enjoy their company in the garden and admire their busyness and apparent indifference to my presence. They have done well out of European colonisation and enjoy blackberries and elderberries, as well as kiwis, figs, and late-season apples. Some even find our suburbs so enticing that they now stick around for the winter.

I have never really understood why daylight saving must end in early April, just when late afternoons begin ceding to darkness far too early. Until I investigated the history of this timekeepers' sleight of hand, I was under the misapprehension that daylight saving was a measure introduced to somehow bring winter's meagre hours of daylight more into line with human affairs—darker evenings being the cost we must all pay for lighter mornings and safer travel to school or work. But short winter days apparently don't enter into it—it's all about those blissfully drawn-out summer evenings. And we have a New Zealand entomologist to thank for getting the idea going in the first place: George Hudson, an avid collector of insects, as he had been since his childhood, in 1870s England.

In New Zealand, Hudson's interests flourished, and he became the leading authority on insects there. He published many books and papers, illustrated with his own exquisite artworks. He was proudly independent and deeply suspicious of government-run science, holding our old friend Robin Tillyard (who described Hardy's hardy nannochoristid) in particular disdain. Hudson also had astronomical interests, even going to the lengths of keeping sidereal time in his home-made observatory—sidereal time is a system that divides the year up into units based on Earth's rate of rotation relative to the March equinox. This allows practitioners to determine the position of stars in the night sky with greater precision than would be possible with the conventional twenty-four hour clock. Such precision is due to a sidereal day being just under four minutes shorter than twenty-four hours. (A full year, therefore, contains one extra sidereal 'day'.)

Perhaps it was Hudson's keen awareness of the artificial nature of conventional timekeeping, as well as his entomological interests, that led him to publish the succinctly titled paper, *On seasonal time*, in the 1898 volume of the *Transactions of the New Zealand Institute.* It was actually a revised version of an address Hudson had given to the Wellington Philosophical Society in 1895, in which he had made a case for putting the clocks back by two hours during summer—so that '*a long period of daylight leisure would be made available in the evening for cricket, gardening, cycling, or any other outdoor leisure pursuit*'. What Hudson didn't mention, at this time, was that his preferred outdoor leisure pursuit was collecting insects.

Hudson's shift work, at the Wellington Post Office, gave him some flexibility; but it didn't always allow him to go collecting, or to work on his illustrations, during the daytime. Clearly, he would always have been keen to make the most of what was left of a day after work hours. Amazingly (or perhaps not), Hudson's ideas caught on all around the world. Port Arthur, in Canada (now known as Thunder Bay), was the first jurisdiction to adopt daylight saving time, in 1908. Tasmania implemented daylight saving measures during both World Wars (to conserve scarce fuel supplies), and again in 1967 (to save energy during a drought that cut hydroelectricity generation). It began to implement daylight saving continuously in 1972, after a fourteen-year parliamentary campaign led by John Steer—that was, again, premised on the benefits of after-work leisure and recreation. Paradoxically (and sadly), New Zealand entomologists didn't get to reap the rewards of Hudson's proposal until 1974—28 years after his death.

Personally, I would prefer to see daylight saving measures applied year-round, and my desire becomes particularly strong in April, when after-work daylight is brought abruptly to an end.

Late-season blues

The community garden in Taroona sits on a gentle, north-facing rise and is, consequently, a great place for growing veggies. On weekends I spend a fair bit of time up there, tending our plot; particularly in April, when it's time to grub out the summer crops, dig in some fresh compost, and plant the winter greens. A few years back I was digging over the plot when I was distracted by a pale violet-blue butterfly that was fluttering around a neighbour's flowering broad bean plants. Yes, it's an odd time of year for broad beans, but they seemed to be doing well enough. I was less interested in the beans than the butterfly, though. There could only be one species that would exhibit such behaviour, and it's a real rarity in southern Tasmania: a pea blue. I parked my spade and hurried over to the bean patch, just in time to see the creature settle on a flower, wings closed. She wasn't after a feed of nectar; instead, she reversed into the flower and arched her abdomen until its tip contacted the base of the flower's style, and there she laid a tiny, mint-green egg. Enchanted, I stayed with the butterfly while she took a brief rest.

Pea blues win my vote (and it's my competition after all) for butterfly of this month. Though rare in southern Tasmania, they are widespread throughout the warmer parts of the Old World; from South Africa, north to southern Europe, and east to India, southeast Asia and much of Australia. At the northern and southern extremities of their range their appearance depends on immigrants, since they are continuously brooded by nature and have no hibernation strategy. They cannot survive cold winters as adults, or as eggs, larvae or pupae. Tasmania is usually beyond their reach; but in years when their numbers build up on the mainland, a few may venture south over Bass Strait. Sometimes they are not uncommon along Tasmania's north coast in late summer, and they may even maintain a toehold there throughout the year; but they are much less frequent visitors further south.

This was the second occasion on which I had seen one in Tasmania. The first was in midsummer, a few years previously, when one exhibiting remarkably similar behaviour had alighted on the flowers of a scarlet runner bean in our own home garden. April is a more unusual time of year for one to turn up, and it suggests that this second one, on the broad beans, might have been locally bred from a migrant much earlier in the summer. Furthermore, given that it was fertile and laying eggs, there must have been more than one around. This individual certainly had no notion of the futility of egg-laying in a broad bean plant in autumn in Tasmania. It might have been better off trying its luck at migrating northwards. I dearly wanted the egg to survive, but I knew it had no hope; and that was regardless of the fact that, only a few days later, the plant was removed completely. (It seems that they had been grown only for their value as nitrogen-rich green manure.)

This particular day was clearly a 'blue' day. Earlier on I had spotted another blue butterfly, that time along the foreshore track only a few hundred metres from the community garden. She, too, was in the mood for oviposition, and her choice of foodplant also left no doubt as to her identity: a saltbush blue. She was a very punctilious little butterfly, spending a good five minutes fluttering around the flower spikes before deciding that none of them was quite what she was after. She headed off over the lip of the low cliff towards the shoreline below, and I didn't see her again.

Unlike the pea blue, the saltbush blue (also known as the chequered blue) is a Tasmanian resident. Indeed, the south-eastern Tasmanian form is considered an endemic and threatened subspecies. On the Australian mainland, the same species ranges far inland, through saline saltbush country. In Tasmania it is entirely coastal because that's where its saltbush food plants grow. This one was interested in a grey saltbush, but she could equally have chosen seaberry saltbush, which is more common locally. I suspect she was a wanderer from elsewhere—perhaps across the Derwent at Ralphs Bay, as I don't think the species is permanently established in Taroona. Saltbush blues seem to have multiple broods throughout the summer; or perhaps their emergence is staggered and represents only a single brood. Either way, they first appear in late October, or November in some spots, and can still be going strong in March. April is really the end of their season, but I did once spot one in early June, in the dunes at Beaumaris.

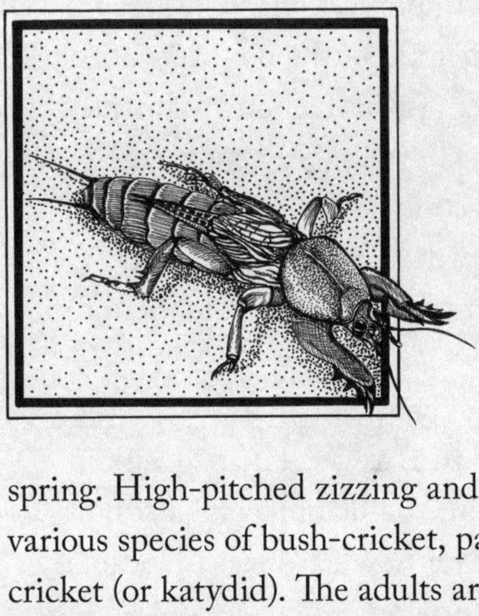

Trilling crickets, silent cicadas

April is also cricket season; that is, the season for trilling and chirping crickets, which have been maturing to adulthood since the spring. High-pitched zizzing and crackling from trees are the work of various species of bush-cricket, particularly the leaf-green garden bush-cricket (or katydid). The adults are almost impossible to spot, since they do very convincing impersonations of wattle phyllodes (the structures—

essentially flattened-out petioles, or leaf stems—that stand in for leaves in blackwood and other similar wattles). For all I know, other arboreal species may also be calling at a frequency higher than I can now hear, rather like bats. Intermittent, stuttering, shrill chirps at ground level, often from under bushes or in cracks in the soil, are usually from black field-crickets—stocky, bullet-shaped creatures, the females of which bear a long, sword-like ovipositor (an egg-laying device) at the rear end. The males' cheerful chirps are, for me, a welcome after-dark fillip on warm April evenings, before the nights turn chilly.

The calls that bring the biggest smile to my face are the continuous trills of mole crickets. These ungainly, soft-bodied crickets with loosely articulated bodies (they remind me of trucks with semitrailers) spend much of their lives underground. Like their mammalian equivalents, they have stout, clawed front legs for tunnel-digging, and much-reduced eyes. They have been feeding up on a diet of grass roots throughout the summer; and, by autumn, will have moulted into their adult form. Come April they will be past their peak, but spells of warm, damp weather will reawaken mate-seeking desires among survivors; and, because April is otherwise rather quiet, their calling becomes more obvious now. I can often hear them from home, trilling their hearts out among the cracked-clay grasslands of Taroona Park. The male mole cricket, whose stubby wings are no good for flight but perfectly designed for rubbing together to produce his trill, digs himself a chamber featuring a funnel-shaped opening to the outside world. There he sits and sings, and sings, and sings, in the hopes of attracting a female—who has the power of flight. His voice is a single-note purr, and though I described the chamber opening as funnel-shaped, it's cleverer than that: it's an exponential horn, designed to maximally amplify and broadcast his song to the outside world. The louder he trills, the more females he is likely to attract.

Mole crickets are thought to be the only insects that modify their environment to amplify their own sounds. If you've taken a walk in the park in the aftermath of autumn rain, you'll appreciate just how penetratingly loud their trilling can be; yet how hard it is to locate the source. The sound resonates and rattles around in your head, defying triangulation. Only when it abruptly stops do you know you're close to the cricket's burrow; but, by then, he has clocked your presence and retreated further underground, so you're unlikely to spot him. At this time of year, a mole cricket's singing is very dependent on temperature: a bit like a battery running out of juice. On chilly days his purring will sound more like a low mumble than a shrill trill.

Mole crickets are more often observed at night, when the females—the ones out searching for males—are attracted to artificial lights. They don't look like something you'd want to pick up, and I'd suggest you go with gut instinct and avoid the prospect of a painful nip and a squirt of cricket faeces. Incidentally, mole crickets are the sole prey of those marvellously metallic, bauble-like bluebottle ants that stalk the grasslands in summer. The 'ant' is the flightless female of a relative of the flower wasps, and she spends her days sniffing out mole cricket nymphs. She, too, has legs designed for digging, and when she senses a mole cricket's presence, she will burrow down with the intention of stinging it and then laying one of her eggs on its now-paralysed body. Her own larva will then gradually consume the cricket, eventually killing it. Mother nature can be callous, but only if we attribute human sensibilities to her.

I *want* to say that it's not only crickets whose calls can drive one to distraction in April, because April is also peak season for Tasmania's hairy-cicadas. And we all know what a din cicadas make, don't we? But, unless you're a female hairy-cicada, the chances are that you'll never hear the serenade of a male hairy-cicada—because rather than sing, he merely taps his feet on a plant stem and hopes that the vibrations will travel along the rest of the plant to an expectant female. Hairy-cicadas clearly haven't read the cicada instruction book. They don't sing; they emerge in the cooler months; they are nocturnal; and, as adults, they only live for about a day. Oh, and they're hairy; all the better to keep them warm on a cold autumn night. It's almost as though they have rejected all cicada traits and decided to be contrarians; but since their lineage pre-dates that of the true cicadas, they're more like a prototype from before cicadas learnt to sing under the heat of a summer sun. At least their nymphs are more conventionally cicada-like, spending years underground, sucking the sap from the roots of trees. And, like true cicadas, the fully grown nymph will one night scramble up the nearest tree trunk using its front legs like grappling hooks, before the integument across the back of its neck splits open and the adult cicada clambers out. Fresh nymphal cases on tree trunks in April are the best clue to the presence of hairy-cicadas, since they last much longer than the adults and don't hide themselves away by day. You can often find these 'exuvia', complete with an attached mass of white threads that look like fungal filaments—they are actually the everted tubes that once lined the nymph's internal air ducts.

The world hosts hundreds of true cicada species, but only two species of hairy-cicada. One is found in the cooler uplands of the southeast Australian mainland; the other is widespread across Tasmania, especially in higher-rainfall areas and at moderate altitude. Ours is as big as our largest true cicada, the red-eye. It has the overall cicada shape, and holds its wings like a true cicada; but its prominent, rounded pronotum ('neck') makes it look more like an oversized leafhopper. Many 'cicadomorph' leafhoppers tap their feet to communicate, and their shared ancestry with cicadas probably explains where our hairy-cicada got the habit. Like moths, hairy-cicadas are attracted to artificial light. Most of those that I have seen have been attracted to the ultraviolet light of a moth trap, but they sometimes congregate under streetlights or other outside lighting. Across most of the state, hairy-cicadas are mottled in various shades of brown; but in rainforest areas, the brown is replaced with beautiful, pale-green marbling. This is, presumably, because it affords them better camouflage amongst the foliose lichens that often bedeck the trunks of rainforest trees. I have only once seen the rainforest version, and that was on an island in a backwater of the Pieman River, just downstream from Corinna (during a Bush Blitz expedition). Transferred from a mossy log to my finger, it sat there looking bemused, and flatly refused to fly away. Eventually I had to prise its legs off, one by one, and return it to its log. It was a magical moment with a creature of great beauty. I would love to see another.

Where? Where? Wedgie! ... There! There! Beetles!

I challenge anyone, naturalist or otherwise, not to be moved by the sight of a soaring eagle. The boundless freedom of the skies, the might, the sheer majesty … I am lost for superlatives. Tasmania's wedge-tailed eagles are, on average, larger than mainland Australian birds and an awesome sight to behold, whether up close, perched on a roadside fencepost, or at a distance, soaring on broad, fingered wings above the skyline and showing off their signature wedge-shaped tail. Yet there are those for whom the sight of an eagle is repugnant and a call to arms. Many eagles fall victim to gunshot, even today, and others are poisoned, deliberately or accidentally. More die in collisions with powerlines or wind turbine blades, or are hit by vehicles while scavenging roadkill. Since so many are dying, it begs the questions: how many are living; and, is the death rate driving the threatened Tasmanian population towards extinction? These questions have been asked frequently, over the years; yet we still don't have answers. One problem is that, although we may know that eagles are present in an area, we don't know how detectable they are (eagles are territorial, so an established pair tends to stay put). This means that we can't readily translate sightings, or lack of sightings, into reliable population estimates.

You wouldn't think that eagles would be hard to spot, or to count. It's rare to drive any of our major highways without spying one or two soaring over some far ridgeline. On quieter roads, I quite commonly come across them in the middle of the road, feeding at roadkill. The Fingal Valley is a top spot for roadside eagles—I once saw a group of five loitering around the remains of a dead ewe. I doubt they were responsible for her death, but they certainly weren't going to pass up the opportunity to help dispose of her corpse. Associating with sheep has caused eagles a lot of trouble since sheep were first introduced, and still does so today—given the number of birds bearing the unmistakable signs of having been shot or poisoned that are delivered for rehabilitation or post-mortem examination every year. Everybody would know that this is illegal, but old habits, and old animosities, die hard; and it is difficult to prosecute the perpetrators. Considering how much the birds seem to like hanging around farming country, it makes me wonder whether European colonisation might have brought some benefits to eagles, notwithstanding the threats. Are their populations denser now than they would have been in times past, before settlers ringbarked, torched, and felled forests for their livestock and crops; and introduced rabbits? I don't know the answer, but even if there are more eagles around than before, there still aren't very many of them, and their fate is firmly in our hands.

In recent years, a citizen science project under the auspices of the Bookend Trust has sought to fill some of the gaps in our knowledge about eagles. *Where? Where? Wedgie!* taps into a growing public appreciation of eagles in order to shed some light on the detectability question. It also increases the proportion of the state being searched for eagles. Over a couple of autumn weekends (after the eagle breeding season), volunteers adopt one or more 4 x 4 kilometre squares across Tasmania and complete a series of ten-minute surveys at different points within their square, noting the location of any wedge-tailed eagles observed. Other birds of prey spotted are also recorded. Observations from the survey become part of a growing body of knowledge about Tasmanian wedge-tailed eagles, including their distribution, territory sizes, number of birds, and so on. We still don't have a definitive answer as to how many birds there are, but projects like this nudge us ever closer.

Chris and I accepted the *Where? Where? Wedgie!* challenge a couple of years back, and picked a square just south of Lime Bay, on the Tasman Peninsula. The corners of the square touched both Lime and Ironstone Bays in the east and Lagoon Beach and Slopen Main in the west. We figured we could survey the square over the course of a single day. Being April, the day was cool, but the weather was set to be fair and we had high hopes for a good day of gentle bushwalking in country that we wouldn't otherwise have thought to visit. Any eagle sightings would be a bonus. Taking the campervan meant that we could brew ourselves a quick cuppa on arrival, while we considered our options. With dry eucalypt woodland blanketing much of our square, our prospects for spotting eagles were going to be slight; other than where there were openings in the canopy, such as along roads and tracks, and where our square took in short sections of coastline. So, using the Lime Bay campsite road for access, we struck out on foot in various directions from it, to try and cover likely spots within our square.

Having surveyed some points in the north (no wedgies, but one brown goshawk, and a glimpse of a sea-eagle), we turned our attention to the southern part of our square, which lay within the Coal Mines Historic Site. Two hundred years ago, convicts from Port Arthur were put to work there, digging coal from a series of deep tunnels beneath our feet. Today, beyond the complex of prison buildings, much of the land is again wooded, with a mixture of wattles and various dry-country eucalypts. Somewhere near the Airshaft (for the tunnel below) we stopped for our first ten-minute eagle observation in this area, selecting a spot where a recent treefall had opened the canopy so that we could look skywards for soaring eagles. Ten minutes came and went. Various birds passed overhead, and many others called from the surrounding bush; but we saw no eagles. With time on our hands, now that our observation period was over, Chris wandered off to read some nearby interpretative panels, while I turned my attention to the fallen eucalypt. It still bore remnants of foliage, and its bark was more or less intact, suggesting that it had probably come down back in spring. I knew from experience that a fallen tree in its first year on the ground can be a magnet for beetles, because the outer wood is still sappy and nutritious, and the bark still offers some protection from the elements. I call it the log's 'Goldilocks period'—not too fresh, not too old. Under the cover of barkness, the nutritious sapwood would be fermenting nicely as yeasts and other microorganisms got stuck into it. They, in turn, would be irresistible to sap beetles; and they would also soften the wood for borers and a range of other beetles, including predatory species intent on feeding on the sap-feeders and borers. All being well, by the end of its first summer on the ground, a fallen tree could be teeming with beetle life.

With these thoughts in mind, I downed my binoculars, dug my closeup glasses out of my rucksack, and began carefully peeling back sections of bark to see who was hiding there. I savoured the characteristic odour of woody fermentation with its hint of vinegar and alcohol. Out tumbled a couple of tiny little sap beetles, almost immediately; and they were soon followed by several others, backing out of the sapwood in which they had buried their heads to feed. I picked them up and tipped them back into a nearby split in the sapwood. The bark was loose enough that it sloughed off in hand-sized chunks, and there were beetles under every chunk. I didn't want to oversample (my euphemism for local habitat obliteration), so I restricted my attentions to just one of the several fallen limbs. Besides the sap beetles, there were three species of curiously flattened silvan bark beetle, as well as various species of rounded fungus beetle, cylindrical bark beetle, lined flat bark beetle, minute hooded beetle, minute brown scavenger beetle, minute bark beetle, silken fungus beetle, rove beetle, longhorn beetle, bark-gnawing beetle, ground beetle; and—the awkward one with no common name—ulodid beetle. You can tell by their names that most of these beetles were exceedingly small; in fact, this is more typical of beetles than you might imagine, as the median size for a beetle, globally, is somewhere between two and three millimetres. I think I counted twenty-two species in the space of my twenty minutes of mindfully poking around in that fallen limb. It more than made up for the lack of eagles.

I do occasionally stop and wonder what it is about poking around for beetles that brings me pleasure. Other than the total absorption of mind, which can be immensely pleasurable in its own right, I think there's pleasure in the discovery, as well as in the knowing. If I hadn't known to expect them, and hadn't known a bit about the beetles that I was finding, I can imagine that it would have seemed like a completely pointless thing to do; indeed, even the thought of looking for beetles probably wouldn't have occurred to me. By the same token, I suppose I wouldn't think of looking for lichens on that same fallen limb, because it's not a subject that I know anything much about and have no way of unlocking the lichenological secrets that, for all I know, might be staring me in the face. I am glad, though, that I can look at a fallen tree and see more than just a fallen tree. I can see beetle food; fungus food. In due course, it might be hollowed out by fire; later it might provide refuge for an echidna, a tiger snake, a bandicoot, quoll, or devil. Maybe the bandicoot might one day fall prey to a wedge-tailed eagle.

Chris and I still had a couple more ten-minute observation periods to complete, though; so we pushed on through the woodland, eventually emerging at the southern end of Lagoon Bay, in time for a late lunch and our next observation. No sooner had we sat down on a rock, and started the timer, than a magnificent immature sea eagle coasted by, right in front of us. It glanced at us disdainfully as it continued on its way up the beach, where it spooked a group of loitering Pacific gulls. We followed the eagle through binoculars, until it disappeared from our sight, then resumed scanning the skies for signs of wedgies. For several minutes there was nothing at all; but then, in the ninth minute, we trained our binoculars on the summit of Sloping Island, some 1500 metres out to sea. There, at last, we glimpsed the silhouettes of not one, but a pair of wedgies. They were hanging in the wind and slowly quartering back and forth over the scrubby grassland, perhaps in search of rabbits. A minute earlier or later we might have drawn a blank, but there was no mistaking them, not even at this distance.

After the thrill of the eagle sightings, we enjoyed a few minutes of quiet contemplation and a spot of wave- and cloud-watching. Then, since we still had one more ten-minute slot to fit in, we retraced our steps back through the woodland to the road, drove around to Slopen Main, and then marched up the beach and back into our square. Pleasingly, the area in question has since been acquired by the Tasmanian Land Conservancy, which will ensure its continued management for nature conservation.

By now the sun was low in the sky, casting long, strangely deformed shadows of the two of us as we scrambled up the dunes behind the beach and stood on the sandy crest, surveying the rushy wetland in front of us. A couple of chestnut teal erupted in panic as we raised our binoculars to our eyes, and I wondered if they were anticipating the sound of gunshot. They landed a bit further away and resumed their quiet dabbling around the fringes of a rushy pool. Small parties of starlings wheeled in from across the bay, performing perfunctory aerial tours of inspection, before settling in lines on the dead branches of a black gum on the edge of the rushes. A distant kookaburra gave a half-hearted chuckle, and a lone pied oystercatcher peeped from the beach behind us. No eagles or other birds of prey upset this scene of evening calm and contentment, and I was almost glad. We, too, appreciated some calm to bookend the day.

By the time we had walked back to the car for the long drive home, the sun had set behind a distant Kunanyi, setting the sky and sea aglow. My thoughts turned to roosting eagles—where would 'our' Sloping Island eagles be roosting tonight? I imagined them perched, side-by-side, on the sturdy branch of a wind-sculpted Monterey cypress, surveying the scrubby grassland in front of them, and its attendant rabbits, and contemplating breakfast.

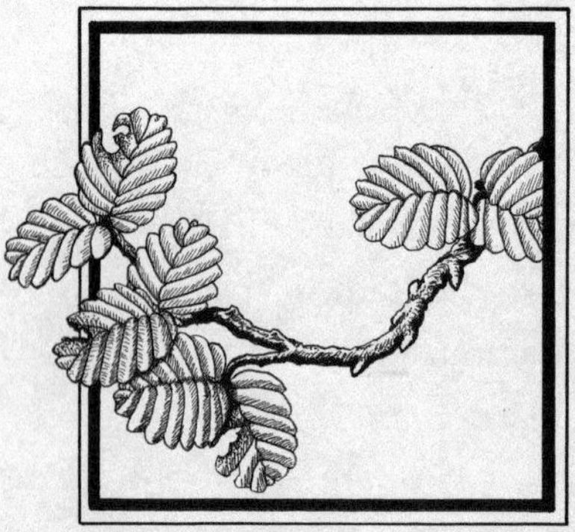

Fagus

Among Tasmanian urbanites (and I include myself here), 'going to see the fagus' in its seasonal finery is an autumn tradition, indeed almost a ritual. The fagus in question is a much-branched montane shrub, or sometimes a small tree, also known as deciduous beech or tanglefoot. *Fagus* is actually the genus to which northern beech trees belong, whereas ours are in *Nothofagus*, which (politely) translates as non-beech, though we tend to call them austral beeches (in fact, the genus has now been split up into more narrowly defined entities based on DNA). The scientific name is more honest than the common, given that the forty-three species of austral beech are in their own family, albeit in the same larger taxonomic grouping as the northern beeches (along with oaks, walnuts, birches, chestnuts and, peculiarly, casuarinas). Most of the austral beeches live in South America, New Caledonia, New Guinea, and New Zealand—a classic Gondwanan distribution. One species, known somewhat incongruously as Antarctic beech, is endemic to northern New South Wales and southern Queensland. Tasmania hosts two species:

myrtle beech (also found in Victoria), and fagus, which is endemic to the island. The latter is our only native deciduous tree, shedding its leaves in autumn and completely shutting up shop until the onset of spring. It's this deciduous habit that draws flocks of visitors to the mountains on fine April days.

There are rather few easily accessible places where you can 'go to see the fagus', but one of the best is Mount Field. Here, you have two options: early in the fagus season you can hike from Lake Dobson up to Tarn Shelf, at an altitude of about 1200 metres, where deciduous beech forms extensive patches of scrub woodland in fire refuges among the dolerite boulders. Or you can wait a couple of weeks for the season to progress, at which point you can walk with the hordes among taller stands of deciduous beech right next to the Lake Dobson Road and around Lake Fenton, at much lower altitude. The roadside atmosphere can be almost carnival-like, complete with gaudily clad and very excited children and an abundance of selfie sticks. If that's your style, who am I to change your mind? But I recommend the Tarn Shelf walk—the reward is so much sweeter for the effort; and it really is a lovely walk, even if you find that you've timed your visit poorly and the fagus hasn't yet 'turned'. The first leg, following the shore of Lake Dobson, passes through enchanting woodlands of snow gum, and groves of giant pandani plants (renowned for being the world's tallest heath plants). The pandani look splendid at any time of year, with their twisted trunks and twirling, strap-like foliage. Incidentally, their common name is a shortening of the scientific name that simply means 'pandanus-leaved', in allusion to the superficially similar foliage of the tropical screw-pines. But to see the fagus on this walk, you need to climb higher, up to the alpine moorland, and then traverse along boardwalked sections of open, rocky, and boggy ground before descending towards a series of tarns along Tarn Shelf. Fagus swathes the steep, glacier-gouged slopes behind the tarns, and if your timing is good, the whole hillside will have erupted in a splendorous, radiant tangle of golden foliage. From the tarns you can either continue and complete a rather long loop (which can, potentially, also take in Lake Fenton); or return the way you came.

Up close, the little corrugated and serrated leaves of fagus may reveal further hues of orange and red, and present endless photographic opportunities. They are undoubtedly beautiful, but my pleasure is tempered by the realisation that they are only putting on this show because the plant is withdrawing life support from the leaves and preparing to discard them. The colours are a last hurrah, once the life-bringing chlorophyll has gone and before the leaves' connection with the living plant is severed once and for all. Seen in such a light, a trip to the fagus sounds rather morbid—a nature equivalent, perhaps, of the dark tourism that brings people to places of past horror and despair, such as Port Arthur. As it happens, Mount Field's name is also on the dark side, in that it honours New South Wales colonial botanist and romantic poet Barron Field. This sounds admirable enough until you learn that the man was also a judge, in which capacity he wrote the concept of *terra nullius* into New South Wales law, setting the stage for two centuries of Aboriginal dispossession and marginalisation. A bracing walk, in cold mountain air, amid splendid alpine scenery and wondrous subalpine woodland, might connect us to something more enduring and uplifting than the veneer of human ideas associated with the mountain's English name. This, for me at least, counts for much, much more than even a 'viewing of the fagus'. And if, while you're there, you get down on your hands and knees among the fallen fagus leaves, who knows, you may even find yourself a southern snow-flea—a hardy little montane minibeast that I will introduce you to next month.

Grounding

Life, even our privileged, First World lives, can be demanding—so many things that need doing, relentlessly one after the other or all at once. Sometimes the pressure is extrinsic: *I have to do this now or I'll feel bad for messing things up for someone else.* Sometimes the pressure is intrinsic: *I want to see this through to the end, even though it doesn't really need doing now.* We all face these pressures, and to a greater or lesser extent, we all succumb to them. But we don't have to be the passive victim of circumstances, since our emotional responses are actually *within*, rather than beyond, our control. The tricky part is giving oneself the mental space to appreciate this. The rest follows more easily.

I'm a latecomer to mindfulness in the formal sense of the word. But I think it's something that I've practised, almost instinctively, for much of my adult life, which is when we all tend to need it the most. If I'm feeling tense, anxious, angry, overwhelmed, or untogether, spending time in nature helps sort me out, reining in the negativity and putting me back together again. It may be a walk in the bush or on the beach, or just sitting quietly somewhere vaguely natural. Initially, my mind is full to bursting with whatever is bothering me; but, before long, I start to notice the sights, sounds and smells of my surroundings. Our sense of vision dominates, and it's usually something sighted out of the corner of one eye that first draws me in. *Are those Tasmanian or brown thornbills skulking in those bushes? Did I just see a cormorant pop up for an instant among the waves over there? Is that mud-dauber wasp making a nest under that rock overhang? Is that a threesome of mating kelpflies on that rock? Was that a meadow argus or a painted lady butterfly flitting over the hedge? Look at all those ants drinking from the extrafloral nectaries on that wattle foliage.* I gradually become aware of the sounds of nature next: the quiet contact calls of the thornbills; the lapping of waves against the rocks; the distant mewing of kelp gulls; the wind in the trees; the buzz of wasps and bees; the whine of a mosquito. And the distinct smells, if not of thornbills and insects, then of seaweed, guano, water rat, fungus, leaf mould, flowers, grass, foliage, decaying wallaby. Observing nature is all about being in the moment—quietly, inquisitively but unobtrusively; not overthinking things; just bathing in the here and now.

Those same tools—which may need to be relearned—can be applied equally to observing oneself. This may or may not take quite some time, but if one has already entered the zone, then time (other than now) has little meaning. Closing my eyes, which helps me avoid distractions from this dominant sense, I begin to notice my own sounds and smells—

internal gurglings, breathing, nasal whistling, sweating; the pulse in my neck; the heartbeat in my chest. The tingling of the cool breeze on my lips; the flow of air over the hairs on my arms; the pressure of uneven ground on the soles of my feet. The tension in my muscles. And then I realise that whatever negativity I had felt has dissipated. I feel grounded again.

As my research career at Forestry gained momentum, there came a point when a walk in nature no longer calmed my chattering mind at night. A newly arrived intern student, who was working with me on modelling the dynamics of dead wood (not as oxymoronic as you might imagine, as we will see, next month), mentioned that he had benefitted greatly from learning and practising Vipassana meditation. As it happened, a ten-day course was shortly to be offered at a retreat just north of Hobart. With the support of my employers, I signed up; and, one calm April afternoon, Chris dropped me off at Dromedary, for the start of the course.

I can honestly say that, at the time, the course was the hardest thing I had done in my adult life. For most of those ten days, we observed 'noble silence'; avoiding eye contact with our peers, and diving deeper and deeper into meditation practice, all the while kneeling on our mats in a darkened hall for several hours each day. It was enough to tip some people over the edge. I stuck it out for the duration (just), by which time my meditation abilities had become good enough to see me through many a dark day (and night) over the years ahead. What got me through, though, was perhaps not what our sincere and well-intentioned instructors had in mind, since we were there to learn how to behave with equanimity towards all internal and external sensations and stimuli. I could never quite tune out the sounds of birds calling in the eucalypts outside the meditation hall; and when they came to me, I welcomed them into my consciousness. Then as now, they grounded me and led me into a safer mental space. I am brought straight back to that time whenever I hear the single-note, plaintive call of a yellow-throated honeyeater, since they were my daily companions and reminders of the world outside the meditation hall; likewise, the wheezy cries of a masked owl, to which I nodded off into a trancelike, hallucinogenic semi-sleep, most nights.

Learning how to meditate to pep up your mental health doesn't have to be as hardcore as a ten-day retreat. Different styles require different approaches and suit different people in different environments. For me, I've learnt that meditative, contemplative, immersive time spent in nature trumps all.

©K.DAVIS 2023

May

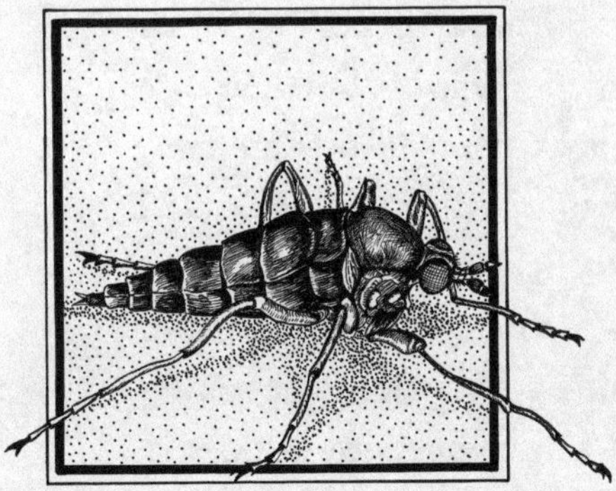

Frosts and darkness return

The thing I like least about winter is the short days. May is when this really hits home. It's dark when I get up to go to work, and it's already dark before I head home again. Weekends, with all their possibilities for nature encounters, are greatly cherished. But nature seems to recognise and accept winter's arrival before I am prepared to do so, and the pickings, when I do get out there, are inevitably lean. Spells of settled weather are not uncommon, though, and a mild May day can be as warm as a cold summer's day. Yet the warmth lasts for only an hour or two, around the middle of the day, and temperatures plummet by nightfall. Frosts are likely by morning, at least towards the end of the month; and the chill persists longer and longer each morning, as the sun's arc lowers in the northern sky.

Frosts and cold kill off the last of autumn's butterflies, so much so that finding a butterfly of the month for May is challenging. But, taking a leaf from September, I thought I would offer you a caterpillar instead. This one belongs to the Tasmanian race of the chaostola skipper, whose common name is merely a repeat of its scientific name. It may have meant something to Edward Meyrick when he named it thus, in 1888, but it appears rather unilluminating to me. The mainland races of the same species go by the name of heath sand-skipper.

I find that name much more informative because, throughout its range, the butterfly is associated with sandy heathland and open woodland, in which its larval foodplants, thatch and slender saw-sedges, abound.

The chaostola skipper is considered threatened with extinction in Tasmania, and is seldom seen during its October-November flight season—not merely because of its overall dark-brown appearance and small size, but also because of its rapid, low-to-the-ground flight, its frequent, extended periods of inactivity, and the general rarity of the adults (even in favourable habitats). However, a few years ago, it was realised that colonies could be discovered much more readily by looking for the caterpillars; or, rather, their silk-lined tubular shelters, which are fabricated from the lower sections of saw-sedge leaves. Furthermore, unusually, the caterpillar stage of chaostola skippers lasts some twenty months, meaning that they may be found at any time of the year. May is as good a time as any, since there are few other distractions.

One classic spot for chaostola skippers is the Peter Murrell Reserve, near Blackmans Bay. It was here, on a training course alongside some forestry and conservation practitioners, that I first honed my skipper-shelter detection skills. Searching requires finding a suitable patch of sandy ground hosting sparse clumps of saw-sedge, then teasing the clumps apart to examine the bases of the sheaths. One that is rolled into a tube is worth investigating further. If there's a skipper caterpillar in residence, it will be head-down. Confirmatory features include white bristles on the head, and a yellowish body, other than the first segment behind the head, which is red. To be honest, it's not a very striking caterpillar, but a find is a find. Once you know what you're looking for, it becomes an unnecessary disturbance to peel open the shelter, and it's best avoided for the sake of the resident caterpillar.

The known range of the chaostola skipper has been greatly extended, thanks to larval shelter searches. From a few fragmented colonies in the Hobart and Channel area, it is now known to occur south to Ranelagh, and north, up the east coast as far as Wukalina-Mount William, and across to Bridport. While still considered threatened, its future seems a little more secure than it once did.

Few insects choose to emerge as adults in May, though there are certain gnats—members of a family wittily named ' winter gnats'—which do so. They look like small mosquitoes, but they don't bite, preferring to dance instead. They're an unpredictable bunch. I've been out looking specifically for them, in what I thought would be high-quality habitats, only to return home and find them in the garden. Likewise, the only one I found on a May camping trip was pummelling its tiny little body against the inside window of the campervan one morning.

Then there is the wingless soldierfly. Only the female lacks wings (she's more a 'walk' than a fly), but because of this trait she's the one most likely to be spotted, dragging her oversized, greyish-brown, egg-filled abdomen across the lawns of suburbia, in search of a suitable place to scatter her eggs. Her larvae feed underground on grass roots, and they're quite happy among those of European grasses; so she is usually spoilt for choice. Somehow, our endemic Tasmanian wingless soldierfly has found its way to New Zealand, despite her limited dispersal abilities. There, European grasses blanket the land from horizon to horizon in many places, offering wingless soldierflies endless possibilities that their ancestors could only have dreamed of.

May is peak season for the Bridgewater Jerry, the affectionate name given to the fog that Hobartians sometimes awake to, on calm autumnal or winter days. The phenomenon depends on a temperature inversion, in which cold, damp air from the upper parts of the Derwent Valley gently sinks, snaking its way down the valley towards Hobart, where it mingles with air that has gained (relative) warmth and extra moisture from the sea. The fog may extend from Bridgewater, on the city's northern flanks, right down to Taroona, and sometimes well into Storm Bay. The fog banks may be only a hundred metres deep, but that's enough to swathe the city and its lower-lying suburbs in chilly dampness for a couple of hours, until it all dissipates under the warmth of the rising sun. A similar phenomenon may envelop the Huon Valley, though they don't call it 'Jerry' in those parts. The origin of the name is still not clear. It appears in James Hardy Vaux's *Dictionary of Flash Language* (convict language used to befuddle their superiors), compiled in 1812. It may once have been Londoner thieves' slang for fog or mist—the kind of weather that was good for a spot of covert thievery.

For me, May is the month to wind down. The garden is already semi-dormant, needing but an occasional weeding; and there is no pressing case for rushing after this bug or that. It's a time for reflection, and for getting on top of indoor jobs. In my working life, too, it has always been a time to take on projects that aren't dependent on summer's warmth and abundance, or to process samples, identify specimens, analyse data, write papers. Some of the vignettes that follow aren't specifically about May, but they reflect the things that I think about and do at this time of year.

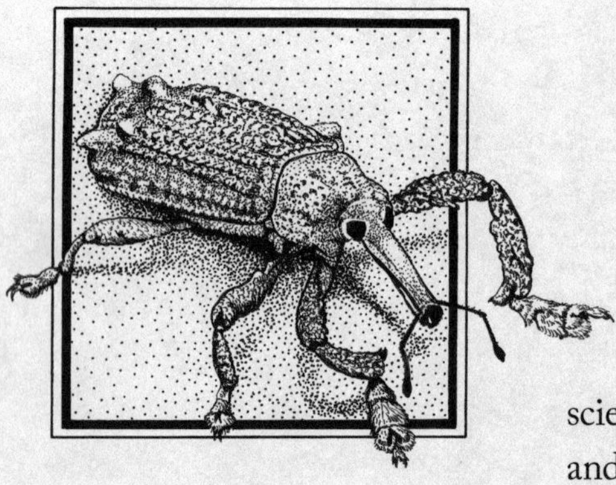

Working in the lee of Lea

Despite the passion of a handful of invertebrophiles, invertebrates aren't generally seen as sexy. Neither is taxonomy—the science of formally describing species and elucidating their evolutionary relationships. Invertebrate taxonomy is, therefore, doubly unsexy. Working out what's what, what's new, what's related to what, what's found where and what has already been described by someone else, requires near-heroic levels of persistence and an almost pathological attention to detail. It almost goes without saying that it doesn't suit most personality types. Taxonomy isn't so much a job as a vocation, a dedication to a very particular way of life. Much may be sacrificed along the way—mental and physical health, opportunities for career progression or a higher salary, and relationships with friends and family. It doesn't have to be this way, but it often is. Yet the legacy of a competent taxonomist may endure for centuries, through species described and the designation of type specimens. Few taxonomic papers are published in today's high-profile journals, because these specialise in cutting-edge discoveries and rapid turnaround. Leading researchers in other disciplines, who publish in the likes of *Nature* or *Science*, are rewarded with kudos, impressive citation indices and successful grant applications. But ask today's groundbreakers how often they rely on research published even a mere decade ago, and many will respond that such research is no longer relevant, and certainly not worth citing.

Taxonomy is different. It's the antithesis of flash-in-the-pan science. A good taxonomic paper will be as relevant in a hundred years as it was on the day it was penned. In the taxonomic literature, every mention of a species binomial (its scientific name) is meant to be followed by a reference to the species authority—that is, the author(s) of the published paper in which the species was first described. Thus, a minuscule species of Tasmanian wood-midge named in my honour by the expert in this group, Mattias Jaschhof, would, correctly, be referred to as '*Peromyia grovei* Jaschhof, 2010'. From 2010 onwards, anyone wanting to learn more about the species, including its description, will know that they can find what they are looking for in a thrilling page-turner by Jaschhof entitled 'New species of *Peromyia* from Tasmania, with a revision of Schiner's and Skuse's types of Australian Micromyinae', published in Volume 60 of the journal *Beiträge zur Entomologie* (Berlin). In any given year the number of people reading Jaschhof's 22-page paper (let alone those citing it) will be minuscule, but the cumulative number of citations could eventually exceed that of a paper in *Nature*, because its content will remain taxonomically relevant forever.

The trope about 'standing on the shoulders of giants' may be more germane in taxonomy than in any other scientific field. I have mentioned a few of those giants, in our Tasmanian context, but there is one whom I regard as the king of them all. What George Hudleston Hurlstone Hardy was to flies, his contemporary, Arthur Mills Lea, was to beetles; only all the more so. In his lifetime, Lea described 5,432 new species of beetle! Nearly all were Australian, and many were Tasmanian. Such is Lea's stature and prodigious taxonomic output that, rather than having any ambition to stand on his shoulders, I am more than content to merely work in his shadow; Lea's lee, you might say. With a couple of beetle-minded colleagues, I recently compiled and published an updated taxonomic checklist of Tasmanian beetles, illustrating it with examples of each beetle family from the museum collections. It was quite an undertaking—the list comprises close to three thousand species, which are either currently known from Tasmania or have been reported from the state in the past. Tellingly, one in five of these was described by Lea. No other beetle taxonomist, before or since, comes near to his contribution.

Unusually, for his time, Lea was a homegrown Australian entomologist, born in Sydney in 1868. His childhood fascination with insects led to a career with successive colonial or state government entities responsible for agricultural entomology, beginning in New South Wales, then Western Australia, then Tasmania, then South Australia. Tasmania is where his taxonomic expertise first came to the fore. Here, from 1899 to 1911, Lea held the position of Government Entomologist. His applied research on codling moths helped to turn around the fortunes of the apple industry (he found that arsenic sprays effectively reduce infestations), before turning his attention to a range of other insect pests of crops. In his spare time, he accumulated a vast and remarkably comprehensive collection of Tasmanian beetles and began describing many of these as new to science, in a series of papers, often underwhelmingly titled (along the lines of '*Some further Tasmanian beetles*'; '*Additional beetles from Tasmania*', and so on). Lea's species descriptions could be rather underwhelming, too—for example: '*smaller and more finely marked than the last*', or, '*like the previous species, but stockier and colouration more pitchy*'. Okay, I'm exaggerating slightly, but, by today's standards, his descriptions were a triumph of brevity and imprecision. If you had specimens of all the related species in front of you, the differences among them might be obvious; but if you were starting from a position of relative ignorance and a paucity of comparative material, then you would probably be stymied.

Regrettably for Tasmania, Lea was headhunted by the South Australian Museum, decamping to Adelaide in 1911. He took much of his collection with him, hindering further local efforts to document the Tasmanian fauna. That said, he left a tantalising series of voucher specimens with the Tasmanian Agriculture Department, whose current incarnation, Biosecurity Tasmania, continues to look after them in their

New Town laboratories. Further specimens were left at the Tasmanian Museum and are currently in my care.

Lea's Adelaide position allowed him to focus more fully on his taxonomic work. His productivity was extraordinary, and it is there that most of his type specimens (the ones that were the basis of his new species descriptions) now reside; even though many represent Tasmanian beetle species collected during his twelve years in the state. Lea was still actively publishing up to the time of his death, in 1932.

Weevils—those comically long-snouted insects that comprise about a third of all known beetle species—were a particular preoccupation for Lea. In South Australia, at the time, there was an entity known as the South Australian Weevil Commission; Lea was, of course, a member. I wonder whether he was its instigator, too, since I can't think of too many other places that have had their own Weevil Commissions. To be fair, the Commission was focused on how to deal with weevils that were a pest of stored grains, rather than on weevil biodiversity *per se*. Perhaps I am not alone in dreaming of a revival of the Weevil Commission, and desiring a more prominent role for weevil taxonomists. A few years back, David Yeates, who heads the Australian National Insect Collection, in Canberra, wrote a satirical online piece entitled '*A dream of Invertebrate Utopia*'—it imagined an alternative Australia, in which every State and Territory had a well-funded 'weevilarium' (much to the chagrin of botanists, who lamented the lack of equivalent institutions—herbaria—for their chosen taxonomic interests). David's point was that there are at least as many weevil species in Australia (and the world) as there are vascular plants, and their economic significance is every bit as great; yet the work of describing and keeping tabs on this immense fauna is relegated to a handful of taxonomists and to collections nestled within collections that receive far less funding and general recognition than our state herbaria. Neither David nor I begrudge the resources allocated to herbaria; it's just that there are many other taxonomic groups equally deserving of our attention—weevils are just one of them.

I have spent a lot of time looking at weevils. They come in all shapes and sizes, and with their long snouts, well-padded feet and bulging eyes, many look rather endearing (at least to one who has gazed at more weevils than hot dinners, over the years). Many weevils are extraordinarily beautiful, adorned with dense swathes of iridescent scales and bristles, which form complex patterns and shapes. In some instances, the scales make certain weevils stand out like sore thumbs; but more usually they break up the weevils' outlines and render them well-nigh invisible in their natural environment. A reliance on scale patterns for distinguishing one species from another has drawbacks though, because scales progressively flake off throughout a weevil's lifetime, when it is out in the elements. Like alopecia or pattern baldness on a human scalp, scales aren't shed entirely randomly, either; and each successive wave of loss may deceptively alter the pattern exhibited by those remaining. Eventually, all the scales may rub off, and a bald weevil of one species, I have found, can look very much like that of half a dozen other related species.

In my Forestry work, weevils were one of three beetle families (along with ground beetles and rounded fungus beetles) that we homed in on as study groups for exploring the ecological implications of silvicultural alternatives to clearfelling in wet eucalypt forest. We chose them because of their sheer diversity, as well as because many are associated with either leaf litter or dead wood on the forest floor, making them collectable using standard techniques such as pitfall trapping. Many weevils are also flightless, meaning that their limited dispersal capabilities might make them vulnerable to local extinction, should we fail to get the management right.

Having settled on weevils for our research, we had to learn how to tell one species from another in thousands of samples; so we built a fine collection of pinned specimens, which eventually amounted to over 450 weevil species and over twenty thousand individual weevils. Even though I made multiple visits to the Australian National Insect Collection in Canberra, as well as to Lea's New Town collection, to try and match our weevils to the identified species held there, we only ever found names for fewer than a third of them. Nearly half of those had been described by Lea. As for the unnamed two-thirds, all we could do, at the time, was allocate codenames to them. Today the Forestry insect collections are incorporated into the Tasmanian Museum collections; and, with the help of some exceedingly talented volunteers, we are gradually working out what some of the codenamed species really are. Most are likely to represent undescribed species. If only Lea were around still to continue the work that he started over a century ago! Perhaps it's time I lobbied for that Tasmanian Weevil Commission.

Our tale of taxonomic intractability and woeful weevil wherewithal eventually drove us to actually apply for, and, in due course, receive a small taxonomic grant. The funding enabled me, and my weevil-minded colleague Lynne (whom we encountered on Ben Lomond, in February, and who is a co-author of our recent beetle checklist), to spend a week at the South Australian Museum, in Adelaide. There, passing a photo portrait of a serious-looking and luxuriantly moustached Arthur Mills Lea, we came to the beetle collections, where we examined hundreds of Lea's type specimens and matched many to Tasmanian Museum specimens that we had brought with us. We felt like kids in a lolly shop, opening drawer after drawer of beetles. Each held the prospect of resolving the identities of one or more of our long-standing mystery species. Each day was spent engrossed in our work, breaking only briefly for lunch. After an early dinner, we would return to the apartment and resume our work, variously processing photos, reading Lea's papers and making notes. When I closed my eyes at the end of each evening, I saw swirling images of franken-beetles: strange chimaeras featuring a jumble of characteristics of the various species that I had examined under the microscope earlier in the day. My dreams, too, frequently remained in worlds of entomology. It was an intense week, but we returned to Hobart replete with scores of updated identifications, hundreds of photos and pages of notes. Yet we had really only scratched the surface, on this initial visit, so we resolved to repeat the process for a further suite of unidentified specimens, at some later date.

Yet, for all of Lea's achievements, the more we got to know his work, during that week, the more we came to rue his rather idiosyncratic approach to taxonomy and the lack of any formal peer-review process for his publications. The more we read them, the less sense they tended to make. We had placed him on a pedestal, but now we found its foundations crumbling. Sometimes Lynne would find that his descriptions contradicted the character states that she could see on his designated type specimens—where she saw a toothed femur, he might describe it as untoothed, for example; or where she saw punctured striae on the elytra (wing-cases), he might describe them as impunctate.

Furthermore, Lea's approach to describing weevils often involved rubbing off most of the scales, to reveal the sculpture of the wing-cases beneath; yet these features were often not as informative at the species level as he had believed, so he had effectively desecrated his own type specimens, rendering them useless for comparison with current, un-rubbed specimens. Lea's specimen labels, too, generally lacked critical details as to where and when an insect was collected (beyond, say 'Tasmania' or 'Queensland'); some merely read 'ditto', which might have made sense when the specimens were all lined up in some configuration known only to Lea, but no longer did, since the specimens had been moved around in the intervening hundred-plus years. To be fair to Lea, he was breaking new ground, and, at times, would have been working at his rather primitive microscope, with his specimens illuminated by candlelight, or by sunlight reflected off a mirror. Nevertheless, what our experiences with his collection demonstrate is why progress in describing new species can be inordinately slow these days, when so much work may need to be given over to trying to interpret the work of earlier taxonomists, or to fixing their errors or inconsistencies. Taxonomists are, undoubtedly, very special people.

Southern snow-fleas

Remember the hardy nannochoristids that we met back in November? They have even hardier cousins that only emerge in autumn and winter. Known to some as snow scorpionflies, I prefer to call then southern snow-fleas, for reasons which I will explain soon. I first heard of them while I was still living in tropical Queensland. I'd attended an entomological conference at which a speaker (Peter McQuillan, I think) waxed lyrical about these strange insects, which hang around on the snow, only on the tops of Tasmanian mountains. Once I had moved to Tasmania, I was reminded of them when I became involved in an ecological survey along an altitudinal transect on the slopes of Mount Weld, overlooking Warra and the southern forests. The pitfall trap samples from high-altitude plots sometimes contained scores of them, but only those traps set in the colder months. I didn't actually see them at the time, because they were snapped up by a doctoral student, who later described the Mount Weld insects as a new species. This brought the total number of snow scorpionfly/flea species to four. Much later, I noticed them in the museum collections, preserved in vials of ethanol.

Several years went by before I went on my own snow-flea hunt, with my young son Ben—his near vision was so much better than mine, and I rated his chances of finding one much higher than my own. I chose a clear afternoon in early May, and we drove to the summit of Kunanyi, since 'Mount Wellington' is the 'type' location from which the species was originally described, back in 1941. It was a mild day down below, but, on the mountaintop, it was only about five degrees. Fortunately, there was little wind chill. The only clouds about drifted by at a comfortable height above our heads, so, in puffer jackets and beanies, we walked a couple of hundred metres along the main track across the plateau, then got down on our hands and knees among the alpine shrubs to begin the hunt. I had thought that it would be easy to find stones to turn and patches of leaf litter to sift, but I was wrong. Most stones were either too deeply embedded or too large; and leaf litter was sparse and intermingled with the prickly shoots of the shrubs. For twenty minutes there were no signs of snow-fleas or any other insect life. A change of tactic was called for, so I reached out and lifted the prostrate branches of a shrub, which had spread out across the surface of a large boulder. Where the outer branches had lain, there was no debris, but nearer the base, a shallow layer of shed leaves and twigs had accumulated over many years. This merged, below, into crumbly humus. The first boulder turned up no insects; but, repeating this technique, I found my first southern snow-flea within a few minutes. I was surprised that Ben had not found one himself, so I called him over to take a look.

Before us was a strange-looking beast—more like a minuscule cricket or a giant, wingless aphid than a flea; purply-brown, with spindly back legs, a bulbous abdomen, downward-pointing mouthparts and long, wispy antennae. Though it didn't seem capable of hopping, it walked off towards the base of the shrub within which I had disturbed it at quite a pace, given how cold the weather was. To prevent it from disappearing, I scooped it up in my hand so that Ben and I could examine it more closely. My hand must have warmed it up, because it became even more determined to flee! Conceding defeat, I introduced it to a sprig of shrub and it clambered about among the branchlets. This won me enough time to take a few photos before returning it to whence it came. We persevered with our search for another twenty minutes or so, but by then the cold had thoroughly sapped our enthusiasm, so we decided that one snow-flea was reward enough.

I have repeated a search of this kind on quite a few occasions over the years. I generally find one or two snow-fleas; never the scores that found their way into those traps on Mount Weld.

But why do I call them snow-fleas, rather than scorpionflies? Although they are part of the scorpionfly insect Order, they resemble neither scorpions nor flies (nor the original scorpionflies, for that matter); indeed, they lack both wings and an obvious tail. But *fleas*? In the northern hemisphere, there is a boreal scorpionfly lineage whose members are widely known as snow-fleas. Like ours, they are flightless, and they are active during the northern autumn and winter. They look rather more flea-like than ours, and they can hop, but in many respects they are similar to ours. In fact, it's quite uncanny how these two flea-like, cold-adapted lineages have evolved independently, at opposite ends of the planet, each from fully winged scorpionfly ancestors. Ours aren't part of the northern snow-flea lineage though, so I think it's sensible to call them southern snow-fleas. Austral snow-fleas might be another option.

As it happens, there has long been discussion in entomological circles about whether the true fleas (the bloodsucking kind) originated as a highly specialised fly that lost its wings and developed a taste for blood, or as an equally specialised sort of scorpionfly. For some, the argument is moot, because scorpionflies and true flies are sister groups, sharing a common ancestor. All this evolutionary divergence happened long ago, so it's hard to reconstruct the order of events, even with the evidence of DNA. But two recently published studies (conducted independently but published simultaneously) have suggested that the true fleas are highly modified scorpionflies. In other words, this means that there are actually three known lineages of scorpionflies that look flea-like: the true fleas, the northern snow-fleas, and the southern snow-fleas. And in case you were tempted to think that the true fleas evolved from the snow-fleas, the DNA evidence (based in part on specimens that I contributed) suggests that this isn't the case! Counterintuitively, the closest living scorpionfly relatives of the true fleas are those lacy-winged nannochoristids. I suspect that this isn't the final word on the tangled tale of flea evolution; but, in the meantime, I'm content to marvel at the ingenuity of ever-inquisitive humans and, all the more, at the unexpected twists and turns that evolution takes through deep time.

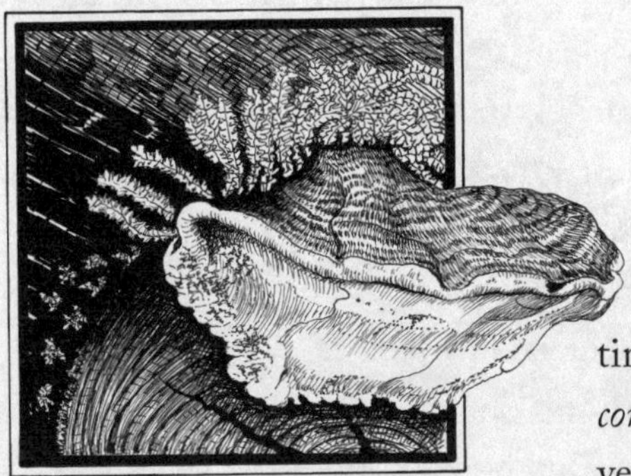

Deadwoodology

I was once told by one of my friends that Forestry Tasmania—my employer at the time—was, in his words, *'rotten to the core'*. Since I spent some of the best years of my working life there you will, I hope, understand why I was reluctant to agree with my friend. However, his words resonated with me in a way that he could never have anticipated. 'Rotten to the core' is, in fact, a good description for the very best sort of log one could find in a forest (in my opinion, at least; hard-core foresters would certainly disagree). Rotting is an amazingly complicated biological process. It involves hundreds or even thousands of species of microorganisms, fungi, and insects. The sage Jimi Hendrix reputedly opined, *'once you're dead, you're made for life'*. He probably didn't have rotting wood in mind, but we will never know for sure.

Regardless of all this, I became a conservation biologist in forestry because I had studied aspects of this phenomenon for my doctoral research. During my decade in the job, I was given almost free rein to develop a broad research program in the discipline that I came to call 'deadwoodology'. My predecessor had got the ball rolling, by setting up a long-term study and a doctoral research project already focussing on dead wood. My role required me to generate ideas for discrete projects under this umbrella, and then to entice researchers to take on those projects, often, but not always, under my supervision. Sometimes I took on the research myself, but much was achieved by postgraduate students, and some by paid contractors. As more and more research began to be carried out, my role extended to synthesising the findings and making them relevant for forest management. To keep things simple, I'll refer to all these researchers, deadwoodologists, as 'we'.

Why, then, is deadwoodology relevant to forest management? During my earliest experiences working in nature conservation, in England, I saw first-hand how that country's forests are highly fragmented and have suffered from centuries of use and abuse. They are a shadow of what nature intended. There is no truly natural woodland left. At best, what remains is seminatural—the product of shifting management practices down the centuries, which have favoured or rejected particular tree species or growth forms. Coppicing, pollarding, grazing, charcoal making, firewood collection, thinning, felling, clearing, planting, hunting—England's forests have had the lot. As a consequence, they are massively simplified versions of nature's postglacial endowment of wildwood. If you restrict your attention to the plants, things don't look too bad; in fact, the remaining forests and woodlands are often visually appealing. But looks can be deceptive, because most English (and European) forests have long been unnaturally deficient in old trees and the dead wood they would normally generate. This appears to make little difference to the flora, and to the herbivorous insects; but I soon learnt that it impacts heavily on the decomposers, as well as everything that depends on the decomposers: those microorganisms, fungi and insects that I mentioned earlier. At least a quarter, and in some cases well over half, of all forest insects have some association with dead wood, so you can see that its loss would be a major conservation issue. And here's the thing: even if you subsequently allow trees in these fragmented woodlands to grow old gracefully, as nature intended, and produce lots of lovely dead wood, the isolation and break in continuity would significantly restrict how much of the original community of organisms will be around to benefit. The damage to the forest's diversity, ecological integrity, productivity and resilience has already been done, and can't readily be fixed.

In Tasmania, forestry is a very recent imposition compared to the situation in Europe. We have the opportunity to keep the decomposer community going in our managed forests, rather than struggling to reinstate it after it has been lost. With more-intensive forest management, including biomass harvesting (for wood pellets for heating or electricity generation), on the horizon, my employers were keen to cover all bases for the public forests in their care. Tasmania's wet forests boast extraordinary amounts of dead wood—some of the highest in the world. If we can't practise forestry in ways that sustain dead wood at functional levels, then we won't be doing sustainable forestry. It's complicated, because doing forestry sustainably will require us to get to grips with scales of space and time that are relevant to both the way forests operate—especially the dynamics of dead wood— and the way forestry operates; that is, how we manage our needs and demands of forests, and how we manage the forests themselves, in response to all this.

Dead wood dynamism might sound like an oxymoron, but it's a fundamental property of forest ecosystems and one that it's critical to understand if we want to protect and preserve the species that depend on it. In most aspects of forest ecology, it's often unproductive and unenlightening to focus purely on the coupe level and restrict our studies to time of harvest; for a forest's cycles of growth and decay, and its flows of species, nutrients and energy, play out over much broader spatial and temporal scales. In deadwoodology, these scales are pertinent, as we shall see.

So, what did we deadwoodologists set out to achieve? First, we needed to establish a baseline, since we were starting almost from scratch—we wanted to ascertain how much dead wood there is in various kinds of Tasmanian forests, and at various stages of their growth and development. We also wanted to learn about its rate of decay; and about the insects and fungi that depend on the dead wood, and how those communities change as the wood decays. Because dead wood originates from living trees, we also wanted to know about the ageing process in those trees, and the role of the decomposer community in tree ageing. Furthermore, because fire plays a dominant role in forest dynamics, especially in generating and consuming dead wood, we wanted to understand its role better. Finally, we needed to make our findings relevant to forest managers, enabling them to plan and implement practices that more closely emulate nature and better cater for the species dependent on dead wood.

May is an appropriate month for introducing you to deadwoodology because much of our fieldwork was carried out in the 'off' season. Dead wood isn't seasonal, unlike most of the other aspects of forest ecology that I was researching at the time; so many of my abiding deadwoodological memories—conducting line-intersect sampling, laying out plots, cutting open logs—date from autumn. Summer fieldwork had ended by then, but the days had not yet become too short or cold to comfortably do what was necessary. It was all deeply engaging, but it was also hard yakka—I must admit that I frequently found myself dreaming of a neat, tidy, English woodland; how much easier our jobs would have been there.

Our aims were ambitious, but breaking the work down into manageable chunks enabled us to get there within little more than a decade; and it was worth the effort. Our findings are too numerous to fully convey here, so I will just focus on some of the most interesting and valuable ones. But first, it's important to note that most of our data concerned eucalypt wood, because other tree species comprised a small proportion of the total volume of wood in the studied areas. Furthermore, our data mostly relate to wet eucalypt forests, because it was believed that these were the forests most in need of management guidelines.

The first thing we found is that dead wood quantities are hugely variable, and no practical field survey technique really captures or overcomes this variability. This is especially the case in wet forests, where we found the largest volumes. Given that a typical line-intersect sample (a standard way of inventorying dead wood) involves a 100 metre transect through the forest, and some logs were easily half this length (because eucalypts can grow to immense heights), there is a clear mismatch of scale. Even when we set up square plots of 100 metres by 100 metres, instead of transects, and measured and mapped every bit of dead wood within them, the variability among plots remained. We also learnt that, counterintuitively, you can't assume that younger stands of trees will have lower volumes of dead wood on the ground than older stands. This is because large, long-lasting logs on the forest floor may be derived from old trees that toppled a century or more in the past, perhaps in the aftermath of a bushfire, rather than from the young trees in the present stand, which probably regenerated in the fire's aftermath.

Things were more straightforward in drier forests, because they aren't typically subject to wildfires as intense as those that occur in wet forests. This may also seem counterintuitive, but dry forests tend to burn more frequently, and, between successive fires, just don't produce as much flammable material as tall, wet forests do, with their typical dense, scrubby understorey. This means that dry forests tend more towards equilibrium in the structure of stands, with more even amounts of dead wood. Since we finished our study, remote sensing using LiDAR has made measuring dead-wood volumes (and the volume of standing timber) much, much easier; but a lack of correlation between stand structure and dead wood quantity remains the case.

The amount of dead wood on the forest floor is related to its rate of decay: the faster it decays, the less dead wood there should be at any point in time, all other things being equal. The problem is that we know that all other things aren't equal, at least in those wet forests where wildfires periodically (but irregularly) kill trees and generate a massive pulse of dead wood. But at least knowing when those fires struck (basically by working out how old the regrowth forest was) gave us some leads into ageing the logs found lying on the ground. Sometimes we could see charring on the logs or on the tree trunks, which offered further insights. Things weren't quite that simple, however, because logs might survive for more than one fire cycle; or they might not have been generated by fire at all. All of our ecological sleuthing skills were needed to try and pin down how long hundreds of individual logs had been on the forest floor. Sometimes we could deduce a minimum time by looking at rainforest trees that had germinated and grown on the log since it had been on the forest floor. Ageing those could be done by counting tree rings from a core, or, sometimes, by counting the number of axial growth whorls. Using these data, we then scored each log's decay class using a subjective scale that indicated how far a log was along the decay trajectory.

Then we had to figure out how long a log typically spent in each decay class (the scale isn't linear), and what proportion of a log's biomass (as opposed to volume) was lost when it decayed from one class to the next. This, in turn, required characterising the density and carbon content of rotting wood in each decay class; but each class may comprise multiple decay *types*, each with its own characteristic density and carbon content, determined by which fungi and insects had been involved in doing the decaying. Then we had to establish the relationship between decay class and decay-type composition, in order to work back and arrive at average values for density and carbon by decay class. Next it was necessary to relate those densities to the measured volumes, bearing in mind that as decay progresses, the volume shrinks and so the original log's volume would have been proportionally greater than what was measured. Then, knowing these two facets of a log (its density and its decay class), we could plot them all on a chart and fit a decay curve to the data, to get an idea of how long logs of a given initial diameter would take on average for, say, 50% or 95% decomposition. In the wet forests, for large-diameter logs (of the sort derived from the toppling of an old-growth tree) the answer is about 70 and 200 years, respectively.

We also looked at how well logs survived fire, including silvicultural regeneration burns. It turns out that they survive remarkably well, with the proportion lost being inversely proportional to their original diameter. That is, the larger the log, the more of the log survives the fire. Very small logs—fine fuels—are usually almost completely combusted; this is, of course, one of the objectives of a regeneration burn, since fine fuels and litter have been found to inhibit the germination of eucalypt seeds, whereas the ash bed arising from their combustion provides the perfect substrate for their germination and establishment.

We didn't set out to answer questions about carbon sequestration and dynamics, but our findings certainly helped shed light on this important topic. It transpires that some of our old-growth wet forests can indeed store vast amounts of carbon, both in the living trees and in the dead wood and soil; but those that do so naturally comprise a rather small proportion of overall forest area. Furthermore, their storage capacity is not indefinite, because either fire will abruptly release into the atmosphere a sizeable fraction of what was stored; or, in the absence of fire, the gradual transition to rainforest will reduce long-term storage capacity by close to half. This reduction comes about because rainforest trees are smaller (and store less carbon) than the eucalypts they replace (and whose logs eventually decompose), and there is little to no compensatory increase in soil carbon.

So much for the mechanics of a decade of deadwoodology. But, looking back, what did it *feel* like? So many words, thoughts and memories come to mind. It was mostly thoroughly rewarding, but also enormously challenging and, frankly, arduous in the extreme. Imagine you're on holiday, walking along a nicely laid-out level track through a tall, wet eucalypt forest, perhaps in one of our national parks. Helpful interpretative signs point out the different tree species and may invite you to ponder their age and the fires that have periodically swept through the land. The going is easy underfoot, and someone has very kindly sawn through any logs lying across the path and cast the cut and upturned billets to the side, where they serve as natural chairs should you fancy a rest. In the cut ends of a typically massive log, you get to see the concentric tree rings in the rot-free, outer portion and to peer into its hollow centre, where (the sign tells you) glow-worms and cave spiders may lurk.

Now imagine leaving the path and trying to head in a straight line for, say, 100 metres, as though you were performing a line-intersect survey to work out how much dead wood there is in this forest. Instantly, your experience challenges. The terrain underfoot is suddenly uneven, with snaking roots, hidden rocks, and fallen branches waiting to trip you up at every step; while head-high tussocks of cutting grass tear at your legs and arms. Before you have travelled 10 metres your route intersects the same log that the formed track simply cuts through. You can either go around or over the log. Going around may mean retracing your steps back to the track; going over, you decide, is harder still. So, your 100 metre walk stretches out into a 200 metre meander, and now you're both exhausted and disorientated; but you dare not stand still for long, since you know that leeches will be inching their way towards your boots; and mosquitoes are already homing in on any exposed skin. Therefore, you soldier on, more alert to the risks and dangers than ever. You want to identify the trees as you go, but because their foliage is high up, doing so means learning to distinguish their trunks on the basis of texture, colour, and diameter. And nobody has told you how to do this, nor how to distinguish which logs came from which tree species, nor how to take an unbiased measurement of a log's diameter, nor to gauge its decay class.

You get the picture. Throw in a hard hat, fluoro vest, steel toe-capped boots, clipboard, compass, diameter tape and hip chain (a distance-measuring device based on a spool of thread that is fed out as you advance), and you have the basics of performing a line-intersect survey. Each time you go out, there are occupational health and safety protocols to follow, plans to be drawn up and adhered to, radio calls to be made. Repeat several hundred times, at scores of different take-off points chosen to conform to stratified random sampling protocols rather than your driving or walking convenience, and you might have a statistically robust set of data from which you can start to look for patterns that, you hope, will offer clues as to how dead-wood dynamics play out across the landscape. But to really get to grips with that, you also need to age the trees in the surrounding forest, and to determine how long a log has been lying on the ground. Nobody has told you how to do these things. You have to take advice, read up, play around with ideas, come up with the protocols and then apply them unswervingly.

You want to learn more about what's going on within a log, so you can better understand where it sits on the decay trajectory; but you've never used a chainsaw before, so you first have to get trained and certified. You need to wear cumbersome protective gear, and lugging this as well as the saw, and fuel, into the depths of the forest is exhausting. For safety, you need to work in pairs. Chainsaws are designed for solid timber. They get clogged up rapidly, cutting into rotten logs, so you do a lot of chain-cleaning. Nobody has delineated or classified all the different decay types, so you have to work these out too and come up with a proforma, so that the classification remains as objective as possible. And you're mindful that each cut you make potentially destroys the habitat that is the driver for this research, so you need to pick your logs carefully to minimise damage.

I'm making deadwoodology sound like mission impossible, but I'm not exaggerating! I wouldn't have done it, though, if I hadn't felt that it was worthwhile; and I trust that my fellow researchers felt the same. For me, the rewards weren't merely academic: they were visceral too. It was a privilege to be licensed to spend so much time in the forests, free of the shackles of urban living and office life; and to delve (literally) into an aspect of the forest's functional ecology that few had previously ventured into, and where so much awaited discovery. Not every excursion involved majestic old-growth—the idea demanded we cover all growth stages—but I particularly enjoyed being in the older forests. These could be experienced at many levels: as my made-up holidaymaker would, certainly—marvelling at the cathedral-like works of nature, and pondering the vastness of time that went into their creation. You could enjoy a spot of birdwatching: green rosellas, yellow-tailed black cockatoos, lyrebirds, scrubwrens, to name but a few. You can get down on your hands and knees and appreciate the smaller plants and animals, the fungi, the mosses and liverworts. I enjoy all these pursuits, but even now I can't stop myself from also engaging in a spot of ecological detective work whenever I find myself in mature wet eucalypt forest. Are all the eucalypts of one age, despite some dominant individuals having twice the girth of others, or do we have more than one cohort here, signifying a non-stand-replacing wildfire? Which age classes of trees bear fire scars or signs of charring on their trunks? How old is that dogwood stand? (Because that tells you when the last fire went through, since dogwood seeds in the soil germinate after fire.) How did that blackwood get to grow so tall and straight, when open-grown blackwoods are short and branchy? (A tall, straight growth form signifies that it grew up in competition with other young trees and shrubs, probably in a gap in the forest, which it eventually came to dominate.) Are those logs of the same tree species as the present stand would indicate? Did they arise as a result of self-thinning of a developing eucalypt stand, or did fire, disease or a windstorm topple the trees from which they were derived? How old are those rainforest saplings that are growing out of that log, and what does that say about how long that log has been lying there? Why is that patch of eucalypts over there so much younger than around here—a past spot fire, perhaps? Why is the foliage on that myrtle branch withered and brown? (Perhaps it's myrtle wilt which entered via a wound caused by a side-strike from a branch falling from one of its neighbours.) Thinking this way helps me to get the measure of a forest, to understand its disturbance history, its trials and tribulations, its achievements in the face of adversity. It engenders a sense of respect, but also of impermanence—come back in five or fifty years and it might look rather different; in five hundred years it would be unrecognisable.

What does all this mean for forestry? Essentially, the better that forestry can emulate natural processes, the more readily the forests will maintain their ecological integrity. Forestry—especially native forestry—is applied ecology. Ecological disturbances don't come much bigger than bushfires, and so we shouldn't be surprised that wet forests cope pretty well with clearfell harvesting if followed by a regeneration burn. Forest management can't fully replicate natural disturbance dynamics, but it can learn from nature. In particular, systems that cater for old trees—and the dead wood that arises from old trees—are ecologically sounder than those that don't. In practice, in wet forests, it's neither ecologically appropriate nor safe to selectively harvest individual trees, when the forest 'expects' a much larger, stand-level disturbance. Better would be to harvest entire stands, but to interlace the harvested area with retained, unharvested forest. These retained areas serve as 'lifeboats' for species that don't survive harvest (or wildfire), yet are able to recolonise afterwards; and also as local sources of ageing trees and dead wood, and the biodiversity that they harbour. This is the basis for the now widely practised alternative to clearfelling mentioned in January: aggregated retention silviculture; as well as for the delineation of a fine-scale patchwork of informal and formal reserves across the managed forest landscape.

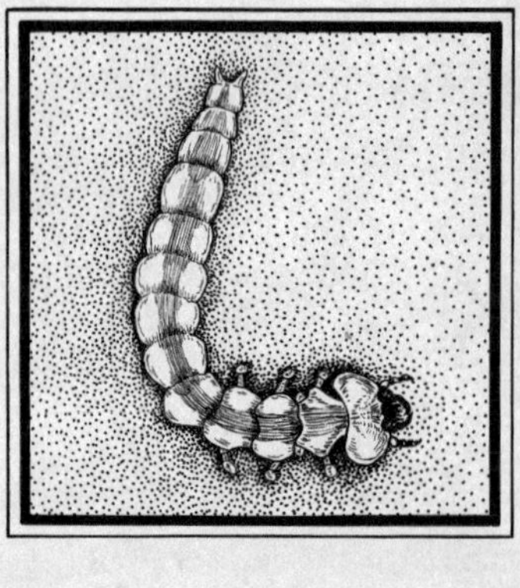

Mudguts

The various rotten-wood types that we deadwoodologists identified in our research on eucalypt logs mostly came without names, so we invented them. We used descriptors based on physical appearance or feel: brown blocky rot, white fibrous rot, jelly rot, and so on. 'Brown' and 'white' signify different biological processes. Brown rot is associated with species of fungi that selectively digest the wood's cellulose, leaving brown lignin behind. White rot is produced by fungi that are able to metabolise the lignin. Small-diameter logs (from younger trees) mostly decay along white-rotting pathways—the white rot often progresses from the outside of the log inwards; whereas larger-diameter logs (from older trees) predominantly follow brown-rotting pathways, often decaying from the inside outwards. One type of brown rot actually had a name, used by loggers, and it soon became my favourite. 'Mudguts' refers to a moist, red-brown, smooth, almost claylike material, with a characteristically earthy smell. It may manifest as narrow seams wedged between other brown rotten-wood types, a bit like mortar between bricks; or it may completely fill the interior of a log. In fact, it may not be an active rot, as most of the other types are; rather, it is probably an endpoint for multiple brown-rotting decay pathways, appearing after all discernible wood fibres have been broken down by a combination of fungi, other microorganisms, termites, and beetle larvae.

Chainsaw operators detest mudguts, because it clogs the chain and foreshadows low-value timber, only suitable for low-grade woodchips. Invertebrate enthusiasts, on the other hand, see mudguts as a goldmine. Inside it, they may expect to find some of the more demanding saproxylic (dead-wood dependent) invertebrate species. A decent seam or pocket of moist mudguts (one completely sealed off from the outside world by other parts of the log) might house several saproxylics, including beetles, velvet worms and springtails. That might not seem like many, compared to the many hundreds of species found in other parts of a decent-sized log; but these particular creatures can be found nowhere else.

Accessing mudguts requires sustained brute force and tools such as chainsaws, wrecking bars and tomahawks. I once referred, in print, to the method as 'THD sampling' (total habitat destruction), which is why I always limited such investigations to a small number of sacrificial logs, or sections of larger logs that were otherwise left intact. I wish I could say the same for all beetle enthusiasts, but there are some who lack scruples—commercial operators in the global stag beetle trade, possibly. Their impacts are all too apparent, when you know what you're looking for, along forest pathways near roads.

In retrospect, I think I was drawn to mudguts because I associated it with rarity and what, in Britain, we referred to as the 'wildwood'—the forest we imagined had once carpeted much of Europe before farmers began clearing it, during the Neolithic, and before whatever remained was transformed into today's isolated patches of highly modified woodland. In Britain—and, indeed, across much of Europe—old trees are now so rare that the specialist saproxylic fauna of mudguts (or its English equivalent: red heart-rot) is much diminished. Saproxylic beetle fragments identified from subfossil wood (often from peat bogs) provide evidence of widespread extinctions over the past few thousand years. Additional losses have been documented from the past couple of centuries. Such is the parlous state of the saproxylic fauna in Britain that these days, THD sampling would, rightly, be considered completely irresponsible. Meanwhile, in Tasmania, old trees and well-aged logs are still commonplace. Long may it remain that way.

In the southern forests of Tasmania, perhaps the most dependable inhabitants of mudguts are Atkinson's jugular-horned beetles. It's a rather clumsy name for a small, red-brown, elongate but dorsoventrally flattened beetle, bearing strange, hornlike projections forwards from its chin. It seems possible that the beetles use these to gouge out bite-sized morsels of mudguts to eat, although they also bear oversized mandibles that would seem equally up to the task. The adult beetles can be found year-round and are usually seen in association with the species' paper-thin, diaphanous larvae. I wonder whether there is some parental care involved. The gouging may be to provide food for the larvae—their intestines, which run the length of their tiny bodies, are usually visibly full of mudguts. Jugular-horned beetles squeeze into the tightest spaces, following thin seams of mudguts right into the heart of large logs. Usually there are damp-wood termites inhabiting the same seams, and perhaps it's their tunnelling or chewing that creates the mudguts, which beetles then feed on. Besides eking out an existence on an impoverished diet, they must also be able to get by on extraordinarily little oxygen; and I can only hope that they don't suffer from claustrophobia.

Stag beetles are stalwart dead-wood feeders, at least in the larval stage, and Tasmania is a stag beetle hotspot, boasting well over thirty species. Many of these get by on a diet of white-rotted wood or even soil humus derived from dead wood; but one threatened species, the flightless Mount Mangana stag beetle, is pickier. The adults are sometimes found under rotting logs; but the chubby, C-shaped larvae are seldom encountered, other than by researchers, since they live deep within logs, usually embedded in mudguts. Mudguts is clearly not a nutritious diet—they may take two years, or more, to mature.

The Mount Mangana stag beetle takes its name from the highest peak on South Bruny Island, but its disjunct distribution, in wet forests, also takes in the Tasman Peninsula, the southern flanks of Kunanyi-Mount Wellington and eastern parts of the southern forests—areas that might have been almost contiguous when sea levels were lower. Being a threatened species, its needs must be considered, by law, when harvesting timber, which is not a simple proposition, given how long it takes to regrow trees capable of giving rise to logs with sufficient proportions to develop mudguts.

Sometimes mudguts is colonised by slime moulds. These otherworldly organisms were once classified as fungi but are now grouped with various amoeba-like microorganisms. Mostly they go about their business as individual, amoeboid cells, feeding invisibly on fungi or on other microorganisms in the rotting wood. Every now and then the individual cells coalesce for mating. They then form anastomosing rivulets of cells that, eventually, converge into a single mass. That's the stage at which 'slime-mould' suits them best. One common species has the delightful name of dog's vomit slime mould. It forms bile-yellow, frothy splotches on old logs and stumps. Others are white, pink, violet or orange, and they may form remarkable columnar structures or mousse-like cushions, which eventually harden to the texture of meringues. If you can bring yourself to look closely at a coalescing slime mould, you may be rewarded by the sight of tiny beetles—imaginatively named cryptic slime-mould beetles. These feed exclusively on the slime mould's cells and spores. Slightly larger, at about two millimetres long, are slime-mould feeding rove beetles. A diet of slime mould is an unusual evolutionary development for rove beetles, since nearly all other species are predators of various invertebrates.

Other inhabitants of slime moulds, more impressive than these tiny beetles, are giant springtails. Springtails are wingless, six-legged relatives of insects, but their origins pre-date the evolution of true insects. They get their name from a feature possessed by many members of the group: a tail-like appendage, tucked under the body, which is held under tension so that, when the animal needs to flee, the 'tail' springs into action and catapults the animal to safety. Giant springtails are quite different beasts, though—they lack springy tails! An order of magnitude larger than the spring-tailed spring-tails, they have pudgy, Michelin Man-like, velvety-grey body segments. The tubercles on their backs may be pillow-like, but in a couple of our local species they taper into arrestingly orange-tipped spines, while in the two-horned giant springtail (my name for a species restricted to the western rainforests) these spines grow to outrageous lengths. Giant springtails spend their days trundling along on stubby, pointy-toed legs, often in 'flocks' like so many minuscule sheep, grazing among the 'fields' of slime moulds carpeting the mudguts. They are very vulnerable to drying out, so they only inhabit the choicest regions of a log, out of reach of sunlight and desiccating warmth. I don't think anyone knows how long they live, but I wouldn't be surprised if it were years.

There is one beetle species, which I shall dub the woody wrinkled bark beetle, which I have only ever found by breaking open large logs to the mudguts within; and even then, I have only found it twice. Once was in Wielangta Forest, between Nugent and Orford. The other occasion was on the Central Plateau, near Bronte Park. My finds, and a smattering of others, demonstrate that it is a widespread beetle (it is also found on the Australian mainland) yet its habitat shields it from all but the most prying eyes. It's a member of the ground beetle family, nearly all species of which are active predators; but a small lineage, known as wrinkled bark beetles, spurned overtly predatory habits and took to feeding on slime moulds in rotting logs, instead. They have adapted their body plan accordingly: in our local species, the body is pitchy purplish-brown, narrow, and about a centimetre long. The wing cases and thorax are etched with deep, longitudinal grooves; the antennae are thick and beaded; and the tiny mouthparts are tucked away beneath the head. If I didn't know better, I'd never call the creature a ground beetle, but the DNA evidence is incontrovertible. Each time I have come across one, it has sat, motionless, among the mudguts, as if stunned by my sudden intrusion into its dark, damp, and otherwise impervious home. How these beetles ever meet one another in such habitat, and disperse from one log to the next, remains a mystery. Presumably, they can fly, but they have never turned up in our flight intercept traps. Incidentally, the grooves and wrinkles in the beetle's integument may have evolved to allow them to carry slime-mould cells when they disperse, ensuring that they will have something to eat once they have settled into a new log. You might call them slime-mould farmers.

Europe's single species of jugular-horned beetle and wrinkled bark beetle are, today, both threatened with extinction and long gone from Britain. We can avert a similar situation in Tasmania, by applying the ecological knowledge we have gained through the deadwoodology research I championed.

©K.DAVIS 2023

June

The heart of darkness

By June, I have adjusted to the idea of darkness. Day length continues on a path of gradual shortening until the solstice, but the rate of change becomes so slow that it is barely discernible. There's no escaping the dark and cold; instead, it becomes necessary to adjust to the new regime, buckle down and work as best one can, with whatever you've got. It's not all bad, though. Days of cold rain, wind and winter bleakness usually remain outnumbered by days of cool but not unpleasant serenity, which often begin with a sharp, sparkly frost beneath a crisp, blue sky. Nature becomes quiet; so, mostly, do I. Often I gaze longingly at the sun-soaked hills on The Other Side, better known as Hobart's Eastern Shore. They may not be particularly green, but that's partly because they are less prone to cloud cover than 'our' side. They also benefit from an extra couple of hours of sunlight in the afternoons. Here, as the sun sinks behind the ranges to our west, and I zip up my puffer jacket and pull my beanie down over my ears, I have to remind myself that it's swings and roundabouts. Looking around the now-dormant fruit and veggie garden that we've nurtured through many seasons, I'm grateful that, in summer, we experience less extreme heat and drought than the other side of the Derwent.

As winter progresses, the Bridgewater Jerry appears less frequently than in autumn, because the air descending overnight from the upper reaches of the Derwent Valley is now both colder and drier. Instead, early-morning

Hobart is subjected to icy catabatic winds, which sometimes howl down the valley at great speed, making me glad that, these days, the Derwent ferry saves me from having to battle bone-chilling sidewinds on the Tasman Bridge, during my bicycle commute to work.

The winter solstice is a turning point of sorts. Though the days begin to lengthen, weather-wise the worst is yet to come. But Chris and I sometimes join other locals in marking the occasion with a sunset drink at the beach. Sometimes we will take the next step and go for a mad, after-dark swim (more a quick splash). The prospect of after-swim mulled wine, homemade mince pies and good company is really what entices us.

June is a difficult month to find a butterfly, but I promised one for every month, and a different one at that. Meadow argus and cabbage whites are prime contenders, since their autumn broods occasionally linger this late; but I've already pressed them into service for other months. Instead, I offer the saltbush blue. It's a butterfly we met back in April, which is usually when its flight season ends. But, one crisp June day, I was on the coast at Beaumaris, and I spotted a saltbush blue fluttering feebly around saltbushes near a beach car park. I doubt it lived more than a day or two longer, as the weather then turned foul, but it brightened up my day and was a vivid demonstration of nature's tenacity, which was all the more unexpected in such a fragile and delicate little butterfly.

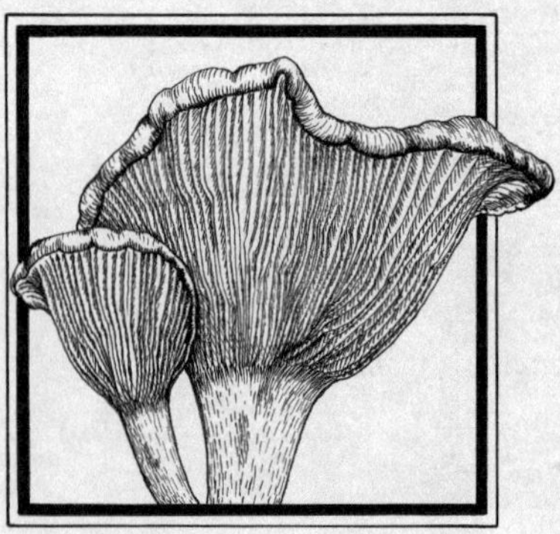

Bioluminescence

Nature doesn't always take darkness lying down, and quite a few organisms produce their own light. On land, against a background of rotting stumps, ghost fungi emit an eerie green glow, no doubt contributing to the widespread reputation of mushrooms for being gateways into other worlds. Photographing ghost fungi has become quite a competitive pastime for some in Tasmania, and my social media channels often feed me sumptuous examples of the photographers' skills. My first putative ghost fungus, spotted on a night walk along the cliff path near my home, turned out to be a little plastic figurine, discarded and lying, glowing, half-buried in the leaf litter. I wondered whether it might be radioactive, since that was the case with the first generation of glowing displays on clocks and watches; so I decided to leave it there, rather than pocket it. I still spot it, occasionally, but only at night. Fungi, too, can be radioactive since they are particularly good bio-accumulators of metals; but their luminescence is due, not to radioactivity, but to a chemical called luciferin, which emits light in the presence of oxygen and an enzyme (called luciferase). Ghost fungi don't glow in the daytime; they save their efforts for after dark, when it is thought that the glow might attract insects to disperse their spores.

If the word 'Lucifer' is familiar to you, perhaps you had a religious upbringing. Since the days of the King James Bible, it has been one of many ways of referring to The Devil without uttering His name. It strikes me as odd, because, in Latin, *lucifer* means 'light-bearer' (which is also what 'phosphorus' means, in ancient Greek). Why should the Devil, Prince of Darkness, be called Light-bearer? Apparently it originates from a translation, or mistranslation, of an ancient Hebrew word that referred to the King of Babylon, who was, by all accounts, a devilishly evil overlord. The original word may have meant 'day star', and this is probably what the King himself liked to be called. But the only sense intended in the Latin translation was 'the evil one'. Not a lot of people know that, but now you do.

Luciferin appears all over the animal kingdom, from the luminous rear ends of fireflies and glow-worms (Tasmanian glow-worms belong to the subgenus *Lucifera*) to the fishing lures of deep-sea anglerfish. Luciferin is also behind the sparkle in sea-sparkle, and that's where I'm heading. If you haven't heard of sea-sparkle, perhaps you have heard of its alter ego: the red tide. Both phenomena are caused by a tiny dinoflagellate whose scientific name translates as 'sparkling night-light'. Dinoflagellates are single-celled creatures that could loosely be called algae because they sit in that branch of the tree of life, and because most are able to photosynthesise. But many dinoflagellates shrug off such simple pigeonholing, as they are also predators, and active ones at that, hurtling through the water, using their whiplike tail to propel them, and hunting down and feeding on other microorganisms. Indeed, our local variety of sea-sparkle is entirely predatory. It feeds on photosynthetic diatoms and the larvae of tiny herbivorous crustaceans called copepods.

Many planktonic algae and microorganisms undergo cycles of boom and bust, because they exist on a nutritional knife-edge; any sudden increase in nutrients must be exploited rapidly before the opportunity passes. Sea-sparkle is no exception, but its boom follows that of the organisms that benefit first from the nutrients, i.e., the diatoms and copepods. A nutrient hike triggers a diatom boom, which soon becomes a bloom, as the dinoflagellates multiply rapidly, sometimes forming slicks where wind and currents concentrate their cells on the sea surface. These slicks of starving and dying cells are tinted red; hence they go by the name of 'red tide', especially when they are washed into shallow bays and onto beaches. Red tides are caused not only by this species, but these days it accounts for most of them.

A red tide by day becomes a concentrated source of phosphorescence at night. Whenever the dinoflagellates are disturbed, whether by waves, raindrops or the passage of fish or shrimps, or by human disturbance, the luciferase reaction is triggered; and each gives off a tiny bluish-white flash of light. Multiply that a million-fold or more and the whole seascape can put on a spectacular lightshow. Phosphorescence caused by organisms other than sea-sparkle used to be glimpsed, primarily, by sailors. It can still be seen this way, and that must be a remarkable sight, casting the ship's bow, and perhaps the outlines of any dolphins travelling alongside, in an eerie, electric-blue glow. But, these days, sea-sparkle is easy to experience from shore, particularly in sheltered bays and inlets along Tasmania's east and south coast. Just follow the red tide and return after dark.

While the dinoflagellates prefer warm waters, they also crave nutrient flushes, and these are more frequent close to shore in the cooler months, when there is more likely to be runoff from the land. Much like with ghost fungi, during this season, my social media feed becomes replete with gorgeous images of Tasmanian sea-sparkle. I love them. They have inspired me to go in search of my own sea-sparkle experiences. At Lindisfarne, on Hobart's eastern shore, a red tide produced a vivid display of sea-sparkle for us, one cold June evening. I dipped my hand in the water and swished it around—the electric-blue evanescence was transfixing. I threw stones in, and delighted in the concentric circles of light radiating from each point of entry, and marvelled at the light-drops that scattered across the sea surface, in every direction. On another occasion, sea-sparkle came to Taroona, and we took off our shoes and socks to paddle and splash in the shallows. The experience was enhanced by the fact that we had been drinking mulled wine to celebrate the winter solstice moments beforehand. At Port Esperance, on a Scouts camp, we brought our kayaks back to shore well after sunset, slicing our craft through a blue glow that was bright enough to aid our navigation.

Even so, I am ambivalent about sea-sparkle in Tasmania. On the one hand, it feels magical, romantic even, to experience this outlandish phenomenon; but, on the other, my experiences are forever tainted with unease, knowing that it does not belong here. Until the 1980s, the organism was only documented in Australian waters from Sydney Harbour. Since then, it has spread around much of the Australian coastline, largely via ships' ballast water and other human-induced means. Sea-sparkle only reached eastern Tasmania as recently as 1994. Now it seems to be resident right around Tasmania, while warmwater eddies have carried it far to Tasmania's south, into the Southern Ocean. This is not an innocuous alga, and its effects on the marine ecosystem are likely to be profound. Furthermore, all those extra nutrients that favour sea-sparkle are usually from an anthropogenic source: stormwater, or agricultural runoff, or, increasingly, faecal matter from fish pens. Sea-sparkle, it seems, is a bit like a starling—beautiful in the right setting, but brutish, invasive and an uncomfortable reminder of our culture's lack of regard for the natural order of things. I wish it no ill, it's very pretty and only doing what sea-sparkle is meant to do; but I wish it weren't here.

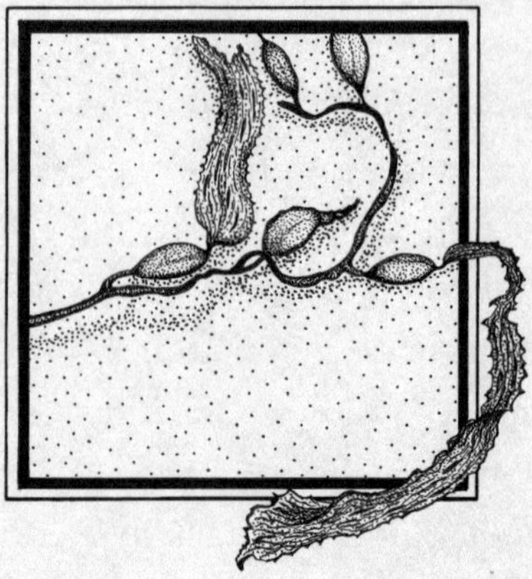

A paddle along the cliffs

I will continue the maritime theme by taking my kayak out for a paddle (or is that taking my paddle out for a kayak)? Either way, placid seas and low sun angles make for a tempting combination on settled winter weekends. I keep my kayak, a sit-on with a pedal-operated rudder, locked up at the beach, so there is little physical effort involved in getting it into the water, as long as there's no swell. Psychologically it's more daunting, facing the prospect of having to wade into the chilly waters, despite layers of insulating neoprene. Once it's done, though, I never regret having made the effort. One particular June morning was especially rewarding, and I will tell you about it, but you will have to forgive the usual digressions along the way.

The air had been still in our garden, leading me to think that it would be a fine day for paddling along the foot of the Alum Cliffs—a stretch of rocky coastline between Taroona and Kingston Beach. Much of Taroona sits on a fan of dolerite rubble embedded in thick clay, which is a product of slumping and weathering of scree emanating from the steep Jurassic dolerite slopes immediately inland. The slumping peaked during the last glacial cycle. It follows one of the ancient fault lines that created the rift occupied by the lower Derwent valley and estuary. It continues to this day, in slow motion, necessitating endless repairs to the Channel Highway. The imposing Alum Cliffs have a different origin, being an exposure of more ancient Permian mudstone.

When we first moved to Taroona, I had assumed that ‘ Alum Cliffs’ referenced yet another place name from the Old Country: the multicoloured sandstone cliffs on the Isle of Wight, which bear the same name. I had first visited those on a school geography excursion, when I was a teenager. Geomorphology had been the lesson of the day.

Viewing the Taroona version through a geomorphological lens, I had been left wondering how anyone could have seen any similarity to the Isle of Wight cliffs. They certainly bore little physical similarity, beyond being sea-cliffs. Then I learned of another Alum Cliffs, lying inland, not far from Mole Creek, in the north of Tasmania. At some point the penny dropped and I realised that 'Alum', in all of these place names is a utilitarian designation—all three are places where you can find alum!

The word 'alum' is the term for a multitude of chemical salts; typically a double sulphate of aluminium and potassium. Alum has been used for a variety of industrial, and even culinary, purposes, and was also a favourite among alchemists. These days it is synthesised, but in earlier times it was mined. Whether this occurred to any great extent at Taroona, or at Mole Creek, I don't know; but at least the place names now make more sense to me.

So here I was, intending to head south along the Derwent's Alum Cliffs, by kayak. Duly dressed neck to toe in neoprene, I arrived at the beach to find the sea completely flat, other than a low southerly swell rolling in, sedately, from some distant, sub-Antarctic storm system. This was promising, as the messy, chaotic, choppy seas that arise from an interplay between local winds and the sea's surface—wind seas—are the enemy of my pleasure-kayaking. Swells are different. They can propagate over hundreds or thousands of kilometres from their source, spreading ever outwards; but as they do so, their amplitude flattens and their frequency diminishes. The rise and fall of their passing may be scarcely discernible on the ocean surface. It makes for altogether more pleasant kayaking. Yet even flattened swells can pack a mighty horizontal punch on reaching the shore—as I found out when I tried to launch my kayak that sunny June morning. I tried to time my entry carefully, but it still took me half a dozen attempts to first wade, and then paddle out through breakers that seemed to arise out of nowhere. Each failure more thoroughly soaked me. *'That's why I wear so much neoprene'*, I reminded myself, through gritted teeth.

Beyond the breaking surf, I was relieved to find the surface calm again. After resting for a few minutes to regain my sense of balance and dignity, I then turned the kayak towards the foot of the cliffs. A few days earlier, we had experienced one of those troughs that periodically dump inordinate quantities of rain on us, swelling creeks, storm drains and rivers. At such times, the waters off Taroona often turn a murky-brown as the Derwent carries away unsightly and, sometimes, unsavoury sediments from Hobart and further upstream. Over the ensuing days, the sea will turn from brown to green as the sediments settle out and nutrients are taken up by rapidly dividing microscopic algae. Today, therefore, I was able to look down upon a luminous, jade-tinged world as my kayak sliced through the sunlit waters; while every stroke of the paddle brought forth the faint but familiar whiff of what I might call 'essence of phytoplankton productivity'—it may well be dimethyl sulphide, or DMS, which is a sometimes-unwelcome product of fermentation that adds straw-like overtones to the taste of beer. In the atmosphere above the world's oceans, the same phytoplankton-produced chemical is a potent seeder of clouds, driving much of our weather.

But I digress, again. In fact, I was on a straight course, parallel to the foot of the cliffs but some twenty metres out to sea, to avoid backwash from the swell after it hit the cliffs. Surveying these from my low vantage point, it was clear that their 'mudstone' designation is a bit of an oversimplification. The near-horizontal strata showed distinct banding, indicating alternating layers of mudstone and silty sandstone. The strata that predominate higher up on the cliffs weather to fawn-brown, while fresh exposures, from intermittent rockfalls, are a gorgeous golden-brown that is especially intense and fetching in low winter sunlight. Towards the base of the cliffs, where the honeyed rock gives way to more sombre, leaden tones, I could discern milky veins of what I took to be alum, occupying fractures and fissures in the rock face, and, in places, spilling out onto the wave-cut rock platform below. It would have been challenging to mine those veins, I mused, with only sea access. But perhaps a hundred or more years ago, when local sea levels were fifteen to twenty centimetres lower, it might have been possible to inch one's way to them without getting wet, from Taroona, at low tide.

My geological reverie was interrupted by a sudden sound, somewhere between a bark and a gulp, emanating from the seaward side of my kayak. Wobbling precariously in my surprise, I turned to look, and found myself staring into the limpid, dark eyes of an equally-astonished seal. Sammy, as the local Australian fur seal is widely known, is an intermittent visitor to the Taroona shoreline. Legend says that he is on his way to or from the Hobart docks, where he enjoys a feed of fish and chips from the punters.

More plausibly, 'he' is several seals that frequent the Derwent. Indeed, the one before me now was a lithe Samantha, with, fortunately, only about a third of the bulk of a Samuel. I felt a mixture of delight and unease. On the one hand, it's delightful being this close to a fur-seal in its element—I could hear her exhaling and watched her nostrils twitching on the end of her pointy snout, while droplets of water rolled along her stout, rearwards-curved whiskers. On the other hand, I knew that my kayak would be no match for even a Samantha if she had attempted any kind of interaction, playful or otherwise. She may have felt uneasy about my presence, too, because she soon ducked her head back down; and, when she next surfaced, she was a good hundred metres away, heading upriver. Perhaps those twitching nostrils had told her 'no fish here'.

I continued my paddle in the opposite direction, and hadn't gone far before something else peeked above the water's surface, before dipping under, only to reappear in the same spot a few seconds later. I advanced cautiously, until I felt something seem to tug on my paddle from below. Imagining more seal encounters, I pulled at the paddle, but it slid gracefully out of the water, draped in the taught, glistening sinew of a giant kelp stem. Olive-brown kelp fronds were now writhing all around me, to the rhythm of the swell, poking up above the waterline, then sinking into the murky green depths. This, too, was unexpected and marvellous. Early navigational charts depicted the Derwent as being choked with giant kelp and whales, but the whales were almost wiped out and the kelp beds are long gone. Giant kelp probably thrived because of the fertilisation effect of whale poo. These days, warmer, nutrient-poor waters pushing ever further south down Tasmania's east coast have transported planktonic larvae of kelp-munching long-spined urchins, which are now established locally. Kelp plants don't like warm waters, either. The loss of kelp must have had a huge impact on local sea and seashore life, from sea dragons to kelp-flies. It also removed the shoreline's living buffer against swells, so these now have unimpeded passage, perhaps transforming a sheltered shoreline teeming with intertidal life into a more exposed and hostile one. All things are connected.

I surveyed the scene. The kelp wasn't dense, but there were plants spaced more or less regularly over a radius of about ten metres. I was used to seeing the odd strand of giant kelp in the waters off Taroona, particularly offshore from the old sewage outlet, because that was where the nutrient levels were elevated—a modern-day replacement for whale poo. It was a pleasant surprise to see it growing here *en masse*. I wondered whether I had inadvertently stumbled on an experiment—I knew that a specially bred, heat-tolerant variety had been planted in various places around eastern Tasmania, to see whether it might recolonise. Or maybe not, since there are still substantial patches of wild-grown kelp further south, towards the Channel (beyond Blackmans Bay). The kelp certainly seemed to be in good health, with its wrinkly, serrated fronds free of encrustations, and its flotation bladders all intact. Those gas-filled, pear-shaped bladders—giant kelp's scientific name translates as 'pear-bearing big bladder'—enable the plants to rise up to life-giving sunlight near the water's surface, from points of attachment on the sea floor, metres below.

Giant kelp grows along cold-water coastlines, generally a few tens of metres offshore, in both the South and North Pacific. It does most of its growing in the winter and spring, before the water warms up. Under optimal conditions, giant kelp stems can lengthen by a metre a day; and they can eventually reach fifty metres or more in length. Sadly, kelps, and subtidal seaweeds of all sorts, are under threat everywhere, usually because urchins are on the rise, but also because of warming waters. This matters, because cold-water 'ocean forests', as they are known, once predominated over tracts of sea-bed around the world that, in total, have a greater extent than the entire Amazon rainforest; and many marine species depend on them for food or shelter. Reversing kelp decline means fixing the urchin problem. In California, the return of once-hunted sea otters (which feed on urchins) has helped to restore giant kelp; although warming waters threaten to send those gains backwards. In Tasmania, it may be down to the rock lobsters, which, locally, are significant predators of urchins. But only the largest, oldest rock lobsters are able to manipulate a long-spined urchin and crush its test to extract the meat within—and fishing pressure means that there are usually too few of those around. A thriving Asian market in urchin roe is giving local divers an incentive to harvest the urchins; but it's a constant battle, because a diver can only reach those living in relatively shallow water, whereas there are urchins in deeper water, that can move in to replace those harvested. Furthermore, a steady stream of urchin recruits is always on hand in the form of larvae settling out from the plankton. Once giant kelp is gone and urchin barrens take hold, it's hard to get it back, because baby kelp plants like to settle in areas already occupied by established plants. All in all, circumstances seemed stacked against giant kelp's comeback, but at least, on this day, and in this place, I could catch a glimpse of its magnificence and its potential. I felt deeply privileged.

Having been bobbing around amongst the kelp for a while now, I realised that I was beginning to feel a bit queasy, so I set my sights on a small cove, not much more than a gap in the rock platform bordering the foot of the cliffs. With a bit of luck, I could haul up there. Once again, the sea conditions proved deceptive, funnelling powerful swells and my kayak through the gap towards a steep and uninviting boulder beach at the back of the cove. With difficulty, I managed to ferry-glide my craft sideways, until I was able to dock next to the rock platform at the side of the cove. I scrambled out, hauling my kayak behind me before the next wave arrived. The platform was more or less level and a good metre and a half above the tide, so I could relax and take in my surroundings. Its surface was pitted and pocked, each shallow impression a mini rockpool encrusted with pink coralline algae and fringed with olive-brown Neptune's necklace and luminous green gutweed. Orange-edged limpets and siphon shells (the latter, limpet-like but from a very different, air-breathing gastropod lineage) clung to exposed surfaces; while dense, glistening beds of mussels crowded every crevice.

I could discern three mussel species. Tiny black flea-mussels predominated, but there were also their larger cousins the beaked mussels, whose purple-tinged, finely ribbed shells protruded from the flea-mussel matrix. Every now and then there was a sprinkling of blue mussels—the only one of the three considered large enough—*muscular* enough—to be worth harvesting for food. 'Mussels' and 'muscles' are etymologically related: both words derive from the Latin for a 'little mouse', supposedly because mussel shells, and some of the body's rippling muscles, bear some resemblance to a small, furry rodent. The shells of some mussel species, known as hairy mussels and horse mussels, are indeed endowed with 'fur' (it's called periostracum), so perhaps that's where the supposed likeness to a mouse originates. Our local (furless) blue mussels have long been a taxonomic enigma. One species is native to temperate Australian waters, but as soon as European sailing ships reached our shores, an interloper from the Mediterranean and North Atlantic (one that readily fouled the hulls of vessels) began to colonise local harbours, docks, and pier piles. In due course it began fouling the hulls of local ships, spreading it further—blue mussels certainly know how to get around. On my *Investigator* voyage I even spotted them a hundred kilometres out in the Southern Ocean, attached to rafts of drifting kelp by their byssal threads (the proteinaceous strands that 'glue' mussels to their substrate).

The North Atlantic and Australian blue mussel species cannot be told apart just by looking at them, and they are so closely related that they readily interbreed. These days, genetic testing shows there are few places in Australia where pure-bred native blue mussels exist, but they still do so along some of Tasmania's remoter, and more exposed, rocky shores.

Some of the shallow pools on the rock platform bore accumulations of shell grit, cast there by storms. These days I could just let the grit be; but I recalled that, in earlier years, I wouldn't have been able to resist scooping some up, since it offered the prospect of rich pickings for micromolluscs. I would always carry resealable plastic bags for just such an occasion. As often as not, when I went to scoop up some of the grit with my hands, I would be dismayed to see the accumulation retract before my eyes and disappear into a crevice. How I came to curse those shell-grit anemones! It often seemed as though they had it in for me, deliberately taking up residence in the spots with the choicest accumulations of micromolluscs, with which to plaster their bodies (presumably for camouflage or physical protection). Still, anemones were applying shell-grit armour to their bodies for millions of years before I came along, so I really shouldn't have begrudged them their share (but I did). On one occasion, not far from this spot, I had managed to find a small patch of grit that hadn't been monopolised by the evil anemones; and later, when I picked through the grit under the microscope, I found that I had collected a Tasmanian air-breather—a marine almost-gone-terrestrial snail in which the aperture's outer lip is internally adorned with tooth-like extensions. Similar extensions borne by their more widespread cousin, the common air-breather, earn that species the colloquial name of 'toothies' among Indigenous makers of shell necklaces. I was delighted with my find, since this was a rarity that I had only collected at one other spot (another rock platform in the far south of the state). Two distant shorelines, forever joined in my mind by their rare molluscs. I love such connections.

Higher up on the rock platform, shards of rock whelk and wavy turban shells told of the local kelp gulls' strategy for extracting the seasnail's flesh. At low tide they would pluck one of the snails from the rocks, fly with it to a great height, and then drop it onto the platform below. With a bit of luck, the shell would smash and the gull would get its dinner. Judging by the sheer number of shell fragments all around me, it was a popular and successful strategy.

Turning my attention to the greyish mudstone at my feet, I metaphorically put on my amateur geologist's hat. Here and there were irregularly shaped stones, clearly of a different geological composition, embedded in the mudstone matrix. Termed 'drop-stones', they are evidence that when the mud was being laid down, the ocean above was awash with icebergs calving off from glaciers, which had emanated from nearby landmasses. These carried with them assorted lumps of ice-ground bedrock, plucked from the base of the glacier, along with finer glacial till. As the glaciers melted, out at sea, those lumps of glacial rubble dropped to the seabed, joining the sediments that would eventually become Permian mudstone, frozen in time rather than ice.

In mudstone mere metres from me, but probably separated from the drop-stone strata by millions of years, there were fossils—delicate, frond-like structures made by colonial animals called lace-corals, and the impressions of clam-like animals called brachiopods. The similarity of these organisms to corals and clams is superficial rather than taxonomic. Both groups of organisms exist today, although contemporary brachiopods are a shadow of their former glory. Back in the Permian, about a quarter of a billion years ago, they formed great reefs similar to those of today's (or yesteryear's) oysters. Conveniently, for geologists, the mix of brachiopod species changed over time, making their fossils valuable for matching up and dating rock strata. These days, the last brachiopod survivors eke out a spartan existence on the seabed. Some live in shallow waters, including around Tasmania; but most are restricted to very deep waters, having been ousted from more favourable spots by the likes of those oysters and other clams.

At this point in my haul-out, I allowed myself a moment's reflection on a much earlier excursion with our young sons, who had been given kid-sized geological hammers and safety goggles and were desperate to go fossil hunting. On that occasion we arrived by rowboat—a rather battered fibreglass one that had been rescued and rehabilitated by a neighbour, before being handed on to us. It was a fun thing to take it out for a row, but only on dead-calm days because, with four of us on board, the waterline came unnervingly close to the gunnels. My recollection is that we found fossils (perhaps the very same ones that I was looking at, now) but that the boys' hammering was no match for the Permian mudstone, which, to their parents' relief (amateur fossil-extraction being generally frowned-upon), refused to yield to their blows.

Judging by the fossil record, life seemed to be going great guns during the Permian. As the millions of years rolled by, the glaciers retreated and vanished. The warmer seas exploded with life, while sail-backed *Dimetrodons* and other synapsids ('proto-mammals') stalked the land; which was, increasingly, swathed in tall forests of horsetails, clubmosses and seed-ferns. But, within another few million years, nearly all life on Earth—in the sea and on land—was wiped out in the aftermath of a runaway greenhouse effect, which may have been triggered by massive volcanic activity. Those reef-forming brachiopods were among the losers.

Crouching on that rock platform, wedged between the crumbling, fossiliferous mudstone cliff and a sea once again teeming with marine life, I found myself contemplating how different the world might have been, today, if there had been no end-Permian mass extinction. Would the brachiopods have continued to dominate and evolve into sea-floor giants or lifeforms capable of colonising the land? Perhaps the mammals would have gained ascendancy over the reptiles a hundred million years earlier than they did. If so, there might have been no dinosaurs; hence no birds. There would almost certainly be no humans, today. Maybe some other animal lineage would have evolved into a physically and mentally dexterous, human-like creature instead, and beaten us to it by millions of years. Luckily for us, we will never find out. I brushed my fingers mindfully over the delicate ridges and troughs outlining the imprints of the ribbed brachiopod fossil, experiencing a brief moment of deep connection with that lost world, which was so distant in time, yet existed right here, in the very spot where I now crouched.

Having pondered the imponderable for long enough, and feeling chilled by the cold June air (and perhaps by thoughts of icebergs), I re-floated my kayak and managed to clamber aboard without capsizing. A westerly breeze had picked up, so I decided to turn homewards, allowing the wind to help propel me towards Taroona. I glided past the cave where, a few years back, a modern-day hermit had taken up residence. His chair still stood vigil at the cave's entrance. These days, the cave is again vacant, its dusty floor thick with the footprints of penguins and water rats. A shrill *keek, keek*, from high on the cliffs, told of a wary peregrine falcon, upset by my presence far below. I couldn't spot it, but a pair of these magnificent birds of prey—reputedly the fastest of all flyers—breeds along here most years; although, at this season, the bird would probably have been prospecting, rather than staking out a nesting territory. I am of a generation—the *Silent Spring* generation, you might call us (alluding to Rachel Carson's seminal 1962 book of that name)—who coincide with the period in which peregrines were very nearly lost to the world. They were victims of now-banned bio-accumulating pesticides, such as DDT, that thinned their eggshells to a point at which few could be brooded long enough to hatch. Though the birds have returned from the brink throughout their near-global range, hearing one still brings me immense pleasure and a sense of relief that we didn't lose them entirely. They are another reminder that nature is resilient, but only up to a point.

The wind continued to push me homewards, so I sat back and enjoyed the ride. Up on the skyline, gnarly, fire-ravaged silver peppermints—arguably the hardiest local eucalypts—somehow found sustenance and anchorage perched atop the skeletal soils typical of mudstone country. I passed the site of a recent rockfall that had sent peppermints and blocks of stone the size of small trucks crashing to the foot of the cliff only a few years previously, after a day and night of heavy rain. I was pleasantly surprised to see tree ferns clinging to a moist crevice half-way up a shaded, south-facing section of cliff; and I was buoyed by the knowledge that a lone clump of invasive pampas grass seemed not to have spread beyond its perch on a slope inaccessible to council weeds officers. A splash of whitewash on the shoreline boulders was also heartening, telling me that penguins were making good use of little concrete houses that had been installed for them among the bower spinach, to encourage them to roost and nest, safe from the threat of predation by feral cats and errant dogs.

Back at the beach, exiting the sea was easier than entering it, as the wind and swell combined to deliver me into ankle-deep water, avoiding a final drenching. There is something deeply satisfying about returning safely from a solo paddle. It leaves me feeling uncommonly at ease with the world. On this occasion, mulling over geological timescales no doubt also helped me to feel at ease with my own fleeting and insignificant place on this planet. Humility, gratitude, exertion, fresh air and nature—what's not to like? That said, the cold winter air was now biting, and I was glad that a hot shower was only a short walk away. Once again, I had cause to reflect on our good fortune in being able to call Taroona, and Tasmania, home.

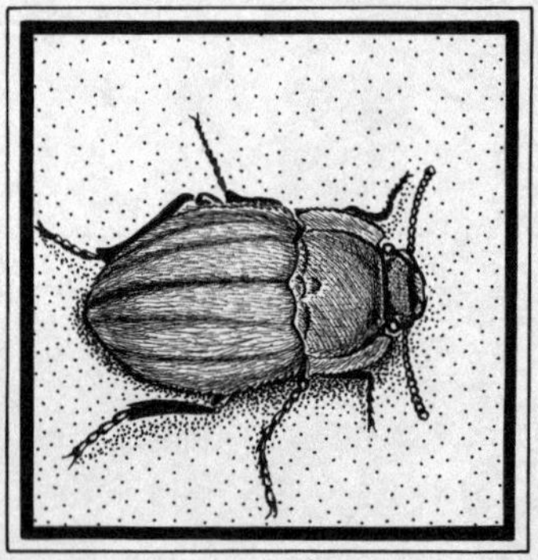

Beach wrack

Even for a hardened malacologist like me, beaches aren't all about sun, sea, surf, and molluscs. They are also home to an exclusive suite of insects found nowhere else. After I had discovered quite a few beach insects on one of our museum expeditions, I mentioned the fact to my entomologist friend Lynne (whom we met back in February, and again in May). Lynne, it turned out, had a fascination with beach insects that ran even deeper than mine, so, one fine but cold June day, we decided to combine forces and go bug-hunting on a few beaches close to Hobart.

If June sounds like an odd time of year to go in search of insects, you'd be right. If June seems like an odd time of year to pack a picnic and head to the beach, you'd also be right; but Lynne assured me that there were beach insects to be found at any time of the year, and the secret to their winter activity was the piles of rotting kelp heaped high on the shore after winter storms. Much as a garden compost heap steams in winter, thanks to the heat of respiration given off by millions of decomposer microorganisms, piles of half-buried seaweed can also rise several critical degrees above the ambient temperature of the surrounding sand. And, where there are decomposers and detritus, there are likely to be detritivores and their predators, too. Lower down the shore, those roles are claimed by various marine crustaceans such as isopods (sea slaters) and amphipods (beach fleas). Some extend to the strandline, and beyond; but, at increasing distances from the water, their roles are usurped by insects, and lots of them.

Our first destination that day was Calverts Beach, which overlooks the mouth of Frederick Henry Bay, on the South Arm Peninsula, just to Hobart's southeast. A high-pressure system in the Bight was directing southerly breezes our way, but these were gentle enough to do little more than induce watery eyes and dewdrops on noses, as we scaled the dunes and jumped down onto the coarse sandy beach. The full magnificence of Frederick Henry Bay was laid out before us, picturesquely framed by Betsey Island, to our right, and the Slopen Main coast of the Tasman Peninsula, across the water to our left. Gannets torpedoed into the lively blue waters immediately in front of us, while a bickering group of silver gulls provided a suitably littoral soundtrack of screeches and whines from further along the beach. I noted, with relief, that the tide was well out, so we were not at risk of a surprise soaking while we had our backs turned. Given the time of year, we could hardly have wished for better conditions.

Scanning the strandline, I could make out many discrete piles of weed, each of which, I imagined, might prove worth investigating. Lynne's experience had taught her that not just any old beach wrack would do, though; so we continued past innumerable dark, wiry amalgamations of crayweed and other algae, until we happened upon a much more massive pile, comprising several tangled stems of giant kelp with their leafy fronds and floats still attached and partially intact. Giant kelp, Lynne assured me, was second only to bull kelp in its suitability for beach insects. Crucially, the kelp appeared to have been sitting on the beach for a good few days, and now lay half-buried in sand. We sank to our knees and began carefully teasing apart the glistening, olive-brown strands, specimen tubes and cameras at the ready. Sure enough, we soon met with our first insects: tiny flies, scarcely bigger than a pinhead. Despite their size, I discerned three different sorts, which I later identified as beach-specific species of lesser dungfly, biting-midge and a uniquely Australian seashore fly bearing the imaginative name of australimyzid, which translates as 'southern fly'. Next, we dug a bit deeper, and some larger flies emerged from the morass. Squat and densely hairy, these were kelpflies. Some were several times the size of others, but I couldn't be sure whether we were seeing more than a single species, because kelpflies vary greatly in stature, depending on whether they experienced feast or famine as kelp-feeding larvae. Their ability to continue to adulthood, regardless of larval deprivation, is a necessity for flies whose larvae live and feed in such patchy and short-lived habitat. It would only take one king tide to snatch away the kelp, leaving the larvae high and dry and with little prospect of finding further sustenance. Kelpflies are found in the temperate zones of both hemispheres, but they reach peak diversity around the kelp-rich shores of Tasmania.

A sudden movement among the kelp had Lynne reaching for a specimen tube. A lithe, dark form about a centimetre and a half long emerged from between her fingers and made a dash for freedom, launching itself into the air and crash-landing a few metres further along the beach, before digging itself into the sand. This was one of the many rove beetle species exclusively found in kelp. As I mentioned in March, when introducing one of their kin (which I had found on a wallaby carcase), rove beetles are unusual in possessing much-reduced wing cases and extra-flexible body segmentation. This enables these beach-dwelling species to weave their way among the kelp strands and through the sand in pursuit of kelpfly larvae, mites, and other prey. It amazes me that they can keep a pair of functional wings tucked under their wing cases. I have watched slow-motion footage of a rove beetle re-folding its wings after landing—the feat is on a par with the best origami artists.

It was now time to really get into the kelp. We dispensed with the niceties and plunged our hands and forearms deep into the heart of the heap, where mucilaginous gloop oozed from the rotting kelp stems and stained the sand a deep, orange-brown. It sounds like a disgusting thing to do, but it was surprisingly pleasant—the gloop was warm, the sand soft, and the odour more 'earthy' than 'seafloor'. More importantly, there were beetles to find down there that we wouldn't have encountered if we had merely picked and plucked delicately at the kelp's surface. Many lived at the interface between the solid kelp and the gloop. Presumably, this is where peak microbial activity is concentrated. A tiny, globular species of beetle that was common in this zone was a beach-specific type of hydrophilid—a word that means 'water-lover', as most other members of the family are water beetles. Slightly smaller, oval, and purplish-brown beetles were beach hister beetles, a family of predatory beetles that specialise in feeding on soft-bodied invertebrates such as those kelpfly larvae. And there were more rove beetles, too—a tiny species with a dark head and tail and a coppery thorax, so small that each grain of sand posed a challenge to their escape from the sudden brightness. Their name loosely translates as the 'southern iota', a term which fits their physical insignificance perfectly.

Having dug down to the bottom of the vein of gloop and found all the beetles we were going to find there, we refilled the hole and repeated the process at several other heaps of kelp, eventually adding a few more fly and beetle species to our list. Then we rinsed our hands as best we could in the water of incoming waves, and enjoyed our beach picnic, warming up with flasks of hot tea. Given the brevity of the winter day, we decided to press on to a couple of other beaches, which differed in their exposure and the composition of wrack left behind by the tides. We rounded the South Arm Peninsula and tried our luck at Opossum Bay, before returning to the relatively sheltered shores of Howrah, in Hobart's eastern suburbs. Though we continued to add more species, it seemed that our first spot was the best, probably because it was the only one with giant kelp. Giant kelp is well known for the marine life it supports while alive, but our beach rummaging demonstrated that its ecological value continues long after death. Given the catastrophic population collapse that has affected giant kelp in recent years, the future for many Tasmanian beach insects may be bleak.

Perhaps it is just as well that not all beach insects are associated with rotting kelp. There are tiny beetles that specialise in scavenging other dead insects that the tide washes up. One such species, Lea's ant-like flower beetle, has a serrated, shield-like extension reminiscent of a miniature hedge-trimmer, emanating from the back of its neck. This may protect its head from competitors or predators. Another, which I shall translate as the beach wrack-seeker, is in a uniquely Australasian sand-dwelling family of beetles, whose bodies are covered in silvery scales and downwards-pointing bristles that may afford protection from windblown sand or ants. Then there are larger beetles in the darkling beetle family, which feast at night on dead crabs, seabirds and fish high on the shore. Some of these burrow into the beach sand by day, while others make nightly excursions from adjacent dunes. Most look nothing like their more terrestrial relatives. One of the finest, with a deep maroon body fringed with silvery hairs (its scientific name translates as 'hoary' or 'greying'), is only found in Tasmania. And there are beach weevils, too, including a couple of intertidal species that emerge from the wet sand at night, probably to feed on algae. Most such beetles are best observed by torchlight, and many are just as prevalent during the winter as in the warmer months. As for flies, in summer at least, there are beach robberflies, beach stiletto-flies, beach marchflies, and many others—most bear silvery hairs, either for camouflage, when perched on the sand, or as protection from windblown sand (or both). We are gradually building up a better picture of which species can be found where, around Tasmania; yet, as in so many areas of natural history in, on and around this island, the field remains wide open for all comers. Why not give it a try?

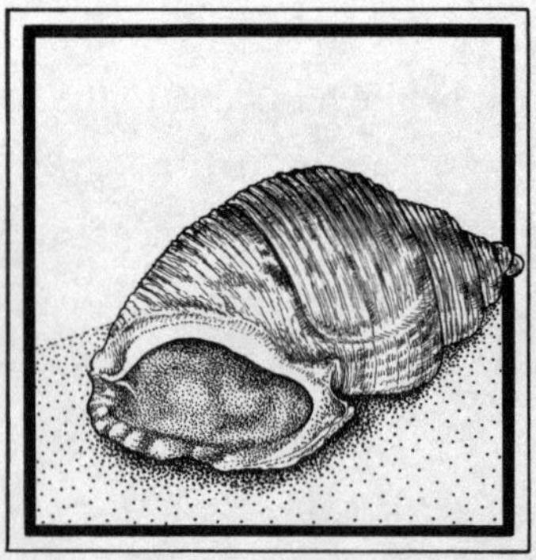

Malacology and mindfulness

With the prospect of a spell of settled weather, I once took a midwinter solo holiday, indulgently quartering the beaches along the east coast, between Orford and St Helens, in my ever-absorbing quest to find, and learn more about, Tasmanian seashells. At the time there was one species that I was particularly keen to find: the southern false-olive. Not only is it a beautiful and rather unusual shell—luminous creamy-orange with a solid, stepped spire—but it's also taxonomically quite distinct (having very few close relatives), and rather rare to boot. Its core range is the coast of New South Wales. Tasmanian records were all from the upper east coast; hence the focus of my trip on this region. It's a species that lives in sand a little offshore, beyond the surf break, where it is a predator, probably of other sand-dwelling molluscs. So, my hopes were raised whenever I found an expansive beach featuring strong surf and a strandline heaped with shells. Beached shells aren't uniformly distributed, though. It takes a particular combination of shoreline geometry, aspect and currents to generate a good shell beach; and, of course, what is good one day may not be so the next—due to the inconsiderate vagaries of tide and wind.

All this I learned first-hand on my perambulations. But I also discovered a couple of spots—already well-known to seasoned beachcombers—where the shells just keep on accumulating, and form massive storm ridges, metres thick in places. One is at Falmouth; the other (imaginatively named Shelly Point) is at Beaumaris. It was to the latter that I was now headed. Swaddled in multiple layers of warm clothing, I walked the few paces from the car park to the beach, savouring the salty, pungent aroma of freshly beached kelp, while my face was pummelled with blasts of icy-cold, dry air. Even in midwinter, with the sun low in the northern sky, all was brightness: the sandy beach stretching into the distance both left and right; the foamy surf breaking on the rocks in front of me; the azure sky merging imperceptibly into the sea beyond the rocks. The surf and wind, too, conspired to serve up white noise. And, at my feet, a broad band of thousands upon thousands of sun-bleached seashells. Each crunching step forward felt like an insult to this place of shells, so I plonked myself down where I was. I estimated that ninety-nine per cent of the bulk of those heaps comprised just one species of dog cockle, but that still left an awful lot of other shells to explore.

So began the now-customary routine: sifting, turning-over, squinting, rolling-over, kneeling, hunching, cramping, attempts at uncramping, endless scrunching. Nature is fractal, I have come to appreciate, like a Mandelbrot set: the closer you look, the more whole new worlds open up in front of you. And the closer I peered at that beach, and the stiller I allowed myself to become, the more I saw. Bottom up and eyes down, my world shrank to the few square centimetres in front of me, and I gave them my utmost attention. My breathing slowed. I no longer felt cold, or discomfort. All extraneous thoughts dissipated. Time stopped. There was only this place, this moment, these shells. I was enchanted by their tiny, intricate forms and diversity. Though I knew their names, they didn't matter now. Textures, colours, patterns. Still tinier particles of sand, crystalline, glinting in the sunlight or glued together with kelp gloop. Beach fleas backing into mini-crevices between the shells. Fragments of penguin quills. Long-legged mites strolling among the discarded skins of beach slaters and barnacles …

I spent hours on that beach, lost in seashells but grounded, contented, found in myself. I did eventually discover a false-olive, but, by then, I'd

© K. DAVIS 2023

July

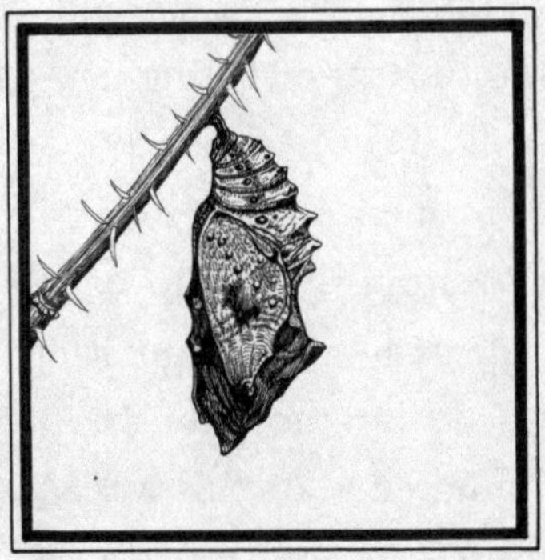

Indoors weather … not

July is a month to endure, rather than relish. I can find myself thinking, '*surely nature intended us to hibernate now*'; and struggle to live up to my own maxims: seize the day; embrace the weather! The days are short, and the weather is often inclement; even capricious. When the sun actually shines, it lacks warmth, and its oblique gaze seems devoid of sympathy. I need coping strategies for such coldness. Those with the wherewithal often head north, and I won't deny that it's a tempting proposition. But there *is* pleasure to be found in all weather, if you're properly prepared. Enjoy your cheeks tingling with a wetting by winter rain, or an upper-body massage courtesy of a pummelling westerly gale; or the bushwalkers' simple satisfaction of a steaming-hot cuppa poured from a thermos. For those who don't mind delayed gratification, there is the satisfying inner glow that follows a brisk walk in the rain, sleet or snow.

It is not easy to come up with a butterfly for July. There are no butterflies that hibernate as adults in Tasmania, so you cannot hope for one to casually pop out to catch some winter sunshine beside your path. I have decided, therefore, to declare July Tasmania's only butterfly-free month. In fact, I take some perverse pleasure from it. Sure, I could have turned to a butterfly egg, caterpillar, or chrysalis; but, as Oscar Wilde once quipped, '*consistency is the last refuge of the unimaginative*'. I wish I'd said that. I'm doing my best to cultivate a little of his imagination.

'Chrysalis', by the way, is derived from Ancient Greek (as usual with these things). It incorporates their word for 'golden'—the chrysalides of many butterfly species are spangled with reflective splotches of gold or silver. I can still recall when I first found the chrysalis of one such butterfly. I must have been about five or six, and had been wandering about with my big brother. It would have been in July—summertime. We were on the quiet, dusty, potholed roads near our house when I spotted it. The chrysalis was attached by its silken pad to the leaf of a stinging nettle, and this is what the caterpillar had been feeding on. It was a thing of such ineffable beauty and wonder that the memory of it lingers still. Before science adopted the word 'chrysalis', naturalists had toyed with 'aurelia' instead—it is a direct Latin equivalent, again incorporating 'golden'. Aurelia didn't stick, but before it vanished from the English lexicon it spawned 'aurelian'—for close to two hundred years this was the accepted term for a butterfly and moth enthusiast. It only went out of fashion after the more scientific-sounding 'lepidopterist' gained currency. Meanwhile, many lepidopterists prefer to use the more general term 'pupa', rather than 'chrysalis'—this is Latin, and means 'doll'. Even the language transforms. Perhaps there will, one day, be butterflies to be found, here in Tasmania, in July.

Darkness and silence

I have trodden the footpaths and pavements of my neighbourhood for more than two decades, so I have come to know them well. On the weekdays when I don't cycle to work (of which there are more in winter than at other times), I like to take a route, for exercise, that follows the foreshore. It starts just a couple of hundred metres from our house. Taking in two sandy beaches, a section of clifftop (with a steep set of steps to get the heart pounding), a small creek crossing, and a rocky headland, it then returns inland, via the oval and park. In summer, it's rare to find myself walking after dark; but in winter it's the norm. You might think that a winter night's walk wouldn't offer many opportunities to encounter creatures. Indeed, the pickings are lean; but there's nearly always something to catch my attention. Most reliable is the bright orange eye-shine of brush-tailed possums, caught in the beam of my headtorch, usually in a tree. Often the possum will just sit there. It seems unfazed by the glare, but perhaps it has been blinded into inaction. Even when not seen, the maniacal hisses and shrieks of possums regularly pierce the silence.

At the far point of my walk, I've occasionally encountered a leucistic possum, that is, one which lacks much of the dark pigment that usually turns a possum's fur deep brown. Normal possums are difficult to discern in the darkness, except by their eyeshine; but this individual's solid, almost golden, furry form is accentuated by torchlight and hard to miss. I find brushies rather cute, from a distance; but that doesn't translate into forgiveness for the havoc that they periodically inflict on our fruit trees and veggies.

Ringtail possums are more unequivocally pleasing. They look even more endearing and are no threat to the garden. Sometimes I only hear them: a wavering, squeaky, almost electronic (or flywheel-like) contact call between family members. It usually comes from within a thicket of understorey shrubs, rather than high in the canopy. Their neat droppings often litter the nature strip outside our house, since they adeptly use overhead power lines as an aerial highway. Once, on returning from a nocturnal winter walk, I looked up to see four ringtails all in a row—their prehensile, white-tipped tails coiled around the cabling as they stared down at me from on high, ears and noses twitching. On another occasion, I heard a scrabbling sound high in some she-oak bushes beside the path, and looked up, anticipating a possum. Instead, my torchlight settled on a black rat, confidently running along a horizontal branch, before leaping across a narrow gap to another, on the other side of the path. We may think of feral rats as ground-dwelling rodents, and that's certainly true of the brown rat; but the lighter-weight black rat is an accomplished and nimble climber and equally at home in the trees, where it is a threat to nesting birds.

Other than the waves, and the occasional indignant, grating shriek of a masked lapwing, the beach is quiet at night. But have you heard of the orange-bellied other-otter? Probably not, since it's my pet name for what others know as the water rat or rakali. It's true that this giant native rodent is essentially an oversized rat that likes water; but it's so much more than that—it is the antipodean equivalent of an otter, in appearance, diet, and foraging behaviour. And it does have an orange belly, along with a thick, white-tipped, well-furred tail. My walk takes in the home range of at least one rakali. Where the creek, intermittent rivulets and storm drains disgorge onto the beach, it's hard to miss the musky, fishy odours that signal (to humans as well as other water rats) that one has scent-marked key points of its territory. Rakalis like to patrol their territory just after dusk, so my walk along the beach sometimes coincides with theirs, although they keep close to the waterline, where their dark, hunched outline is barely discernible against the rippling wavelets. By day, the broken, hollowed-out tests (shells) of urchins, or a pile of crab legs on the beach, are signs of a rakali's successful night of fishing in the shallows.

Returning home via the park, I often catch a glimpse of a frogmouth out of the corner of my eye. More precisely, I sense a ghostly blur, vaguely the shape of a frogmouth, or some silent movement. These owl-like hunters of moth and mouse are masters of the art of not being seen. Their deep, upwards-inflected, repetitive hooting is almost subliminally quiet, as though they don't want anyone to hear them either. By day, frogmouths are even harder to spot. Their intricately mottled, grey-brown plumage and motionless pose render them near-invisible on a chosen tree limb, hiding in plain sight. Bandicoots, on the other hand, secrete themselves away by day in thickets of low vegetation such as tussock grasses and blackberries. I am pleasantly surprised by bandicoots' resilience in the suburbs, given that they seem as though they would make excellent quarry for dogs and cats. On the Australian mainland, foxes have more or less eliminated them. Long may Tasmania remain fox-free.

Both southern brown and eastern barred bandicoots frequent my local park, and it's unusual not to see one or the other during an evening walk. Brown bandicoots are rather portly, almost barrel-shaped animals, and usually seem unconcerned at my approach; whereas barreds are lither and sprightlier, but more skittish. They are prone to leap off in random directions when spooked. At one time I would see more browns than barreds; but these days, for reasons unknown, it's the other way around. The snuffle-holes that they create in the grassland, as they search for tasty roots, tubers or insect grubs with their pointy snouts, are a characteristic sight by day, not only in the park, but also along the nature strips and even in our lawn. In natural settings, these holes are valuable places for seedlings to establish themselves in and thrive, so the bandicoots' gardening services help boost plant diversity and ecological succession, to the benefit of all.

There are also invertebrates that may be spotted on winter evening walks. Wearing a headtorch on a beach at night is asking for trouble—the light source and adjacent parts of one's face can attract legions of confused and, apparently, deranged kelpflies. Much as I like kelpflies, it's often more pleasant to walk on the beach with the torch at hand-level, although one then misses out on the enchanting twinkle of mica and quartz crystals in the sand.

Millipede-killers—a large, flightless, rather grotesquely proportioned species of assassin bug—are often out in the open after dark, sitting around on paths and awaiting the passage of prey. Despite their eminent collectability for study, there aren't many taxonomists in the world with expertise in assassin bugs (as with so many invertebrate groups); and our local species remains scientifically undescribed and unnamed. In the park, not all eyeshine comes from possums, bandicoots or frogmouths. Lower down there is likely to be bluish-white eyeshine, like tiny diamonds in the grass. It belongs to wolf spiders, which leave their burrows at night to go a-hunting, even in winter. The intensity of the reflected light sometimes suggests truly monstrous spiders, but I am constantly surprised when I home in on the source. It is, inevitably, a tiny spider with a leg span of mere millimetres.

One moonless evening, last winter, I decided that I knew my route well enough to attempt the entire walk without using my torch. The streetlamp-lit sections were a doddle, but I wanted to challenge all my senses on the unlit sections of foreshore, clifftop path and parkland. The transition from lit to unlit was abrupt, and I had to give my eyes time to adjust before continuing on my way. There was enough reflected light on the sea for the shoreline sections to be relatively straightforward, bar the odd unanticipated boulder. The clifftop section was in near-total darkness, though, and I was reduced to inching my way forward by tiny steps. Without the benefit of eyesight, I noticed how my hearing came to the fore.

We think of bats as masters of echolocation, but even humans have basic abilities that enable us to generate sound-derived mind maps. Echolocation essentially amounts to triangulating one's position, and the position of potential obstacles, by registering slight differences in the timing of sounds detected by each ear, especially when those sounds are echoes of noises we generate, ourselves. I can't claim to have mastered the technique on my night walk, but I was intrigued by how my body seemed intuitively able to 'feel' where the path ahead lay, based largely on auditory input and feedback: the sounds of my footsteps and breathing, and a gentle breeze in the path-side bushes. No doubt this revelation is old news—I imagine our ancestors having taken these abilities for granted over most of our evolution. It's probably only since artificial lighting has become so ubiquitous, that we have begun to pause long enough to give this innate sense our conscious attention.

My nonvisual senses now heightened, I sat for a while on the bench at the top of the steps. There didn't seem to be much to hear, to be honest. I struggled to distinguish real sounds from internally generated white noise. It's strange where, and how, claustrophobia can strike, even out in the open. A warm draft of air brushing my cheeks put me on high alert. Goosepimples rose to the occasion, and I felt a sudden shiver down my spine. Part of me wanted to up sticks and flee from the grotesque, predatory monster whose breath I might just have felt on my face. Or maybe it was a walking, grinning human skeleton (my usual imaginary ghoul of choice; perhaps I have spent too long in museums). I strained to see or hear anything that might calm my nerves, but there was nothing. Sensibly, and against all my instincts, I decided to sit tight until the feeling passed, and it eventually did. Once my heart had stopped pounding, I resumed my walk, with a slightly quicker descent to the beach, where I knew there lay the prospect of starlight, open space, and safety.

I have given up trying to walk in the pitch dark. We're not meant to be nocturnal, I've learnt. It's why we have torches.

Aurora australis

Some nights, the sound of traffic intrudes upon my dreams. It's unusual, since our street is normally quiet, but it does lead to a beach. And the beach faces due south, making it a popular spot for aurora-watching. Auroras (or the aurora australis, to give the phenomenon its full name) peak around the equinoxes, when the sun's 'north-south' magnetic field coincides with, but is magnetically opposite to, Earth's; but the long nights of winter enhance the chances of an aurora coinciding with darkness. When it does, and if the forecast is for clear skies to the south, the word soon gets out. I receive aurora alerts on my phone, as would many of those who arrive by car; but if it's past bedtime I won't know about them until morning; or until traffic noise awakens me. If I hear a marked increase in traffic earlier in the evening, I will sometimes wander down to the beach to see what all the fuss is about. A good aurora will attract scores of observers, sometimes more than the car park can accommodate. There are all sorts. Many are what I might call 'casual hopefuls', who have seen some of the spectacular photos that abound online, and turn up expecting to see something similar with their own eyes. Then there are the serious, sometimes fanatical aurora chasers—photographers for the most part, who come with all the gear and know how to maximise their chances of capturing that perfect shot. Some will bring bivvy bags, swags or camping chairs and set themselves up for the night; others will take one knowing look at the sky and decide to cut their losses and leave.

The thing about auroras at our latitude is that they tend to be very faint. Usually, they only manifest across a small part of the sky, close to the southern horizon. To the naked eye, they typically appear as an eerie green glow. At their peak there may be other colours and, very occasionally, vertical bands of light, termed 'beams' and 'picket gates' by afficionados. These may wax and wane over mere minutes, or hang about for hours. We're lucky to be able to see them at all. At a similar latitude in the northern hemisphere—in a European context, think Marseille or Barcelona—there would be no expectation of seeing auroras. Fortunately, for us, the magnetic south pole, over which the auroras blaze, wanders around over the years and is, presently, a long way north of the geographic south pole. Furthermore, it is north in the direction of Tasmania and New Zealand, rather than towards South America or South Africa. This means that our latitude relative to the *magnetic* south pole is a good twenty degrees further south than our actual position.

The interplay between solar wind and Earth's magnetic field choreographs our auroras. The sun constantly throws charged particles and radiation in all directions; but sometimes a solar flare ('coronal mass ejection', to give it its proper name) will fling more than the usual quantity in our direction, constituting a 'solar storm'. When the storm reaches Earth, the planet's magnetic field shields us from the worst impacts. This occurs, partially, by deflecting the wind towards the magnetic poles. The stronger the solar wind, the more it will penetrate our shield around the poles and interact with gases towards the outer limits of our atmosphere, ionising them. In simple terms, collisions with oxygen molecules produce green light, nitrogen molecules produce pink and deep red, and hydrogen and helium produce blue and purple. Auroras can also induce a low-frequency crackling sound, though it may take an exceptional solar storm, and extra-sensitive hearing, for these to be detected at ground level.

Until the advent of digital cameras with ultra-sensitive charged couple devices (the digital equivalent of photographic film, for those who remember the old ways), there really would have been little to see here. If you had wanted to photograph the aurora australis, you would have needed to head much closer to Antarctica, preferably to *terra firme* on a subantarctic island, or the icy continent itself, so that you could steady your camera. (You might have arrived aboard the Antarctic Division's vessel, *Aurora Australis*, affectionately known as 'Orange Roughy'. Its modern-day replacement, *Nuyina*, is named with the Palawa kani term for the aurora.) Today's cameras have democratised aurora photography. With the help of a tripod (allowing for long exposure times), a decent digital camera (and even some phone cameras), can gather the faint light of a distant aurora and build up a picture that reveals the richness of its colours and luminescence. Follow up with a bit of digital post-processing and you potentially have some stunning photos. That's why serious aurora chasers keep coming back to places such as Taroona.

My own half-hearted attempts at aurora photography have generally not ended well, but to be honest, I'm not really that bothered. It's a phenomenon that I like to witness from time to time, as much for the shared sense of excitement among watchers as for the aurora itself (or should I say 'herself', since she is often referred to as 'lady aurora'). Although the lady may sometimes dance, her green, pink, and blue dress undulating gracefully in the solar wind, an aurora is mostly quite clearly inanimate. I think that's the issue for me—auroras are works of nature, to be sure, but of physics, rather than biology. They may generate a sense of awe, but, at heart, they are entirely explicable by the laws of physics, leaving little room for wonder. Contrast that with almost any encounter with the living world, and the faint glow of a distant aurora on the edge of space pales towards insignificance; for this naturalist, at least.

I am more awestruck by the sight of a starry sky. We in the south are fortunate that we look straight out into the Milky Way, viewing the galaxy's disc more or less end-on. We are also blessed with some of the clearest skies on the planet, relatively untainted by light, soot and other pollutants. The stars are crisp and twinkly, rather than a smudge of dim light, as experienced by many in the north. At different times of the night, one can witness an arc of stars and luminous dust (the 'milkiness' that gives our galaxy its name; and is the root of the word 'galaxy') stretching from one horizon to the other, or arching directly overhead.

The best aurora photos, in my opinion, are those which capture this celestial majesty too, and remind us of the vastness of space and our little place in it—'*far out in the uncharted backwaters of the unfashionable end of the western spiral arm of the galaxy*', as Douglas Adams wrote, in *The Hitchhiker's Guide to the Galaxy.*

As I wander home, having taken in aurora happenings at the beach, I spot a bandicoot foraging on the nature strip. I wonder what she has found, and where she has been hiding by day. I wonder if she has youngsters in her pouch. And I wonder about her chances of dodging the near-constant stream of cars heading to and from the beach tonight.

Molluscs for the masses

One of the projects that kept me 'sane' (note the inverted commas), as I recovered from my first bout of leukaemia, was writing and publishing my book, the *Seashells of Tasmania: A Comprehensive Guide.* I was motivated to do this, as I was learning about my seashells, because I had reached the limits of what I could identify using existing books. Back in October, I mentioned Margaret Richmond's two-volume set on Tasmanian seashells. Hers was a valiant attempt, and she put her heart and soul (and savings) into the venture; yet it still only covered a small fraction of the seashell species that I was finding. Furthermore, many of the names were outdated, and, occasionally, incorrect. (There's nothing shameful in that; it's just how knowledge advances.)

I had the pleasure of meeting Margaret, towards the end of her life. She had a wicked sense of humour but was struggling with constant arthritic pain—a condition that she assured me could be somewhat eased by a glass of good port. I made sure to bring her a bottle whenever I visited. Gone were the days, though, when she would head out on multi-day fieldtrips to various parts of the Tasmanian coastline (often in the company of young protégés, whom she took under her wing), to record which seashell species could be found, and where. During our conversations, Margaret expressed lingering doubts about the wisdom of having spent so much of her later years on her seashell quest; but I assured her that her efforts were widely appreciated and would be part of her enduring legacy. After Margaret passed away, her collection was donated to the museum. Many of the smaller shells arrived in cut-down cardboard boxes. The luxuriant labelling on these made it clear that their original occupant had been a fine bottle of port.

The silver lining to my first leukaemia episode was a period of several months, over winter and spring, when I was at home, recuperating. July was one of those months, which is my rather tenuous justification for placing this vignette here. What better way to pass wintry days than photographing shells, and using these as the basis for writing a successor to Margaret's books? I bought myself a digital single lens reflex camera, attached my old manual macro lens and ring flash, and set up shop at the dining table. My collection, at the time, occupied the entirety of a six-shelf bookcase in the study, with each species occupying a single plastic bag, placed within a brown paper party bag dedicated to its family, with all families arranged taxonomically. I started at the beginning and kept going until I got to the end – all up, it amounted to some 350 species (just those over about ten millimetres in length). Processing the photos to my satisfaction took more time; but eventually I had accumulated a set of acceptable images of all the species that I thought one could reasonably hope to find on a Tasmanian beach. Next, I arranged these into a series of thirty family-level colour plates. I had already compiled a database of my Tasmanian shells, and I extended this to include data-fields covering such vital statistics as size, the species' distribution within and beyond Tasmania, habitat, depth range and so on. These data then formed the basis for the text that I compiled to accompany the book's plates. Everything somehow fell into place and, before I knew it, with the help of the Tasmanian Field Naturalists Club, I picked up a consignment of beautiful books from the printers. I was thrilled.

Not long after we had delivered most of the boxes to the Club, I was leafing through one of the shiny new books when I spotted a glaring error: two of the cuttlebone images had been juxtaposed, and one was misidentified! This was entirely my fault. I was mortified. There was only one thing for it: a corrigendum. I hastily prepared a sticky label template with the corrected text and retrieved the boxes. Chris and I then spent a number of hours affixing a label to the inside back cover of every book. It wasn't ideal, but it was an acceptable outcome and a valuable lesson in humility for yours truly.

The book sold well, and within a few short years the Club was out of stock. Thus, it wasn't long before I was contemplating the need for a second edition, which I would again compile in my spare time, updating species names and many of the photos. It duly came out in late 2019, and has continued to sell well, prompting another reprint in 2023. I'm always proud, if a little bashful, when I bump into people I don't know and there's a moment of realisation, '*oh, you're* that *Simon Grove! We love your shell book! We keep it next to the toilet.*' What higher praise, what better measure of one's worth to society, could one hope for in life?

Middens in our midst

Taroona High School is blessed with what must be one of the most scenic locations in the country. I feel sure that its commanding, waterfront position, perched on a prominence overlooking the western shore of the Derwent, is a potent factor in the emotional wellbeing and academic achievements of many of its students; our sons included. From how many other classrooms can a student gaze out across sports fields to the sea, and see dolphins, seabirds, seals, and even the occasional whale? How many other schools give their students the opportunity to walk down to the beach and poke around among the seashore rocks during lunch hour? Humanity's appreciation of commanding views seems universal, and I'm sure this appreciation was shared by Lutruwita's First Peoples. They would have had more practical reasons to spend time in the School's location, though; their longstanding occupation is evident on the foreshore. Beyond the earthworks demarcating the levelled ground of the sports field, and in the shadows of a gnarled Monterey pine, is an irregular, grassy mound. In places, it has been worn bare by the passage of generations of schoolchildren, revealing the shelly fragments within. In a part of the island where many shell middens were plundered for lime in the early days of the colony, this one, at Taroona, is among the most intact remaining.

A midden is an archaeological term commonly understood to mean 'an ancient rubbish dump', but it would be more accurate to think of it as a living place. If you've ever fossicked in a former house site, you are likely to have found piles of bricks and perhaps some bottles, old cutlery and other metal objects, plastic and more. Such sites differ little from today's rubbish dumps, except in scale: the latter are fast becoming vast middens of plastic and twisted metal. By contrast, pre-colonial middens are almost entirely organic. Bar the odd blackened or varnished fireplace pebble, discarded or lost stone tool, or fragment of charcoal, they comprise shells and bones—the long-lived leftovers of innumerable meals of seafood and animal meat, accumulated over hundreds or thousands of years. Eating seafood generates a lot of shell waste, which helps explain why most known middens are coastal. They can be seen all around the Tasmanian coastline, from sheltered estuaries to wind-swept dunes, especially in the west and south, where sand-blows sometimes leave piles of ancient shells and stone tools improbably perched on dune crests. Middens are where my interests in seashells intersect, or perhaps collide, with some of my remorse over the colonists' injustice towards Lutruwita's First Peoples.

During the years when our sons attended school in Taroona, I came to take the foreshore middens here for granted. They were just something that was there, when we walked to or from school or weekend soccer matches. I recently decided that I should take a fresh look at them, so, one cold, wet July day, I rugged up and extended my usual foreshore walk northwards a few hundred metres, to the school. Even in the space of a mere decade or two, I noticed how storms and rising seas had eaten into the informal path we had once used to approach the school. I found myself gingerly picking my way among rocks, accumulations of sun-bleached shells, and fresh, slippery weed, before coming to what we knew as the 'school beach'. With the school doubling as a polling station, Chris and I would sometimes row our boat here from Taroona Beach on election days. 'Boating to voting' became our private, ludicrous, ritual.

From the school beach, the main midden is but a short stroll away. It sits some three or four metres above an inflection point in the coastline, where it abruptly switches from east-facing to north-facing. Visually, it may not seem like much, as is common with middens. Everything is covered by a protective layer of earth or vegetation; but, where that's been eroded, hundreds of shell fragments become visible. Here, most were from mussels and wavy turban shells, whose solid, circular opercula, or 'front doors', were scattered like chalky-white coins across the ground. Middens usually comprise only species large and common enough to have been worth harvesting for food. The only other species I could discern, among the fragments, were Venus clams and lined whelks. Most were off-white and partially demineralised, lacking the glazing and solidity of fresh shells; as though they might have once been plunged into hot water or embers; or, at the very least, been exposed to sun, air and rain for a very long time. Interestingly, I could see no oysters among the midden shells, yet these are locally common, as beach-washed shells, today. Indeed, an artificial reef is being constructed in shallow waters within fifty metres of this living place. It will be seeded with spat as part of a wider program to revive Tasmania's native oyster reefs and the vast network of marine species they support.

I clambered down to the north-facing section of beach below the midden. Here, it is hard to discern where midden might end and beach-deposited shells begin. Most likely, the shells that I was seeing at the top of the beach were a mixture of recently deposited ones, and ancient ones from the midden, above. I wondered how much longer this evidence of enduring human habitation would survive, with sea levels clearly rising.

I scrambled up the slope again, through well-established, protective plantings of pigface, saltbush and she-oak, and turned to face the river, trying to imagine what it must have been like in pre-European times, when the midden was actively being added to. Was there ceremony or reverence involved, perhaps recognising that the shells represented a visible manifestation of inter-generational continuity; or were the discards cast aside as casually as we would do, today? I conjured up the sounds of children playing, perhaps skimming stones on the water, while adults contentedly stoked the fire and cracked open the freshly-steamed turbans (or warreners); sharing stories or jokes as they worked, played and ate. Perhaps women were crafting shell necklaces—rainbow kelp-shells (or maireeners), a favourite, are still common locally. Or maybe they were arguing, irritable, sitting there half-starved, with teeth chattering in the cold. Life has never been all fun and games.

From the top of the slope, it was clear that the dolerite boulders littering the nearby shoreline had been rearranged to form what I had once assumed was an ancient tidal fish trap, akin to those known from the north-west coast and Victoria. It would have been a perfect place for one, giving the locals access to fishy alternatives to their shellfish meals. I have since learnt that the arrangement is, in fact, nothing more than a makeshift swimming pool for the school. It dates from the early 1970s, when the adjacent shoreline was infilled with builders' rubble (a modern-day midden).

Off to the right, there are more intertidal boulders, and more excellent foraging grounds for anyone seeking winkles, whelks and warreners. As I watched, a noisy flock of crested terns alighted atop a prominent group of guano-capped boulders just offshore. A group of black-faced cormorants, wings outstretched in a vain attempt to dry them in the rain, were clearly unimpressed with the new arrivals. With low groans and croaks, they were soon rock-hopping to lower-lying, but quieter, boulders nearby. I recalled that from my kayak I had once seen a seal hauled out on the same rocks, but it would have been invisible from the land.

Staring out through the rain to a grey smudge on the far side of the Derwent—the South Arm Peninsula—I pondered the inevitability of all these former living places becoming submerged. Others already had; markers from an even earlier time in human prehistory, when sea levels were much lower. At that time, an icy Derwent disgorged into the ocean somewhere much further south, beyond the southern tip of Bruny Island. Still further back in time, before the most recent glacial maximum, there was a period with a climate as warm as, or even warmer than, today's, known as the 'Eemian', or, more simply, the 'Last Interglacial'. The Derwent's course, then, was probably similar to today's, as it is dictated by ancient fault lines. Across the river, at Mary Ann Bay (near the northern end of South Arm), there are shell deposits a few metres above the tideline. These can be distinguished from a midden by the species they contain, including many small shells that, broadly, would not be out of place on a local beach today. The strange thing is that those shells are also scattered throughout surface deposits on flat ground above the bay, some twenty metres above today's tideline. An earlier interpretation of this phenomenon invoked tectonic uplift, but more recent work, to which I contributed, reassessed the evidence and concluded that those shells were eroded out of the lower-lying deposits by the strong westerly winds that prevailed during the ensuing glacial period—winds persistent enough to push the shells uphill, to their present position.

Mary Ann Bay is a mysterious place. Many winters ago, we visited it with our small children, with the express aim of casting a message-in-a-bottle to the waves. It seemed like a good idea at the time, though I would no longer engage in such wanton littering of the oceans. Nor would I lead young children down a steep shoreline, onto a rocky outcrop jutting just above the waterline; even *if* it offered the prospect of a perfect casting-off point for our bottle. The sea gods were clearly unamused—we all nearly drowned. No sooner had we cast the bottle into the swirling waters than ever-increasing swells rose around us, isolating us on top of the outcrop. We were within seconds of deciding that we would have to swim for our lives—and our sons couldn't swim at this stage—when the swell mysteriously diminished, reconnecting our rock with the shoreline and allowing us to flee. Shaken, we felt duly chastised. Now, when we gaze out towards that bay from Taroona, memories of Neptune's wrath are never far from mind. As for our message-in-a-bottle, perhaps it still rides the ocean waves, in the company of countless discards from humanity's throw-away societies. Or maybe it travelled mere metres and added one more item of rubbish to the strandline at Opossum Bay.

Chilled by the memory and the weather, I decided it was time to head homewards. I trudged through the unrelenting rain, pondering whether the fact that middens remain the most tangible and commonly encountered evidence of Aboriginal occupation and stewardship of the land is something to rue or to rejoice over. On one hand, the subtlety of other physical evidence of occupation undoubtedly made colonial attempts to annihilate Indigenous people and their culture easier, and, perhaps, the memory of what was, harder to hold on to. On the other hand, a culture that, for thousands of years, leaves few traces but the remains of meals, is surely one to celebrate and learn from. Of course, nothing is simple—a discerning eye would see that entire landscapes bear the hallmarks of past Aboriginal management; but for the most part, it seems, we have chosen not to discern, nor to learn.

©K. DAVIS 2023

August

Spring's promise, often broken

August invites mixed outlooks. The optimist in me notes the lengthening days, and dreams, wistfully, of better times ahead. The pessimist counters that increasing day length isn't the only trigger for a natural renaissance, and reminds me that, in the air, coldness is likely to prevail for many weeks; and it will linger even longer in the soil and water. Then the realist chips in, pointing out that, as with so much in the world, there is truth and meaning in both perspectives. It is still officially winter, so we can't expect anything else; but sometimes, when the sun shines a little more strongly, and the land warms, and fluffy little cumulus clouds billow up for the first time in months, it *does* feel like spring is on its way. Pottering around in the garden, I am often accompanied by a cock blackbird. He shows little fear, and is standing by for the moment when my scratchings will expose a worm or grub. Awaiting the next titbit, he practises his territorial song. It's barely audible because he doesn't project the sound. In fact, he is whispering, without even opening his beak. He even practises alarm calls this way, and I wonder if all songbirds do this. I imagine it's a bit like humming a tune in your head, so that you can be sure of yourself when you come to sing it aloud for others to hear.

Starlings do all their practising out loud, it seems. The local gang members, who hang out on the wires in front of our house, chatter incessantly among themselves, mostly through signature, rather unmelodic, squelches, peeps and whistles. At this time of year, they seem to anticipate the arrival of spring migrants; I hear them incorporating mimicking calls into their repertoire. Presumably, they have memorised these from the previous spring's soundscape. They aren't perfect renditions, but sometimes they are enough to fool me into thinking that the fan-tailed cuckoos have returned a full week, or more, before they really do.

Returning cuckoos, real or imagined, are no concern for masked lapwings; but, in parks and on patches of grass throughout the suburbs, and even on the flat rooves of buildings, these highly-strung birds are in a state of alert. They are guarding new-born, stilt-legged fluffballs from all comers. A pair invariably tries to nest on the flat roof of one of our museum buildings, at Rosny, and their antics have become my standard audio and visual accompaniment to early spring.

Tasmanian lapwings are far more likely to dive-bomb passers-by than are Tasmanian magpies—the latter seem less prone to swooping than their mainland counterparts. Brooding lapwings have other tricks in their toolbox, too. If they sense you heading their way, it is likely that one parent will wander off, tangentially to your approach, with a wing held limply at its side—feigning injury. Should you be taken in, it will lead you safely away from its chicks, then miraculously cure itself, flying off amid a lot of fuss and noise. If you stay your course instead, and get too close to the chicks, the guarding parent may push its 'elbows' forwards menacingly, showing off the upturned, yellow spurs that are normally hidden from view. These give the birds their colloquial name: spur-winged plovers. The display is no idle threat—the spurs can seriously injure a predator, and the tactic must save plenty of chicks from an untimely death. Sadly, spurs don't work with cars—many a pair of roadside-nesting lapwings loses their offspring to traffic, especially when the naïve chicks totter over the kerbs and cannot climb back up. At this time of year, I can more or less guarantee passing one or two flattened victims on my cycle commute. On one occasion, I arrived in time to rescue a hapless little bundle, which had obediently sat still, in accordance with its parents' command, and remained motionless as I approached. I was able to scoop its almost weightless body into my hand and liberate it on the grass verge, nearby. All the while I was subjected to a tirade of grating shrieks and threatening advances from its parents. These only subsided once I had released the chick and retreated a good twenty metres. You would think that the parents might have shown some gratitude, but I guess it was not part of their behavioural repertoire. Their take on the situation was probably that only their heroic threats had allowed their chick to survive death by monstrous, bipedal predator. Still, over the ensuing days and weeks, I enjoyed seeing the chick thrive and, eventually, morph into a gangly, feisty, young-adult lapwing.

Lapwings may enjoy the warming ground, but August's rising air temperatures also prompt butterflies to take to the skies again. One early August day, I spotted an unseasonably precocious common grass-blue. It was sunning itself on a daisy in Taroona Park, as I was walking home from an afternoon visit to the community garden. It was a one-off; more usually a cabbage white constitutes my first sighting, at the end of the first week or beginning of the second. For this, it qualifies as my butterfly of the month. I know it's an invader and a pest, as I discussed in March, but at this time of year one must take what is on offer and show gratitude for small mercies. My usual spot for observing this portent of spring is the office tearoom. It has a balcony that overlooks a garden and strip of lawn, on which, if it's a sunny day, my eyes are likely to be fixed as I enjoy a cuppa or my lunch. Leaves, feathers, and stray tissues blowing by on the breeze cause frequent false alarms; but, sooner or later, a lone male cabbage white will put in a cameo appearance, skirting the flowerbeds in search of nectar sources among early-flowering oxalis. Bumblebees, honeybees, bluebottle blowflies ('blue-arsed flies') and European droneflies (more invaders, I know) boost my balcony insect species list. With them, I can finally declare the new season open.

Chilly chafers and buzzie beetles

I have made quite a few references, in this book, to the benefits of being 'in the moment' in nature, in tune with bodily sensations arising from within our own bodies and our immediate surroundings. I imagine that this is how most organisms exist, unencumbered by the 'higher' thoughts and emotions that usually rule our lives and can place a barrier between us and the natural world. But living entirely 'in the moment' must have its downsides. For instance, I sometimes wonder whether life as an insect might be one of continuous sensory overload; even assault. Imagine inhabiting a tiny body whose chitinous exterior literally bristles with sensors, yet there is only room inside your head for a minuscule brain with which to process all the sensory inputs. It's hard, isn't it, from our privileged, big-brained position? The mind of an insect must, surely, have evolved to follow rules of thumb that keep the need for mental processing to a bare minimum and, in most situations, work—*'Temperature or light levels creeping up? Must be time to get going for the day.' 'Pheromones detected? Must be a potential mate, move towards the source.' 'Dark shadow moving across my field of view? Potential predator! Drop to the ground or fly away.' 'Green stuff? Vegetables.' 'Bright light against a dark background? That'll be the moon—great for navigation.' 'Continuous sheet of blue or white? Sky. Up we go.'* We do this too, of course: a bustle in the hedgerow won't, as a rule, lead us to assume the May Queen is having a spring clean. We are more likely to be instinctively alarmed: our bodies will automatically galvanise us into fight-or-flight mode, assuming that a predator or rival is about to leap out of the shrubbery and accost us. If the assumption is correct, we'll be well-placed to deal with it. If not, there's no real downside; it's simply better to be safe than sorry. But, while we have the mental capacity to override such knee-jerk reactions, I suspect that insects don't.

This isn't about hedgerows, or shrubberies, though; it's just a roundabout way to introduce a couple of beetle species which commonly cause puzzlement at this time of year, especially if encountered *en masse* and in unlikely situations. I have seen these spectacles many times now, but I, too, found them intriguing on my first encounters, so I will share them with you.

When I worked at Forestry, it was always a pleasure to enter and exit the building via the beautiful glass-roofed dome at the front entrance, and pass through the lush forest planted within it. Breathing in the heady aromas of eucalypt and myrtle first thing in the morning would set me up well for a day in the office. But the dome lacked animal life. There were no birds (other than the odd sparrow), and I saw no insects. Until, that is, one August morning, when I arrived at work and the reception staff surprised me with a takeaway container full of dead and dying black beetles. They had been swept up that morning, from the pathway that led through the dome. I was, of course, known as a 'bug person' by then, so I was expected to know what was going on with all these 'bugs' that had suddenly appeared. But I was stumped. It seems strange now, in fact. All I could think of was the most obvious thing: that the beetles, whatever species they might be, had been attracted to the little lights that lined the path (which had been left on overnight, apparently); they had found their way in under the sliding doors, and here they were. I couldn't identify them, nor tell their backstory.

Nowadays, many residents of Hobart and other towns will have been greeted by accumulations of these beetles, either on their front porch, or at a shop entrance, or beneath a streetlight. We frequently receive enquiries about them at the museum. But, back then, they were still a novelty.

As it happened, I had noticed these mystery beetles, myself, only the previous weekend. Chris and I had been for our usual stroll along the Taroona foreshore, and the receding tide had left a thick, dark ribbon on the sand, comprising thousands of dead and dying beetles. The local silver gulls were enjoying an unexpected bonanza, but even they couldn't dent the numbers. The phenomenon has since become an almost annual occurrence, always during this season.

I now know that the beetles were red-headed cockchafers. If that term sounds a little indelicate, you could call them red-headed pasture-scarabs. Etymologically, 'cockchafer' is quite innocent though. In British English, a chafer is a sort of scarab beetle, and putting 'cock' at the front merely signifies that it is a particularly big one (as a cockerel is to a hen); hence the cockchafer is the largest chafer species in Britain. The word 'chafer' is related to German and Dutch equivalents that, today, merely mean 'beetle', but once they meant 'gnawer'—it was used for beetles that would damage plants by chewing. (A 'beetle', by the way, is a 'little biter', etymologically.)

Our red-headed cockchafers are modestly proportioned, begging the question: why aren't they simply called chafers? To that I have no answer; but I can tell you that they are about a centimetre long and solidly built, with stout spurs on their legs, and compact, lamellar clubs to their antennae. Despite their name, they are entirely black, or at least deep brown. They take the 'red-headed' part of their common name from the colouration of their larvae, in which the possession of a reddish-brown head serves to distinguish these otherwise off-white creatures from larvae of black-headed pasture-scarabs.

It's unusual for insects to be named for their larvae. In this case, the names were coined by agricultural entomologists, since the beetles' larvae, which are also known as curl-grubs, on account of their C-shaped bodies, are agricultural pests. They consume the roots of grasses, sometimes giving rise to bare patches in paddocks, lawns and cereal crops. They have long been a problem for farmers on the Australian mainland; but their rapid expansion southwards, across the pastures, parks and gardens of Tasmania, is a recent phenomenon and probably related to a warming climate. These days, they are just another local marker of the seasons for suburbanites, since the adults time their emergence for the lengthening days (or shortening nights) of late winter. They probably don't feed, so in the few days (or nights) of their existence, they must reproduce and disperse. A fatal attraction to artificial lights seems to be the consequence of one of those mental shortcuts that I mentioned earlier. It probably started out as a reliable rule for flying in a straight line at night, by the light of a distant moon. Being active at night rather than by day is sensible, to avoid being eaten by birds; but it backfires spectacularly in urban environments. That said, plenty of beetles still succeed in reproducing, and they set the stage for further beetle accumulations in the years ahead.

But what of those drifts of beetles along the strandline? I think these are also the consequence of a maladaptive mental shortcut, for a beetle species that has had no need to evolve a strategy for coping with the sight of open water, nor with water's ability to reflect light from overhead. For such an insect, flying over water at night must be very confusing, especially in modern times, near brightly lit coastal settlements. '*Which is the correct moon, the one up above or the one down below?*' '*What about all those other 'moons', some of them up above, some of them down below?*' A beetle that flies headlong into the shallows is not likely to live to tell the tale, but it is likely to get washed up on the next tide (if not gobbled up by fish first).

It's not only red-headed cockchafers that are fooled by open water. I have found that patrolling the tideline is rather a good way to collect insects, and (as I recounted in June) there are beach-dwelling insects that make a living scavenging unfortunates' corpses. A little later in spring, the shoreline can be thick with dead and dying eucalypt leaf-beetles (gum beetles). On fine spring days, the leaf-beetles' dispersal mechanism is 'ballooning'—allowing updrafts to carry them high into the sky over their forest home, where breezes may carry them many kilometres. You can readily witness this 'aerial plankton' from any vantage point in eucalypt country, such as a rocky knoll, once you get the hang of focusing your gaze on the airspace rather than the trees. My guess is that the leaf-beetles and other insects gauge up from down partly by where the bright blue or white light is coming from, perhaps, additionally, relying on sunlight's polarisation. But confusion reigns if they get blown out to sea, where their entire field of view is bathed in light. Indeed, the sparkling sea may serve as a super-stimulus, overriding the beetles' other senses, and causing them to plough into the water. Either that, or they simply get exhausted and end up crash-landing into the sea.

My second beetle marker of the new season also brings the museum many enquiries from the public. I first came across the species in question on a Forestry field visit to the Florentine Valley. It was on one of our annual research exposé days, which was, for some reason, being held in late winter. These outings were planned so that we scientists could present some of our research findings to the Executive, whose members were then to judge their relevance to the business of forestry and guide where they thought the research should be taken next. I can't actually recall what specific issue was discussed at this particular stop, perhaps because I was transfixed by a seething mass of small, metallic-blue beetles at my feet. Each was only a few millimetres long, and a deep cobalt blue. There must have been hundreds of them, swarming over the swathe of low vegetation along the middle of the dirt road. It was as though they were trying to escape some hidden menace at ground-level, climbing over one another to reach the highest spikes of the rushes and sedges. As often as not, having reached the high point, they would be jostled by others and leap into the air, propelled by their thickened and, apparently, very muscular back legs. Crash-landing on the ground nearby, they would then resume their frantic clamber. At the time I didn't know what they were called, but I now know that the common name for these metallic-blue, flea-like beetles is … (you'll never guess): 'metallic-blue flea-beetles'.

The captivating behaviour that I observed that late winter's day was entirely typical for the species. It is commonly found along trackways and in rough, overgrown paddocks—wherever there is an abundance of the beetles' larval foodplants: those buzzies with the annoyingly tenacious seedheads that I talked about back in January. That's why the beetles are also widely known as 'buzzie beetles'. Later in the season, their black larvae can be spotted feeding on the buzzies' leaves, and by late summer they will have pupated, giving rise to a new generation of adult beetles in autumn. It seems that these hibernate over winter, but are quick to awaken with the tentative return of the sun's warmth in August. The ground remains cold and damp at this time of year, though, and that is my explanation for why the cold-blooded beetles climb upwards—essentially to warm up in the sun. But why the rush to warm up, and why do they all choose the same sprigs of vegetation on which to sunbathe? My guess is that each beetle gives off aggregation pheromones that encourage others of their kind to come and join them, with their reflective blue bodies also acting like beacons in the sunlight. The more beetles in the aggregation, the greater the power of attraction produced by their combined pheromone output and blueness, and so the aggregation grows still larger, until all the local beetles are drawn in. Presumably, the ultimate purpose of aggregating is for finding other beetles that are now pleasantly warm and in the mood for mating (something they seem to spend a lot of time doing). Mating this early in the season might give the next generation the best chance of growing to maturity over the ensuing warmer months, before temperatures again plummet in autumn.

Watching the antics of the flea-beetles induced a state of near-hypnosis in me, and I had little awareness of time passing. But, soon enough, I sensed a general murmuring and shuffling of feet among my fellow foresters, signalling that the discussion (to which I should probably have paid attention) had reached its conclusion. We had been standing around in the cold for long enough, but before we returned to our utes for onward travel, I pointed out the beetles to the assembled group, and was pleased to see the interest that they sparked. When not desk-bound, foresters seem to spend a lot of their time looking upwards, seeing beauty in the straightness of the boles and the symmetry of the crowns of the trees in their care. These beetles, which were now exerting their power to aggregate over humans, too, were a reminder that, sometimes, beautiful things can come in small packages, right at your feet.

The lament of the swift parrot

After a trip away, or even just a week at work, I look forward to spending time in the garden. August starts off quietly, but soon becomes busy, as the need to prepare veggie beds and get the upper hand in the (endless) war on oxalis, twitch, and other weeds escalates. The garden is my happy place, though: it offers time and space for quiet reflection, while also giving me a physical workout. The nearby park is within earshot of home, which means there's always something to listen to, other than the usual suburban sounds of lawnmowers and barking dogs, and the fluty, dawn-to-dusk songs of cock blackbirds broadcasting from rooflines and fenceposts. At this time of year, galahs are very vocal in their enjoyment of the fresh green pick offered by the park's grasses and capeweed; wattlebirds scrap among the blue gums; and common bronzewing pigeons, hiding among the wattles, ramp up their monotonous, far-carrying, hooting calls, which have been mistaken for strangely diurnal owls.

My thoughts are especially liable to be interrupted by the calls of one bird species, which I have come to love above all others: the swift parrot. Although they breed only in Tasmania, they spend the colder months as nomads, wandering through the south-eastern corner of the Australian mainland. They generally turn up in Taroona around the end of August, or at the beginning of September, usually when a spell of anticyclonic weather and gentle northerly breezes helps them cross Bass Strait. A first encounter is typically fleeting—they're not called 'swifties' for nothing. A brief, querulous, upwards-inflected call, overhead, is enough to get my attention. If I'm lucky I'll catch a glimpse of a party of a dozen long-tailed, pointy-winged green birds zipping across the sky, heading to or from some flowering blue gum in the park or along the foreshore. Their arrival is a milestone in spring's progression, or winter's departure, and I have come to imbue it with great significance—as much for what it portends for the coming summer as for the pleasure of reuniting with these long-absent friends. As more arrive, the frequency with which I hear their overhead calls increases. To the bemusement of friends and neighbours, I find myself breaking off midway through a conversation and pointing skywards, uttering '*Swifties!*', followed, shortly afterwards, by: '*There! Over there! Gone now.*' If you're not tuned into them, they're so easily missed.

Swifties are less easy to miss once they have begun to make themselves at home again, and have sussed out the best blooming gumtrees. Blue gums are sporadic bloomers—they may burst into flower at pretty much any time of the year; but flowering seems to be contagious: once one starts, others in the neighbourhood soon follow. This means that the many birds which depend on blue gums are well-attuned to homing in on them, wherever they happen to be flowering. While for me it's the 'cat pee' scent of flowering blue gums that assails my senses and draws attention to the source, it's the trees' showy, globular, off-white flowers that probably act as a visual beacon for swift parrots. They primarily feed on the nectar and pollen of blue gums and have special brushes on their tongues for licking up pollen. Their predominantly green plumage makes them hard to spot amid foliage; but they're sociable and gregarious birds, and once they've found a good food source, they chatter constantly to one another in an amicable, sing-song sort of way. Every now and then, the chattering will pause; then the birds will simultaneously erupt and, explosively flashing their red underwings, they will dash to the next tree on their rounds, '*peep-peep-peeping*' excitedly as they go.

The early settlers of Hobart Town reported vast flocks of swift parrots, but since those days their numbers have steadily diminished, and at an accelerating rate, it seems. Much of their breeding range—the grassy blue gum woodland that once blanketed large swathes of south-eastern Tasmania—was converted to sheep-grazing country long ago, or has given way to sprawling suburbs. Consequently, the birds have been squeezed into increasingly marginal, wetter forests along the periphery of their former breeding range, where they run up against other land uses. Forestry is commonly blamed for the parrots' rapid decline, and it is certainly not blame-free. However, close inspection of the data behind the published research on swift parrots leads me to unambiguously point the finger at a greater threat: sugar gliders.

Sugar gliders (or, more accurately, Krefft's gliders, because the original sugar glider has recently been found to comprise three sibling species) are almost unbearably cute-looking arboreal marsupials. Their soulful, forward-facing eyes, accentuated by dark-furred rims, seem purpose-made to pull on human heartstrings. It's believed that they were brought from near Melbourne to Tasmania, in the mid- to late nineteenth century—as pets. Inevitably, some escaped or were released, and they established self-sustaining populations in the Launceston area. Over the intervening hundred-plus years, their range has expanded into all but the wettest, coldest, or most isolated tracts of Tasmanian forest. Warmer and drier conditions during recent decades may have aided this territorial expansion.

Since gliders are nocturnal and arboreal, we seldom see them; but native fauna certainly feel their impact. They largely subsist on a diet of insects, which is supplemented with sugary tree sap lapped from wounds they've made in the trunks and branches of gumtrees, with their sharp teeth. That isn't all they do with those teeth: they are great opportunists, and are partial to birds' eggs and baby birds—so you can see how they have become a threat to swift parrots. The parrots nest in small hollows in old eucalypts, and the female does most of the brooding. They haven't evolved any behavioural mechanisms to cope with the new threat of predation by sugar gliders. When a sugar glider sniffs out an occupied hollow, the sitting female is soon dispatched by the glider in its bid to reach the eggs or nestlings below. Not only is the entire brood lost, but the adult female is, too. In recent years it has become clear that the swift parrot population is grossly imbalanced, with many more males than females, and relatively few young birds of either sex. A recent population viability analysis (a form of population modelling) accounted for extremely high levels of predation by sugar gliders, as observed at various study sites around Tasmania, and concluded that the birds' numbers would decline catastrophically—close to the point of functional extinction—over just three generations. Recent genetic analyses seem to concur with this grim prediction.

It seems like clutching at straws, but there may be some hope for swift parrots. Sugar gliders haven't reached Bruny or Maria islands, and, in some years, when the blue gums are flowering there, or in adjacent mainland parts of Tasmania within parrot commuting distance, most of the population can reproduce in relative security. But these conditions only eventuate in around two years out of every five. In other years, it seems the birds would be better off not even attempting to breed, such is the rate of population attrition inflicted by sugar gliders. If only we could warn them! In the meantime, it is hoped that specially designed nest boxes and localised culling of gliders may give the parrots enough breathing space to hang in there, until a longer-term solution is found (if, indeed, one can be found in time). Even that strategy is problematic though—not only is it difficult and dangerous to work in the crowns of old eucalypts, but, with the birds' inconveniently peripatetic breeding strategy, it is difficult to accurately predict where they will nest in any particular year.

Invaders such as these sugar gliders, and the blackbirds and starlings that I mentioned earlier, are both troublesome and troubling: they have become (unwitting) colonisers of the territories of Lutruwita's indigenous inhabitants, and ousters thereof. So were my people, the English—wittingly so. As a first-generation immigrant from the other side of the world, I am acutely aware that I am living on land which belonged to others, and that my impact on this land—and, indeed, the planet—is great. Lutruwita's people learnt how to tread so lightly that, today, we view the cultural landscape that they honed, which the rest of us inherited or usurped, as the gold standard of 'naturalness'. We even refer to swathes of it as 'wilderness', as though those First Peoples had no role in its creation or maintenance. Since colonisation, we have left enormous, heavy footprints across their land and along their shores, though ours haven't entirely erased theirs. I have a heavy ecological footprint, myself, with regular intercontinental air travel to stay in touch with my family, often timed so that I can swap cold southern seasons for warm northern ones. My health journey has produced vast amounts of medical waste: bales of chemo gowns, heaps of nitrile gloves, great lengths of plastic tubing, and mounds of blood bags, syringes and other materials. My research has involved the felling of trees, and perhaps even forests, and the destruction of rotting logs. My collecting trips have relied on polluting four-wheel drives, helicopters, boats and ships; as well as the killing of countless small animals (I do not miss the irony that I am 'collecting for posterity'). Then there are my frequent escapes into the quietness and clean air of the natural world, undertaken by means of a noisy, diesel-fuelled campervan. I am a hypocrite.

Maybe we are all hypocrites. Perhaps it is an unavoidable accompaniment to life in our present culture. Perhaps, too, developing an awareness of one's hypocrisy, and confessing to it—as a reforming alcoholic does—might be the first steps towards righting matters. To extend the footprint analogy, the only footprints over which we have some control are those that we are yet to leave behind—those we might leave for posterity. 'Posterity' means 'that which will be behind us'. The word came into English from Latin via Norman French, and like those languages, appears to have fallen out of favour. This seems a shame; to me, it conveys a future of *intent*, one shaped by choices we make today. Similarly, 'treading lightly on the Earth' signifies, if not an intent, at least a desire to impact less, to leave a smaller footprint, and to disappear sooner into the background. Perhaps we can imagine ourselves leaving lighter traces when we think only of our own footprints; but, if we regard ourselves within the scope of our societies and our species, it rapidly seems apparent that reducing our collective footprint will be, and is, the greatest challenge of our times.

Throughout our individual lives, we are all on intersecting journeys of discovery that have a bearing on our relationships with the natural world, and hence on the ecological footprints that we leave behind. The 'discovery' part may be limited to a growing realisation concerning our impacts on nature, our (delusions of) control over it, and our dependence upon it. The most formative journeys of discovery are ones of growing wisdom and enlightenment arising from this realisation. Such journeys enable us to project our footprints forwards in time with a sense of pride, hope, and optimism—knowing that we will have learnt from our own past, from those around us, and from those who came before us. There is no better time to embark on such journeys than while winter prevails but portents of spring can be glimpsed—a metaphorical August, poised for new beginnings. Perhaps there is no better place, either, than here, in Lutruwita-Tasmania, where nature is still all around us.

Except it increasingly isn't. Gardening in August isn't the same without swift parrots. Last year I heard them flying overhead just twice, and they numbered fewer than half a dozen birds each time. Our lives will be the poorer for their passing.

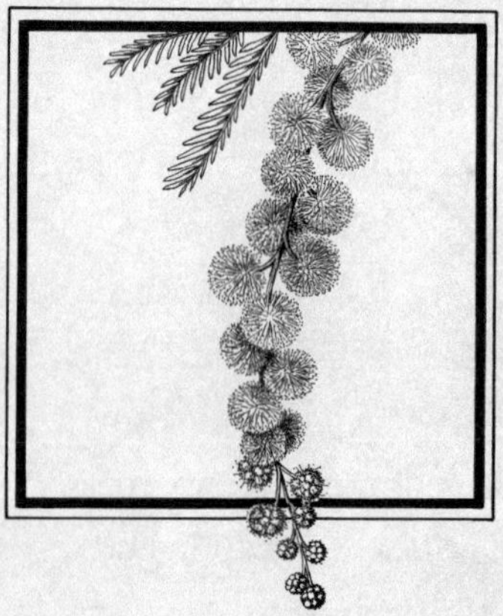

Wattles that heal

Months after I arrived in Tasmania, all those years ago, I felt I was still floundering in my Forestry job. I had seen a lot of forest, begun to crunch data, met numerous knowledgeable professionals; but I lacked contextual perspective. When you live and grow up in a place, you develop a broad, deep knowledge of that place. Here in Tasmania, I was far removed from my own upbringing. This was also clear to Tim, my affable and supportive boss, so we decided to do something about it. First, I was sent on a series of road trips with longer-serving colleagues to the far corners of the state. Talking to them, and seeing things first hand, taught me much about forest ecology and management. The process was like a jigsaw: you pick up a piece and get to know it well, then other pieces; then they click together, and a picture begins to emerge. The process didn't stop when the trips came to an end, of course. I continued learning for the duration of my tenure. If I hadn't worked in the sector for all those years, I might never have realised how long it takes to even begin to understand the complexities of forests and forestry.

Some eleven months into my job, a special opportunity arose. I drove to Launceston airport for a rendezvous with a charter aeroplane company, and for the next couple of hours, we flew a wide circle over the forests and hills of northeast Tasmania. I had inherited several long-term research projects in this region, and the expedition would offer me a bird's-eye perspective of the landscapes in which they were embedded, so that I could better understand the context of our findings. It was a near-perfect day for it, with a gentle breeze and dispersing cloud. Our little twin-propeller two-seater banked low over the hills around Evandale, before striking east-northeast towards Mount Barrow and the northern slopes of a freshly snow-clad Turbunna-Ben Lomond. Once my initial vertigo had subsided, I relaxed and peered with interest through the windows. We passed directly over Jacob's Ladder, and continued eastwards, bumping along below the clouds all the way to the coast near Scamander. Turning northwards, we passed over the granite country of The Gardens, before meandering westwards over the hills of Blue Tier and Ringarooma, and then Mounts Victoria and Arthur, before descending again to the airport. Two hours from take-off we were back on solid ground again, but what a fascinating two hours!

I learnt much about the forests of the northeast that August day, and about my own naivety. From the time we approached Mount Barrow, to the time we began our descent, I had been peering with consternation at the lemon-yellow crowns of great forest trees in full bloom, wondering what on earth they were. Some rainforest tree, no doubt, since they towered over the tree ferns that formed the understorey. Sassafras? Myrtle? I really was plumbing the depths of my ignorance. Back on the ground, my pilot, whose local knowledge seemed much deeper than his youthful appearance had suggested, put me right: they were silver wattles. I had to mentally backpedal to accommodate this information, because, even at this early stage in my Tasmanian journey, I regarded silver wattles as a 'weed' tree—one that was particularly good at colonising disturbed ground. If a fair percentage of the forests that I had flown over were dominated by silver wattles, what did that say about their history?

Had I taken more interest in Tasmanian history, I would have known already, but *human* history was never my thing. Others will know that Tasmania's northeast was once gripped by gold and, especially, tin fever. In the late nineteenth century, and early twentieth, the region attracted legions of panners, diggers, prospectors and miners; as well as all the operators and others needed for pop-up shops, pubs, hotels, houses, brothels and banks in the middle of nowhere. People came from far and wide, including a sizeable contingent of Chinese entrepreneurs and labourers (their legacy is celebrated, today, with the 'Trail of the Tin Dragon' touring route). There were no environmental regulations in those days—no Environmental Protection Agency, no Forest Practices Authority, no Permanent Native Forest Estate Policy, no Forestry Commission, no water and sewerage authority, no local council planning zones. The boom was effectively a government-sanctioned free-for-all: dig and cut now, we'll pay later. Not only were outcrops blown up and alluvial land scraped bare in search of ore, but water races were dug, dams built, cow paddocks and turnip fields established, charcoal pits, sawpits and brickmaking kilns constructed. Fires were lit, too, eating into the forests. And countless buildings were made of timber, hewn from what must have previously been, predominantly, majestic wet eucalypt forest and rainforest.

Mining booms always seem to end in busts. The prospectors moved on, mines closed; infrastructure and whole towns were abandoned. Nature had been severely browbeaten, but it was not defeated. Over the ensuing century, nearly all of the visible scars have healed, with one of the principal agents of healing being silver wattle. Wattles are exemplary fixers of our environmental transgressions, kick-starting ecological succession whenever they are given the opportunity. I mentioned their role back in November, so I won't repeat; suffice to say that the preponderance of silver wattles in Tasmania's northeast can be traced to the events I have just outlined.

Silver wattles are short-lived trees, so several generations have come and gone, each contributing incrementally towards the land's recovery by fixing soil nitrogen and providing food and shelter for other plant and animal species, which, in turn, play their part. Today, many of the remaining wattles are tall forest trees. They are reaching the ends of their natural lives but still put on a spectacular floral display each August, as I had witnessed during that flight. What happens next in the succession will depend on which other trees and shrubs survive in their vicinity. Eucalypts will be slower to colonise than rainforest species, because they suffered greater losses during the mining era, and because their seed is more-or-less gravity-dispersed, so it is not carried long distances. Eucalypt seed also needs to land in full sunlight on mineral earth—preferably an ash bed—for the seedlings to stand a good chance of establishing. It is, therefore, unlikely that wet eucalypt forest will re-establish where wattles now stand. A hundred years hence, a future version of me flying over these forests might still see the occasional silver wattle, but much of the landscape will probably be a sea of rainforest green.

It's fitting that our national floral emblem is a wattle, given that our nation's prosperity has long been dependent on massive vegetation disturbance to benefit sheep, cattle, wheat, mining—and wattles. Nearly everybody in Australia knows that wattles are acacias, and *vice versa*, but fewer will know that there are about 950 species of *Acacia* across Australia. Even less known is that there are just a handful of *Acacia* species elsewhere in the world. Those iconic, giraffe-browsed, flat-topped acacias on the plains of the Serengeti? They are no longer *Acacia*s, even though the first member of the *Acacia* genus described was from Africa. In the early 2000s, it became widely accepted that the botanical concept of what constituted an *Acacia* needed an overhaul, since the African, American, and Australian/Pacific lineages had been found to be distinct entities. Under the taxonomic rules of precedence, the African lineage should have retained the *Acacia* name, but Australian botanists successfully argued that wattles' prominent role in Australian culture, as well as their much greater diversity here, ought to trump this. Delegates at the 2005 International Botanical Congress agreed, and the decision was upheld at a later Congress. So now African acacias can still be called acacias with a small 'a', but botanically speaking they are *Vachellia*s.

Calling silver wattles weeds was a *faux pas*, if we take 'weeds' to mean 'plants that have established themselves in the wrong place, ousting some other plant that we wanted there'. In the context of Tasmania's northeast forests, silver wattles are facilitative and restorative: no wattles means no rapid healing of scars and no return of more mixed forest. However, weeds in my backyard can also restore bare soil; it's just that I resent and resist their advances. Left to their own devices, chickweed and oxalis would give way to twitch grass and dandelion, and later, perhaps, to blackberries. Ivy, mirror bush, cotoneaster, and elder seedlings, sprouting after passing as seeds through the guts of blackbirds, would take root in the shaded thickets. Before long we would have a young (and exotic) woodland. It's not what I would want for my backyard, but, ecologically, it would be much more of a restored and functional system than regularly dug-over bare ground supporting broccoli or zucchini plants for a few months of each year. Perhaps we would do well to heed the words of nineteenth-century American essayist Ralph Waldo Emerson, who defined a weed as '*a plant whose virtues have not yet been discovered*'.

What's my point? It's that nature isn't a values-based system; it's our culture that imposes values on it, favouring one species or one successional stage over another. In the garden, we favour early-successional species, but only some of them (broccoli and zucchini rather than chickweed and dandelion). In forests, we tolerate early-successional species (wattles) if they pave the way for later-successional ones (rainforest trees). Left to its own devices, nature exists in a state of dynamic equilibrium—species come and go across the landscape, while the whole remains whole. There is no profligacy, yet there is exuberance, built on a premise of frugality where the unwritten rule is that matter, nutrients, and energy cannot be wasted, only redeployed, and recycled. How different this is, compared with the way we humans have conducted our affairs. Where we see weeds, perhaps we should see ecosystems under reconstruction—nature doing what it can to right our wrongs.

What, then, of rewilding? Across Europe and North America, increasingly ambitious projects are returning vast tracts of former agricultural land to nature, employing varying degrees of human intervention to set the direction of ecological recovery. Long-gone species, such as lynx, bison and beaver, are being reintroduced; and, where this is impractical, ecologically analogous species (such as the tougher breeds of domestic cattle, pig, and horse) are put to work on ecosystem reconstruction. The results are impressive, and the rewards for nature and people multiply over time, often in unexpected ways. It requires a big shift in mindset in these parts of the world, where farming has been going on for so long that it's viewed as the natural order of things.

Does rewilding have relevance to Australia, and Tasmania in particular, where modern farming is a more recent imposition? I think it does, but it won't be straightforward. Our ecology is different. Abandoned agricultural land won't necessarily revert to the kind of native bushland that was there before it was cleared (or before colonisation). It would certainly *function* better—in terms of nutrient cycling and all the other important things that ecosystems do—compared with your average farm paddock. And it would be wild if we let it be so; but it wouldn't be 'natural'. More likely, it would evolve into some novel ecosystem with a mixture of non-native and native plant and animal species. Are we ready for this? Is this what we would want our descendants to experience as 'nature'?

Rewilding would be most likely to work to nature's advantage in areas that retain sufficient remnants of 'naturalness'—for simplicity: what was there before we impacted upon it—to act as sources of material for expansion, recolonisation and reclamation. A negligible degree of 'unnaturalness' (such as a proliferation of exotic woody weeds or feral animals) also helps. Logged-over, partially cleared, grazed, or burnt forests and woodlands would be prime candidates. Even then, as my silver wattle example demonstrates, a return to some sort of 'original' forest type might be a long time coming, and may never eventuate; so 'success' also depends on there being nearby reservoirs of the native species that we want to recolonise. This begs further questions: does it matter how recolonisation plays out? What are we trying to achieve? What does 'original' or 'natural' even mean? Given that our megafauna is long extinct, what we imagine as the 'natural state' is just that: a figment of our imagination. In what ways would those creatures—giant echidnas, marsupial lions, giant devils, gargantuan wombats and so on; and, of course, thylacines—have influenced Lutruwita's ecosystems? We may never know. What we do know is that the human footprint on this island stretches back tens of thousands of years. Now that the climate is rapidly changing, even the 'natural' ecosystems of the past few hundred years will come to be seen as increasingly anachronistic.

There's an unresolved tension between the rewilding ideal of letting nature do its own thing, and more interventionist approaches towards regenerating what has been lost or creating some idealised version of what should be; and there's also a fine line between 'gardening' and facilitation. Should we scatter seeds, or plant trees even? (Perhaps, if they can't recolonise of their own accord.) Should we import leaf litter and rotting logs to help invertebrates and fungi recolonise? (Arguably yes, since these will do much to restore ecosystem function, even if we seldom see them at work.) Which approach is best in an era of rapid change? Do we even have to choose? Alone, I don't have the answers, because it's largely a matter of our cultural values, and these will continue to evolve over time just as surely as will any patch of bushland. If you need convincing of this, consider how the early settlers thought it would be a good idea to bring European songbirds (including pest and nuisance species such as sparrows, blackbirds and starlings) into the colony, something that today most of us rue. It wasn't long ago that conservationists thought it would be wise to liberate non-native lyrebirds in Tasmania (at Hastings and Mount Field), or Flinders Island wombats on Maria Island, though such introductions would be frowned upon today. And let's not forget that feral deer are still given legal protection across Tasmania, despite their rapidly increasing ecological impact.

When we ask ourselves what we should do, or what is best, the one thing we *cannot* do is start from some ideal time in the past, when 'nature' was more 'natural'. We can only begin here and now, where we are; and 'where we are' is within a complex mosaic of indigenous species (and people) and immigrants; and of recently altered landscapes and communities, and ancient ones, barely changed for millennia; and everything in between. We are enmeshed within numerous systems and ecosystems—familial, social, cultural and economic; as well as in more-than-human biological communities, and the systems that regulate the physical and chemical aspects of our planet. Furthermore, we are diverse: each of us is like a different species in the forest—some may like to cluster together, like the extrovert starlings or buzzie beetles; others may be quiet, and prefer to be alone or away from human company more. Some of us tend to look upwards, at the trees or the sky; while others look at what's at their feet, or listen, or perhaps scent the air—relying on other senses to read the world. Some of us bear indigenous heritage, and some carry knowledge and ways of knowing that have been acquired over many years, perhaps thousands of years; others have little knowledge of things beyond an ambit bounded by their interests, or what they have been exposed to.

Within all this complexity, there's *me*. I have my own identity and values, which have come into being through my own particular ways of experiencing the world: my senses, my learnings, and my inheritances. I assert my individuality where I may; yet we are entangled with each other and the world. It is of this entanglement—our relationships—that we make the world. A thorny issue that arises from my entanglement derives from my former profession, in the field of forestry. It is my past, and part of my making, so it has, naturally, been a thread running throughout this book ('like a running sore', some might feel). To be fair, and specific, my profession was as a conservation biologist within the broader discipline of forestry. But this is no confession; I'm fully at ease with my lived experiences, which have afforded me a particular set of insights. Lively debate continues in these circles—and more widely in society—as to wherein lies the optimal trade-off between a *smaller* physical footprint for forestry and a *shallower* one (assuming we have to have one at all). A smaller physical footprint may polarise the landscape into expanses of essentially unmanaged forest, on the one hand, and pockets of intensively managed forest, on the other. A shallower footprint integrates the same level of production across the wider landscape, spreading any impacts more thinly. This is known as the 'land-sparing versus land-sharing' conundrum. Conservationists of different persuasions may claim the moral high ground at any point along this spectrum. My own leanings, based on my personal journey of discovery, are towards land-sharing, which is in ascendancy elsewhere in the world; but the Tasmanian, and Australian, *zeitgeist* favours land-sparing (which, it's worth remembering, goes hand-in-hand with its alter ego, land-sacrificing).

It's a conundrum, to be sure, and one that can be deeply divisive; though it need not be so if, like the way natural ecosystems run themselves, we can allow ourselves and our societies the tolerance to accommodate different ways of living in the world, and accept that these will change over both space and time, sometimes in unexpected ways. This particular conundrum can be viewed through a classical economic lens, too: that is, regarding the extent to which we rely on our forests to supply products to meet demands (this is, after all, the main *purpose* of forestry). With ever-increasing human populations and levels of consumption, it's hard to constrain *demand*-driven production to sustainable levels; whereas *supply*-driven production is more readily rooted in sustainability. Limit supplies to those which can be sustained, and (notwithstanding differing definitions of sustainability) the rest follows.

Sustainability itself may be best viewed as a journey, rather than a destination, because our ideas as to what constitutes sustainability change over time just as surely as the forests do. Changing ideas and practices ('adaptive management') can, then, be a sign of a healthy level of engagement in the issue, rather than as continual policy failure. But trees grow slowly, so it could be said that forestry is about delayed, rather than instant, gratification—it is in it for the long haul. In forestry there is, therefore, a natural resistance to frequent change. This doesn't sit well within our rapidly changing political or cultural landscapes, where a perception of conservatism or obstinacy can frustrate, disgust or galvanise those whose life journeys have offered up different lenses, different ways of seeing or being.

On being in nature

Like trees, we all live and grow; we all change over our lifetimes. Sooner or later, we must all, surely, come to the realisation that the gift of *living* is incredibly special. It should never be taken for granted.

Life doesn't last forever; it may not last as long as you have anticipated. But what is life for? The Grove school of thought on this matter might appear nihilistic, for I think that our lives have no intrinsic meaning. This means, though, that we should put our efforts into giving our lives—ourselves— extrinsic meaning. When we are embedded within a supportive social network and a tolerant society we have meaning, and make meaning, through our relationships with others around us. It's a reciprocal, if not always equal, relationship. Ideally, we grow into the meaning granted to us by those who surround us (and through the meaning we may have given to others). We feel a sense of belonging, are comforted by this, and are motivated to maintain or enhance it. However, not everyone has such relationships; nor would everyone agree with this view, admittedly. This is merely the Grove school of thought.

What's this got to do with nature? Well, our relationships aren't just with humans; they are also with the other lives with whom we share the world. We don't just gain meaning from people, but through contact with nature, too. It's not (usually) an either/or, people/nature thing. Too much time alone in nature generally makes a person crave human company and society; too much time in society, without contact with nature, can make one want that contact. The secret may be in knowing what balance works best for oneself. I suspect that many people aren't very aware that nature could bring additional meaning to their existence, since so many of us live lives far removed from nature—for them, the opportunity scarcely exists. I find this the biggest tragedy of modern civilisation, and it seems to be increasing. It seems as though we must, often, choose whether to live according to material imperatives or, broadly speaking, 'spiritual' or 'naturalistic' ones; it's hard to do both. The pursuit of happiness through materialism can feel like running to stand still. With nature, you don't really need to pursue anything to be rewarded. And, in Tasmania, nature is all around us—we are incredibly lucky—and we can choose whether we engage with nature, or not.

I am convinced of the essential role of nature in our lives, but such conviction comes, of course, from my own personal experiences. I cannot imagine living a mentally well life—indeed, any life at all—without access to nature. When circumstances deny me the opportunity to get out into nature, I really struggle, as I did when suffering from leukaemia and its aftermath. Nature seems like the master key to a meaningful life. Just passing through nature is good, but the more I am in nature, and *with* nature—really looking, listening, and thinking with nature—the better my mood. Up to a point, at least: even nature nerds can love humans and crave their company, at times. Yet nature is a circuit breaker if I am dwelling on maudlin thoughts; it frees my mind to enjoy life. It helps bring the good things life can offer back into focus. The nineteenth-century American naturalist and essayist, Henry David Thoreau, expressed the sensation it gives well: '*I took a walk in the woods and came out taller than the trees.*'

An innate attraction to nature appears to be deep within us all. In the mid-1980s, the late and great American entomologist, evolutionary biologist and freethinker, Ed Wilson, introduced the world to the concept of "biophilia". The term had been coined two decades previously by psychoanalyst Erich Fromm, who defined it as 'the passionate love of life and of all that is alive'. Naming it enabled many of us to recognise that deep-grained feeling within ourselves—I remember experiencing an aha! moment when I first came across it, and a shift within myself, that made me feel a little more comfortable: '*Aha! So that's why I love being in nature! I'm not such a freak after all!*' It's only a word, though; others have been thinking, feeling, talking about, practising, and preaching the concept for centuries, and probably since the dawn of humanity. In Western culture, where, arguably, the need is greatest, countless books have dissected the concept from countless perspectives: physiological, psychological, spiritual, and so on—I don't need to dwell on them. It's actually simple, I believe: we evolved as part of, and immersed within, the natural world, and we are hardwired to sense belonging through, and derive meaning from, ongoing contact with plants, animals, and the wider natural world. So how, then, can we expect to live healthy, meaningful modern lives entirely apart from—*divorced* from—nature? If your response is along the lines of '*no worries, I'm fine thanks*', then, either you have more connection to nature than you realise, or it might surprise you, and wake you up, should you go for a walk and smell the roses (or, preferably, the flowers of leatherwood, or some other native plant).

The concept of biophilia is all well and good, but it's all about how nature can help *us*, as individuals or societies. It's a paradigm that sees nature as a backdrop (albeit essential) to our paramount lives, and a prop for our ambitions—our incessant desire for improvement: 'progress'. However, if our children are also to benefit from nature as we have, and, moreover, if nature is to be able to continue to unfurl for its own sake, we need to go beyond a quasi-utilitarian view of nature and understand that a truly healthy relationship is reciprocal. Why should the Golden Rule only apply to humans?—Remember that I said I didn't believe in our lives having intrinsic meaning? For me it's all about relationships: if we are kind to nature, nature may reward us, and *vice versa*. Our track record is appalling, though. Pope Francis once said that '*God always forgives. Human beings sometimes forgive. But nature never forgives*'. I beg to disagree. Nature may sometimes deliver shocks and catastrophes that affect millions; yet, as a whole, the planet only remains habitable because nature keeps it so. In this regard, nature is the epitome of forgivefulness and benevolence. Nature cares little whether we are optimistic or pessimistic about its future. But it does matter, materially, to nature, whether we feel love and care for it, and grief, guilt, and fear at its loss.

The physical landscapes that surround us are emotional landscapes, too. They are more than simply the backdrop for the lives we and all the creatures around us lead for the short extent of our existence. They resonate with meaning; with something much bigger than us. They must be cherished. This means that they ought not to be kept at a safe distance from our touch, but integrated into our lives in ways that enable us to appreciate, value, enjoy and be awed by them, and to respect, cherish and properly care for them. Tasmania's wild hills and intimate valleys have borne witness to my own highs and lows, and, without doubt, to those of many others. The hidden coves and nooks have assuaged my sorrows and anguish; the rugged peaks have heightened my moments of euphoria. I have left my footprints on many Tasmanian beaches; they, too, have tales to tell. In a sense these are landscapes of time, as well as space, or place: they bridge the past, present and future. Only our narrow and, perhaps, aberrant conception of time leads us to value the here-and-now of our own existence above all that came before and all that is to come. My memories, and my trace, linger in Lutruwita now, joining those of innumerable others. There is sadness in the memory of what has been lost, and shame in the memory of what has been unfairly taken, and in how it was taken. There is also joy in the ceaseless bustle of plant and animal life, and peace in knowing that most of the actors are largely indifferent to all the indignities that we have inflicted on their kind. The trees of the forest bear the memories of fire, wind, and snow. Some bear scars from the stripping of bark hundreds of years ago, for shelters or watercraft. Axes and saws have left their mark, too, and influence the forest regenerating in their aftermath. The grasslands and woodlands are shaped not only by fire, but also by the boom-and-bust cycles of wallabies whose keen senses remain alert to predators long since condemned to oblivion by our ignorance and arrogance. Moss bugs and others are living memories of deep time: of ice ages and vicariant continents. Viewed in a big-picture way, the biosphere, atmosphere, and geosphere are all one system, one global landscape, working together across vast expanses of space and time in a way that makes Earth fit for the purpose of sustaining life, *our* life.

At a philosophical level, humanity continues to argue about whether we are part of nature, or apart from nature. In my view, we are both. We are a 'natural' species like any other—we are expressions of nature, and deeply tied to all its other manifestations on this planet. At the same time, we are nature's *enfant terrible.* As a society we have chosen to live as though we are something separate, something monstrously different from that which gave us life. This has enabled me to use the word 'nature' to signify the world *other* than humans, so the distinction is not without value. Indeed, if we regard ourselves as part of nature, then we seem to be able to justify doing whatever we like with the rest of the natural world—because our actions are 'natural'. But what a dud deal that would be for the rest of the natural world, and, ultimately, for us. Better would be to apply our sense of *natural* justice and extend it to our relationship with the natural world. Perhaps that might be easier said than done, but what's stopping us? We have little to lose and much to gain; while 'business as usual' would be more like a Faustian pact: an illusion of being in charge that ultimately sees us lose everything while gaining nothing of real consequence.

On being alone

For better or worse, I'm one of those for whom there is sometimes no place lonelier than a crowded room; except, perhaps, a crowded city. It's not that I don't like people, it's more that I'm one of those introverts at heart, and being overtly social is not my forte. I struggle, too, with noise and with multiple conversations happening at the same time. It feels like acoustic assault. Fortunately, these days we are learning to be more aware, and accepting, of each other's differences, and I no longer feel the need to apologise for what would once have been considered character deficiencies or abnormalities. Perhaps, one day, we'll see that there are many more aspects of our personalities and outlooks that are non-binary: we are multi-dimensional, and sit at various positions along a multitude of spectra; and there's nothing wrong with that.

But where was I? Yes—I am comfortable in my own company; furthermore, I generally have no difficulty striking up a conversation with myself (being alone and being lonely clearly aren't one and the same thing; although, oddly, some languages lack this distinction). Where my self-talk takes me is context-dependent: my thoughts can become maudlin when I'm alone, tipping me into a full-blown funk—a downward spiral of woe, angst, regret, and despair. But in the 'right' context, being alone can nourish and repair, raising my mood to the heights of happiness, joy, and euphoria. It is only now, on the cusp of my seventh decade of existence, that it's begun to dawn on me that the key to a more contented life is setting oneself up for less of the wrong and more of the right contexts, and retaining a degree of equanimity, or even detachment, towards either eventuality.

It will probably be clear to you, by now, that the way I set myself up for self-nourishment and repair is by being alone in nature. It isn't that being alone in nature is qualitatively better than being in nature with friends; but these are different ways of being in nature and they engender different experiences. The journey from home to my chosen natural destination can be an emotionally fraught period of transition. I'm alone—it's when I'm most likely to sink into a funk. There's no logic to it, it just happens, when my mind is free of its usual occupations and in a holding pattern, with no direction or guidance, no absorbing distraction. The demons that have been lurking somewhere over my left ear since my transplant seem to enjoy tapping me on the shoulder at times like these. They usually appear, not as concrete recollections of past traumas, but as increasing unease with the world and my place in it. I have flashbacks, not to the traumatic days I spent in hospital, but to random, usually quite pleasant events in my life; yet each is accompanied by regret that those moments have passed and can never be regained. Indeed, sometimes it seems as though the happier the memory, the sharper the pain. If I don't keep the demons in check, they become increasingly bold, tapping, prodding, and heckling to the point where I wonder if I can continue with the journey, or even with life. It's not that I actively want to kill myself, normally; it's more a sense that I wouldn't mind waking up dead one day. *'To be or not to be, that is the question.'* However, I know myself well enough to appreciate that these are just feelings; that '*this too shall pass*'—the equanimity bit is important. Being able to view one's emotions impartially and from a distance helps tame them. I am thankful for the Vipassana meditation course I did, years ago—it taught me this. As the saying goes: in life, pain is inevitable; suffering is optional.

These mental tricks help, but a better solution is to take a break from the journey and go for a leg stretch somewhere at least vaguely natural; it will draw me in. A quiet country lane with skylarks ascending over the adjacent fields and cabbage white butterflies laying eggs on roadside brassicas will do. Better still: a patch of native bushland or beach where I can listen to birdsong, photograph insects, or fossick for seashells. Before long, the natural world will have worked its wonders and my mood will lift. A good indicator is when

I catch myself thinking out loud, or maybe striking up conversations with the plants and animals. *'You're a lovely spinebill, aren't you?' 'Sorry to disturb you, Mr Slater.' 'Ooh, you're a bug I've not seen before, where did you come from?'* You get the picture. I prefer to take this as a sign of improving mental wellbeing, rather than deepening madness; though the view might not be shared by occasional passers-by, who stumble innocently into my little world.

Once I get to my ultimate destination, in nature, I'll be back in my element. True nourishment begins. I am both lost and found, and I couldn't be happier. Time seems to stand still, and I feel a sort of foreverness and togetherness rolled into one, as though I am touching infinity. Everything seems connected, full of purpose and meaning.

Lockdown ... and beyond

Although I like being alone, the lockdowns imposed when the COVID-19 pandemic arrived in Tasmania were another matter. Tasmania had always felt like a sleepy backwater, far from earth-shattering events (like 9/11)—it had seemed as though I had closed the door on these, when we arrived here, all those years ago. Long may that be so; but a pandemic, by definition, goes everywhere. When histories of the pandemic are written, my story won't be part of it; neither, most likely, will yours; yet our experiences during it will remain deeply etched in our memories, and in the traces of our lives that remain after us. In their detail, they are profoundly personal, yet their emotional context may be common to many of us. Most likely, there might be grief, fear, anguish, and anger; but there may also be love, passion, release, reawakening, grounding, and reconnection. My experiences contain some of each.

As the virus spread around the world and global case numbers skyrocketed, my initial stance was one of bravado. If the disease is going to affect so many of us, I reasoned, better to get it sooner rather than later, before the health system collapses. I mentioned this to Rosie, my trusty oncologist, who was swift to advise me of the dangerous foolhardiness of my stance, considering my unorthodox and demonstrably sub-par immune system. I took her advice and retreated from my workplace with a microscope and boxes of micromollusc and insect samples, which I could continue to work on at home, for months if necessary. My days fell into a simple routine. Breakfast; a covid-safe solitary walk to the beach and back; a few hours of microscope, database, or website work; lunch; a quick solo bike ride to the top of the hill and back; more work; another quick walk to the beach; dinner; chat with family, and online chats with distant relatives; telly; and bed. It wasn't a particularly exciting existence, but it was productive, sustainable, and relatively stress-free. I was anxious about relatives in the UK, where the pandemic raged early; but I held few personal health worries, figuring that I was currently in a safer place than almost everyone else on the planet.

What I missed, though, was nature. I was lucky not to be living in the middle of a big city. We had the beach, garden, and views to wooded hills. I could watch birds flying over; I could hear them in the park across the road. Yet I yearned for a decent bushwalk or camping trip, despite it being winter. I took to closing my eyes and imagining myself up on the mountain on a hot summer's day. My meditation training and practice paid off. To a soundtrack of zizzing tinnitus, I imagined mountain cicadas calling across the alpine shrublands. I conjured images of parties of black currawongs tumbling through the skies, and I could hear their calls. I could smell the flowers, the gum leaves, the mosses and the bogs, and I could feel the warm breeze on my cheeks. It wasn't the same as being there, but it was the next best thing.

Weeks passed, then months. Winter turned to spring, and there was always something new to see, hear, smell, or feel in the garden. On the television, vaccines began to be spoken of, at last. I received mine sooner than most, and felt a little bolder. I began returning to work once restrictions were eased. News came from England that many of my vaccinated relatives had caught Covid, but none had suffered greatly, and all had recovered. Once I had received my second dose of the vaccine, I gave up working from home entirely. Commuting by bike was pleasantly traffic-free for a while; and because it was deemed to be vigorous exercise, I required no mask. Meanwhile there were very few outbreaks in Tasmania, and still I felt safe. I had my first booster, being given priority again. Restrictions on travel lifted; then cases in Tasmania soared. But we could now go camping and bushwalking, and it was summer. Chris and I took advantage of these new freedoms as often as we could and breathed in nature, deeply.

At work, I kept myself to myself as much as possible. It wasn't hard because I had the collections halls to work in when not in my office. I had my second booster early, too, and I continued to feel as safe as anyone could expect to feel in the 'new normal'. Inevitably, Covid eventually caught up with me, and when I was least expecting it—at a moment that turned out to be the beginning of another wave of infection across Tasmania, with one of the variant strains. Fortunately, it was summertime again; so, after spending three days holed up in bed, I was able to pass a few more days recuperating in the garden while enjoying the sun's warmth, the twittering birds and the buzzing insects.

Yet people spoke of a 'new normal'. Buoyed by the re-emergence of birdsong in unusually quiet cities, by sightings of whales and dolphins in bays usually buzzing with watercraft, and by starry skies startlingly devoid of aircraft contrails and the haze of pollution, many of us talked of reawakening and reconnection with nature; of how we need to reset our relationships with the planet; and how the economy is not the sole measure of our collective wellbeing. Sadly, the prospect of some new way of living, one that is more accommodating of nature and its role in our wellbeing, seems to have foundered. The 'new normal' feels increasingly like the old one: on television, the News covers the same old issues; politicians play the same old games. Road, air, and sea traffic has returned to pre-pandemic levels. Contrails and air pollution are back. Birdsong is again absent from our cities and our collective consciousness.

As I write these words, I am reminded of Joni Mitchell's 1970 hit single 'Big Yellow Taxi'. Her lyrics, now playing in my head, famously decried our tendency not to appreciate what we have until it's gone. The song resonated with the nascent environmental movement, and its galvanising influence may even have given rise to some better outcomes for nature, in the intervening decades, than might otherwise have been the case. Even so, much, much more of the natural world has gone during the intervening years. Joni's words continue to resound as the attrition carries on. It has all happened during our lifetimes, and on our watch.

Most traditional cultures have lived by the ethos that we don't inherit natural wealth from our ancestors so much as borrow it from our descendants. It's those future footprints, again. Today, in our pursuit of endless economic growth, we are on a track that would see us borrow so much, from a finite planet, that it can be no more than a one-way street to the End Times. Little of real worth will remain for posterity.

Not long ago, we arrogantly named our species *Homo sapiens*, the Wise Human; then it was suggested we call this pivotal period in Earth's history the Anthropocene, the Human Age. That name sounds a little too positive to me. A more suitable term might be Eremocene—the Age of Loneliness. Coined by Ed Wilson, who had also introduced us to 'biophilia', this name references the rate at which we are emptying the planet of life. Most geological Ages lasted millions of years; the Eremocene will be lucky to endure past a thousand. Perhaps our species might be capable of displaying individual wisdom, but, collectively, our behaviour has been anything but wise. Somehow, we need to turn things around: to degrow, to focus our collective energies and abilities on quality of life for *all* rather than on the quantity of products that some of us can consume.

We can start by spreading the word that there's more to life on Earth than the concerns of people and the material trappings of humanity. Biophilia is within us all. Some of us already know this; others may need to be shown, nudged, encouraged, enticed, or cajoled. Nature has no voice in our world without us; so, let nature speak through us. Speak kindly and with honesty and integrity, even if driven by anger, frustration, or fear. For the moss bugs and the humans. We're all in this together.

Given how 'the new normal', post-Covid, looks disturbingly like the old normal, it might seem as though the more things change, the more they stay the same. *'Plus ça change, plus c'est la même chose'* is a quote I introduced early in my writings, right back in September, and it seems appropriate that I repeat it here. Given the book's insecty theme, it seems fitting that this aphorism, by French author Alphonse Karr, appeared in a journal called, in English, '*The Wasp*'. Contrary to the way this saying is usually interpreted, Karr was actually making a point (in referring to 1840s French society) about how much better it might be if people were prepared to shrug off old attitudes that hold us back, and rise to the challenge of the new.

I agree with Karr. It's high time for *change*, and rather less of the *même chose*. After all, nature has always been about change, and we are inextricably bound to nature. If we choose to, we can be the new normal that so many of us desire, the future of treading lightly on the Earth. We can nudge the 'naturalness' back into our species, and put *ourselves* back in touch with nature.

Index of English and scientific names

I have restricted this index to Tasmanian species of mollusc, insect, spider and crustacean whose common names are not readily searchable online—sometimes because I made them up especially for this book, or because I have used a generally available term to refer to something more specific in the Tasmanian context. Species are listed in their major taxonomic groups, and then alphabetically, by the common name used in the book.

Following convention, each species binomial (the unique combination of a species' generic and specific name) is accompanied by its authority (wherever it has been described)—essentially, the person/people who described the species, and when. Where the species has been assigned to a different genus since its original description, the authority appears in parentheses.

As an aid to taxonomic pigeonholing, the species' family is also provided – it's the last word, in parentheses, ending in 'idae', which is the standard family-level designation in zoology. For reference, the basic taxonomic hierarchy follows the order:

Kingdom – Phylum – Class – Order – Family – Genus – Species

Molluscs

Airbreathing seaslug: *Onchidella nigricans* (Quoy & Gaimard, 1832) (Onchidiidae)
Bicoloured native slug: *Cystopelta bicolor* Petterd & Hedley, 1909 (Cystopeltidae)
Big-navel nautilus: *Nautilus macromphalus* Sowerby, 1849 (Nautilidae)
Broad-hinged venus clam: *Sunetta vaginalis* (Menke, 1843) (Veneridae)
May's margin-shell: *Dentimargo mayii* (Tate & May, 1900) (Marginellidae)
Ken Graham's whelk: *Buccipagoda kengrahami* (Ponder, 1982) (Buccinidae)
King Island whelk: *Tasmeuthria kingicola* (Tate & May, 1900) (Buccinidae)
Milligan's carnivorous snail: *Victaphanta milligani* (Pfeiffer, 1853) (Rhytididae)
Ponder's sundial-shell: *Heliacus ponderi* Garrard, 1977 (Architectonicidae)
Port Phillip topshell: *Spectamen philippensis* (Watson, 1880) (Solariellidae)
Pumice-snail: *Litiopa limnophysa* Melvill & Standen, 1896 (Litiopidae)
Southern false-olive: *Zemira australis* (Sowerby, 1833) (Pseudolividae)
Spicer's spindle shell: *Dolicholatirus spiceri* (Tenison-Woods, 1876) (Fasciolariidae)

Insects

Atkinson's jugular-horned beetle: *Prostomis atkinsoni* Waterhouse, 1877 (Prostomidae)
Beach hister beetle: *Halacritus lividus* (Lea, 1925) (Histeridae)
Beach hydrophilid beetle: *Ercycodes* species (Hydrophilidae)
Beach lesser dungfly: *Thoracochaeta* species (Sphaeroceridae)
Beach marchfly: *Cydistomorpha vetusta* (Walker, 1848) (Tabanidae)
Beach robberfly: *Stichopogon maritima* (Hardy, 1934) (Asilidae)
Beach stiletto-fly: *Anabarhynchus* species (Therevidae)
Beach wrack-seeker beetle: *Phycosecis litoralis* Pascoe, 1875 (Phycosecidae)
Beautiful little southern dronefly: *Austalis pulchella* (Macquart, 1846) (Syrphidae)
Black-and-yellow flower wasp: *Catocheilus apterus* (Olivier, 1811) (Thynnidae)
Bone-skipper fly: *Piophilosoma palpatum* Hendel, 1917 (Piophilidae)
Brown-winged soldierfly: *Octarthria brunnipennis* (Fuller, 1934) (Stratiomyidae)
Crested jewel beetle: *Castiarina jubata* (Blackburn, 1890) (Buprestidae)

Cuckoo spider-hunting wasp: *Psoropempula* species (Pompilidae)
Cuckoo bee: *Coelioxys tasmaniana* Rocha-Filho, 2016 (Megachilidae)
D'Albertis' grasshopper: *Russalpia albertisi* (Bolívar, 1898) (Acrididae)
Damp-wood termite: *Porotermes adamsoni* (Froggatt, 1897) (Termopsidae)
Dronefly-like axe-snouted fly: *Pelecorhynchus eristaloides* (Walker, 1848) (Pelecorhynchidae)
Faithful moss bug: *Hemiodoecellus fidelis* (Evans, 1937) (Peloridiidae)
Gnat-sized gnatbug: *Systelloderes* species (Enicocephalidae)
Golden bee-fly: *Sisyromya aurata* (Walker, 1849) (Bombyliidae)
Greasy straight-faced hoverfly: *Orthoprosopa grisea* (Walker, 1835) (Syprhidae)
Great Lake sun-cockroach: *Polyzosteria Yingina* Henry, 2021 (Blattidae)
Green frog-eating ground beetle: *Catadromus lacordairei* Boisduval, 1835 (Carabidae)
Green-sided soldierfly: *Odontomyia* species (Stratiomyidae)
Griffith's moss beetle: *Notolioon griffithi* (Lea, 1907) (Byrrhidae)
Grove's wood-midge: *Peromyia grovei* Jaschhof, 2010 (Cecidomyiidae)
Hardy's nannochoristid: *Nannochorista dipteroides* Tillyard, 1917 (Nannochoristidae)
Hoary darkling beetle: *Edylius canescens* Champion, 1894 (Tenebrionidae)
Inland jackjumper ant: *Myrmecia haskinsorum* Taylor, 2015 (Formicidae)
Intertidal weevil: *Aphela* species (Curculionidae)
Lea's ant-like flower beetle: *Mecynotarsus leai* Pic, 1942 (Anthicidae)
Long lauxaniid fly: *Paranomina longa* McAlpine, 2019 (Lauxaniidae)
Lowland flightless longhorn beetle: *Dorcadida bilocularis* White, 1846 (Cerambycidae)
Marchfly-shaped biting snipefly: *Spaniopsis tabaniformis* White, 1914 (Rhagionidae)
McQuillan's jewel beetle: *Castiarina macquillani* (Barker, 1988) (Buprestidae)
Mount Weld southern snow-flea: *Apteropanorpa warra* Palmer, Trueman & Yeates, 2007 (Apteropanorpidae)
Nunamara bald-headed fly: *Clesthentia crassioccipitis* (Nagatomi, 1991) (Apsilocephalidae)
Scratchy jewel beetle: *Castiarina rudis* Carter, 1934 (Buprestidae)
Semi-glowing hoverfly: *Hemilampra* species (Syrphidae)

Sieve-necked leaf beetle: *Palaeomela cribricollis* (Lea, 1929) (Chrysomelidae)
Southern iota rove beetle: *Iotarphia australis* Cameron, 1943 (Staphylinidae)
Spotted-winged water snipefly: *Dasyomma maculipenne* Hardy, 1920 (Athericidae)
Spot-winged nannochoristid: *Nannochorista maculipennis* Tillyard, 1917 (Nannochoristidae)
Tanyderid cranefly: *Nothoderus australiensis* (Alexander, 1922) (Tanyderidae)
Tasmanian gnatbug: *Oncylocotis tasmanicus* (Westwood, 1837) (Enicocephalidae)
Tasmanian Kosciuszko-dwelling grasshopper: *Kosciuscola tasmanicus* Rehn, 1957 (Acrididae)
Tasmanian mountain earwig: *Gonolabis tasmanica* (Bormans, 1880) (Anisolabididae)
Tasmanian painted marsh beetle: *Pictacara tasmanica* (Blackburn, 1892) (Scirtidae)
Tasmanian southern snow-flea: *Apteropanorpa tasmanica* Carpenter, 1941 (Apteropanorpidae)
Toothed lacebug: *Ceratocader dentatus* (Hacker, 1928) (Tingidae)
Two-horned giant springtail: *Womersleymeria bicornis* (Womersley, 1940) (Neanuridae)
Two-streaked darkling beetle: *Coripera bistriata* Carter, 1908 (Tenebrionidae)
Walker's flightless longhorn beetle: *Dorcadida walkeri* Gahan, 1893 (Cerambycidae)
Woody wrinkled bark beetle: *Leoglymmius lignarius* (Olliff, 1885) (Carabidae)

Spiders

Harris' peacock spider: *Maratus harrisi* Otto & Hill, 2011 (Salticidae)

Crustaceans

Brine shrimp: *Parartemia zietziana* Sayce, 1903 (Parartemiidae)
Clam shrimp: *Eulimnadia rivolensis* Brady, 1886 (Eulimnadiidae)
Freshwater prawn: *Paratya tasmaniensis* Riek, 1953 (Atyidae)

Acknowledgements

This book owes its completion to the encouragement of friends and family, who at times recognised its worth more than I did. I extend heartfelt thanks to them all. In particular, a very big thankyou to Jocelyn (Jo) Bornemissza, whose deep interest and unfaltering faith in my writing abilities, and diligence in reading and commenting on every page I wrote, spurred me on. My wife, Chris, encouraged my writing endeavours from the outset, accepting my frequent weekend writing retreats into the wilds, or into my study, with grace and goodwill, and reading through the entire manuscript. My mother, Pauline, also kindly read an early draft, and encouraged me to see the process through to publishing, while Karen Richards provided thoughtful feedback on a later draft. Tony Brown and Mary Mulcahy checked over my writings on First Peoples and museum matters respectively, and I thank them for their helpful feedback. Joy Lawn, Fred Duncan, Sarah Tassell, Steve Turton, Mark Neyland and Simon Stephens also checked over particular sections to ensure that my writing accorded with their experiences or knowledge. Andi Magnusson cleverly modified the AI-created feature illustration for the Interlude chapter, adding in the figure in the hospital gown and doctoring the chook on the bed so that the figure was no longer staring at the bird's rear end.

I noticed Keith Davis's marvellous nature sketches posted online, so I approached him about whether he would be willing to illustrate this book. I am delighted that he accepted, especially given the number of illustrations involved, his generosity, and his ability to complete them in such a short timeframe. Thank you, Keith.

I dropped the first names of various people into the book at various points, so now I'd like to acknowledge these friends and colleagues with their full names, listed here under their respective months of first mention. September: Dr David (Dave) Green, Bill and Jill Chilvers; October: Dr Sarah Tassell, Dr Karen Richards; December: Chris Spencer, Dr Dmitry Shcherbakov, Dr Kirill Eskov, Rob de Little, Don

Hurd, Dr Kirrily Moore; January: Dr Jeffrey (Jeff) Skevington, Dr Marie Yee, Dr Susan (Sue) Baker; February: Dr Catherine (Cathy) Byrne, Dr Lynette (Lynne) Forster, Charlotte King, Dr Mark Neyland, Gabrielle (Gabes) Balon; Interlude: Pamela Lovell, Dr Rosemary (Rosie) Harrup, Fred Duncan, Mercedes Villarreal (now Duncan), Dr Guillermo Martinez Pastur, Dr (María) Vanessa Lencinas, Dr Elena Lukashevich; July: Simon Stephens; August: Dr Timothy (Tim) Wardlaw.

For the first print edition, I also thank the very professional publishing team at Forty South, and my editor, Daniela Brozek, at Bright South, whose depth of engagement, attention to detail, and creation of the map (incorporating some of Keith's images) contributed greatly to the book's final form. And for the second print and ebook editions, I extend my thanks to Dr David Reiter at Interactive Publications for his patience and for going above and beyond.

Made in the USA
Monee, IL
13 September 2026

62975572R00227